Microchip
Fabrication

Other McGraw-Hill Reference Books of Interest

Handbooks

BENSON • *Audio Engineering Handbook*

BENSON • *Television Engineering Handbook*

COOMBS • *Printed Circuits Handbook*

COOMBS • *Basic Electronic Instrument Handbook*

CROFT AND SUMMERS • *American Electricians' Handbook*

DI GIACOMO • *VLSI Handbook*

FINK AND BEATY • *Standard Handbook for Electrical Engineers*

FINK AND CHRISTIANSEN • *Electronic Engineers' Handbook*

HARPER • *Handbook of Electronic Systems Design*

HARPER • *Handbook of Thick Film Hybrid Microelectronics*

HARPER • *Handbook of Wiring, Cabling, and Interconnecting for Electronics*

HICKS • *Standard Handbook of Engineering Calculations*

INGLIS • *Electronic Communications Handbook*

JURAN AND GRYNA • *Quality Control Handbook*

KAUFMAN AND SEIDMAN • *Handbook of Electronics Calculations*

KURTZ • *Handbook of Engineering Economics*

STOUT • *Microprocessor Applications Handbook*

STOUT AND KAUFMAN • *Handbook of Microcircuit Design and Application*

STOUT AND KAUFMAN • *Handbook of Operational Amplifier Circuit Design*

TUMA • *Engineering Mathematics Handbook*

WILLIAMS • *Designer's Handbook of Integrated Circuits*

WILLIAMS AND TAYLOR • *Electronic Filter Design Handbook*

Other

ANTOGNETTI AND MASSOBRIO • *Semiconductor Device Modeling with SPICE*

ANTOGNETTI • *Power Integrated Circuits*

ELLIOTT • *Integrated Circuits Fabrication Technology*

HECHT • *The Laser Guidebook*

MUN • *GaAs Integrated Circuits*

SILICONIX • *Designing with Field-Effect Transistors*

SZE • *VLSI Technology*

TSUI • *LSI/VLSI Testability Design*

Microchip Fabrication

A Practical Guide to Semiconductor Processing

Peter Van Zant

Second Edition

McGraw-Hill Publishing Company

New York   St. Louis   San Francisco   Auckland   Bogotá
Caracas   Hamburg   Lisbon   London   Madrid   Mexico
Milan   Montreal   New Delhi   Paris
San Juan   São Paulo   Singapore
Sydney   Tokyo   Toronto

Library of Congress Cataloging-in-Publication Data

Van Zant, Peter.
   Microchip fabrication : a practical guide to semiconductor
processing / by Peter Van Zant. — 2nd ed.

      p.   cm.
   ISBN 0-07-067194-X
   1. Semiconductors—Design and construction.   I. Title.
TK7871.85.V36   1990
621.381'52—dc20

89-36002
CIP

1 2 3 4 5 6 7 8 9 0   DOCDOC   9 5 4 3 2 1 0

ISBN 0-07-067194-X

*The sponsoring editor for this book was Daniel A. Gonneau, the
editing supervisor was David E. Fogarty, the designer was Naomi
Auerbach, and the production supervisor was Suzanne W. Babeuf. This
book was set in Century Schoolbook. It was composed by the McGraw-
Hill Professional and Reference Division composition unit.*

Printed and bound by R. R. Donnelley & Sons Company.

*For more information about other McGraw-Hill materials,
call 1-800-2-MCGRAW in the United States. In other
countries, call your nearest McGraw-Hill office.*

To my wife Mary DeWitt.

It is hard to avoid cliches in describing the support from a loving spouse, so I will just say thanks to a lovely lady and good friend.

Contents

Preface

When *Microchip Fabrication* was first published in 1984 we knew that, like the semiconductor processes it described, it would suffer technological obsolescence. Now a second edition has become necessary. The book has been reorganized, updated, and rewritten. Feedback from a diverse spectrum of readers has contributed greatly to this edition. We thank all those that have purchased the first edition and shared with us their comments.

The second edition has the same goal as the first; a comprehensive first text in semiconductor processing, not requiring a math or science background. As such, it should prove valuable to a wide range of readers, from production workers to personnel serving the industry with equipment and materials. New material has been added to make the text more appropriate as a teaching text for college courses or in-house programs.

To the second edition have been added overview, objectives, expanded key terms, and review question sections, for each chapter. These sections are included to facilitate use of the text in a self-study program. A new chapter, "Manufacturing Technology," examines the economics and production control issues of a semiconductor fabrication production line. The reader may get annoyed at the many forward and reverse references. We apologize. One problem with writing about this technology is that some process steps and some types of equipment are used in a number of different processes. In the interest of efficiency those process steps are described once and referenced where they are used in other processes.

Another problem is the intimate relationship of semiconductor device and circuit structures and processes. Understanding the importance and control of a particular process is aided by an understanding of the benefits of the process in a particular device structure or material and vice versa. We suggest that the reader read the process chapters and reference forward to the device and circuit descriptions in the last two chapters.

Many people have made valuable suggestions to this edition. I am particularly indebted to Anne Miller, who has distributed the first edition and taught semiconductor training programs using *Microchip Fabrication* as a text. Her feedback and advice has been most appreciated. Jim Hayes also taught from the text and provided appropriate suggestions and feedback. Bill Kiba generously contributed information on gallium arsenide processing. On the production side I thank Anne Thurston for producing the new illustrations and Terri Oropeza of Copy Ink for help with the use of the Microsoft Word program and the shake down of our Macintosh SE.

Peter Van Zant

1

The Semiconductor Industry

Overview

In this chapter you will be introduced to the history, development, and trends of the semiconductor industry. The major product types and transistor building structures, along with integration levels, will be explained.

Objectives

Upon completion of this chapter, you should be able to:

1. Describe the difference between discrete devices and integrated circuits.
2. Write the definitions of the terms *solid state, planar processing*, and N-type and P-type semiconducting materials.
3. List the four major stages of semiconductor processing.
4. Explain the *integration scale* and the implications of processing circuits of different levels of integration.
5. List the major process and device trends in semiconductor processing.

Birth of an Industry

The computational needs of society outgrew manual manipulation of data in the late nineteenth century. A crisis developed in the 1880s when it was predicted that the 1890 census would take 10 years to condense and correlate. This length of time was unacceptable and led the U.S. Census Bureau to sponsor a contest for a new method of processing the data. The winner was Herman Hollerith, who invented a mechanical tabulating machine that completed the census data-crunching in an amazing six weeks.

His machine was a mechanical system driven by an electrical motor and introduced, for the first time, the punch card. (Hollerith went on to form the Tabulating Machine Company, which eventually evolved into IBM, the International Business Machines Corporation.) This tabulating machine was one of the earliest computers, a machine designed to perform automatic calculations on data. The technology of computers continued to progress with the development in the 1930s of computers using electromechanical switches.

During World War Two, a considerable amount of effort went into the development of intelligent and fast machines to crack secret codes. One result of these efforts was the world's first electronic computer, named the Electronic Numeric Integrator and Calculator or ENIAC for short. The ENIAC was first demonstrated at the Moore School of Engineering in Pennsylvania in 1947.

This first effort hardly fits the modern picture of a computer. It occupied some 3,000 ft^2, weighed 50 tons, generated large quantities of heat, required the services of a small power station to operate, and cost \$400,000 in 1940 dollars. The ENIAC was based on 19,000 vacuum tubes, along with thousands of resistors and capacitors (Fig. 1.1). Despite its mammoth size and component count, its computing power was less than that of a modern programmable engineering calculator.

However crude this early computer now seems, it was the system that gave birth to the electronics era with all its famous progeny. The revolution it launched led in the mid-1970s to the end of the industrial age and the dawn of the information age. In that decade, more workers became employed in the processing of information than in manufacturing.

Many inventions have been made along the way from the monstrous ENIAC to lap-top computers. A measure of this progress was the duplication, in the mid-1970s, of the entire electronics of the ENIAC on

Size, ft	30 × 50
Weight, tons	30
Vacuum Tubes	18,000
Resistors	70,000
Capacitors	10,000
Switches	6000
Power Requirements, W	150,000
Cost (in 1940)	\$400,000

Figure 1.1 Eniac statistics. *(Foundations of Computector Technology, J. G. Giarratano, Howard W. Sams & Co., Indianapolis, Ind., 1983.)*

a $\frac{3}{8}$-in^2 piece of silicon, drawing less power than a light bulb and costing less than $20!

The Solid-State Transistor

Vacuum tubes were the primary electronic devices (or components) of electrical products, from their invention in the early 1900s until the 1950s. A vacuum tube consists of three elements—two electrodes separated by a grid in a glass enclosure (Fig. 1.2). Inside the enclosure is a vacuum, required to prevent the elements from burning up and to allow the low-resistance transfer of electrons.

Tubes perform two important electrical functions: switching and amplification. *Switching* refers to the ability of an electrical device to turn a current on or off. *Amplification* is a little more complicated. It is the ability of a device to receive a small signal (or current) and amplify it while retaining its electrical characteristics.

Vacuum tubes, while providing the required electrical functions, suffer from a number of drawbacks. They are bulky, prone to loose connections and vacuum leaks, fragile, require a relatively large amount of power to operate, and their elements deteriorate rather rapidly. One of the major drawbacks to the ENIAC and other tube-based computers is a short operating time before one or more tubes burn out and require replacement.

These problems were the impetus leading many laboratories around the country to seek a replacement for the vacuum tube. That effort came to fruition on December 23, 1947, when three Bell Lab scientists demonstrated an electrical amplifier formed from the semiconducting material *germanium* (Fig. 1.3).

This device offered the electrical functioning of a vacuum tube, but with the advantages of the solid state, no vacuum, small and light-

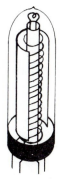

Figure 1.2   Vacuum tube.

Figure 1.3   The first transistor.

weight, low power requirements, and a long lifetime. First named a "transfer resistor," the new device soon became known as the *transistor*. The three scientists, John Bardeen, Walter Brattin, and William Shockley, were awarded the Nobel Prize in 1956 for their invention of the transistor.

Solid-State Devices

The transistor, like the vacuum tube, is a three-element device. But instead of the electrons *jumping* between the elements as in a vacuum, the electrons *cross* electrical junctions formed in the solid-state material. These junctions are the interfaces between regions in the semiconducting material that are rich in either electrons or holes (Fig. 1.4). The electron-rich regions are called *N type* (for negative) and the hole-rich regions are called *P type* (for positive). The electrical junctions are more formally called N-P or P-N junctions. The concept of holes is explained in Chap. 3 and junctions are explained in Chap. 12.

Besides transistors, solid-state technology is also used to create diodes, resistors, and capacitors. *Diodes* are two-element devices that function in a circuit as a switch. *Resistors* are monoelement devices that serve to limit current flow. *Capacitors* are two-element devices that store charge in a circuit. In some circuits, solid-state technology is used to create fuses. (Refer to Chap. 14 for an explanation of these concepts and how these devices work.)

Discrete devices

The 1947 announcement of the first solid-state transistor launched the solid-state electronics age and the microelectronics industry. The industry is divided into two major segments, the semiconductor industry and the electronics industry. The semiconductor segment refers to the many firms and their suppliers that produce the solid-state devices. The electronics segment encompasses the industry that designs and produces the vast number of semiconductor-device-based products, from consumer electronics to computers. The electronics industry includes the manufacturers of printed circuit boards.

The semiconductor industry was in full swing by the early 1950s,

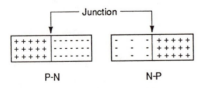

Figure 1.4 P-N and N-P junctions.

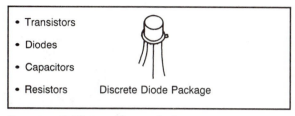

- Transistors
- Diodes
- Capacitors
- Resistors   Discrete Diode Package

Figure 1.5 Solid-state discrete devices.

supplying devices for transistor radios and transistor-based computers. In this early stage, the semiconductor industry products were separate transistors, diodes, resistors, and capacitors (Fig. 1.5). In the
wafer-fabrication process, only one device was formed in each chip
area.

These devices, containing only one device per chip, are called *discrete devices*. Generally, discrete devices have less demanding operational and fabrication requirements than integrated circuits. In general, discrete devices are not considered leading-edge products. Yet,
they are required in most sophisticated circuits and traditionally have
accounted for 20 to 30 percent of the dollar volume of all semiconductor devices sold.

Integrated circuits (ICs)

The dominance of discrete devices in solid-state circuits came to an
end in 1959. In that year, Jack Kilby, a new engineer at Texas Instruments in Dallas, Texas, formed a complete circuit on a single piece of
the semiconducting material germanium. His invention combined
several transistors, diodes, and capacitors (a total of five components)
and used the natural resistance of the germanium chip (called a "bar"
by Texas Instruments) as a circuit resistor. This invention was the integrated circuit, the first successful *integration* of a complete circuit in
and on the same piece of a semiconducting substrate (Fig. 1.6).

The Kilby circuit did not have the form that is prevalent today. It
took Robert Noyce, then at Fairchild Camera, to furnish the final
piece of the puzzle. In Fig. 1.6 is a drawing of the Kilby circuit. Note
that the devices are connected with individual wires. Noyce applied
the technique of using a thin, evaporated surface film of aluminum
and the surface patterning process of photolithography to connect the
devices. Noyce's integrated circuit was in silicon.

Earlier, Jean Horni, also at Fairchild Camera, had built a diffused
junction transistor on a silicon chip to create a solid-state transistor
with a flat profile (Fig. 1.7). The flattened profile was the outcome of

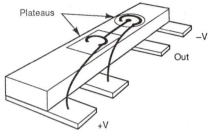

Figure 1.6 Kilby integrated circuit from his notebook. *(Courtesy of Texas Instruments.)*

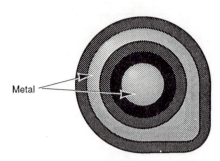

Figure 1.7 Horni "teardrop" transistor.

taking advantage of the easily formed natural oxide of silicon (silicon dioxide) that also happened to be a dielectric. Horni's transistor used a layer of evaporated aluminum that was patterned into the proper shape to serve as wiring for the device. This technique is called *planar technology*.

The Noyce integrated circuit became *the* model for all integrated circuits. The techniques used not only met the needs of that era but contained the seeds for all the miniaturization and cost-effective manufacturing that still drives the industry. Kilby and Noyce shared the patent for the integrated circuit.

Starting with the simple Kilby integrated circuit, the industry has never looked back. Year after year has seen the inclusion of more and more devices in a single circuit. In 1964, Gordon Moore, a founder of Intel, predicted that integrated-circuit density would double every year. This prediction became known as Moore's law and has proven to be very accurate (Fig. 1.8).

The industry keeps track of these integration levels by a scale (Fig. 1.9), ranging from small-scale integration (SSI) to ultra-large-scale in-

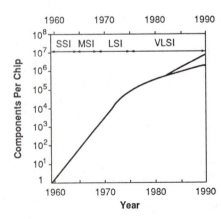

Figure 1.8 Growth in IC density *[After Moore, "VLSI, What Does the Future Hold?", Electronic News, 42(14) (1980).]*

Level	Abbreviation	# Components per Chip
Small Scale Integration	SSI	2 - 50
Medium Scale Integration	MSI	50 - 5000
Large Scale Integration	LSI	5000 - 100,000
Very Large Scale Integration	VLSI	Over 100,000 - 1,000,000
Ultra Large Scale Integration	ULSI	> 1,000,000

Figure 1.9   IC integration scale.

tegration (ULSI). ULSI chips are also referred to as very very large-scale integration (VVLSI). At some point in the near future, probably at around 4-million devices per circuit, a new designation will come into usage.

Stages of Manufacturing

Solid-state devices are manufactured, many at a time, on disks of semiconducting materials called *wafers* or *slices* (Fig. 1.10). The first stage of manufacturing is *material preparation* (see Chap. 2). In this stage, the raw semiconducting materials are mined and purified to meet semiconductor standards (Fig. 1.11).

In stage two (Fig. 1.12), the material is formed into wafers in a process called *crystal growth* and *wafer preparation* (Chap. 3). This stage is either performed by a separate department within the semiconductor company or supplied by vendors.

It is in stage three, *wafer fabrication* (Fig. 1.13), that the devices or

1. Material Preparation
2. Crystal Growth and Wafer Preparation
3. Wafer Fabrication
4. Packaging

Figure 1.10   Stages of semiconductor manufacturing.

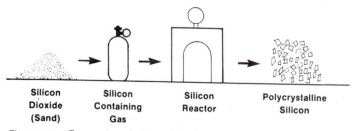

| Silicon Dioxide (Sand) | Silicon Containing Gas | Silicon Reactor | Polycrystalline Silicon |

Figure 1.11   Conversion of silicon dioxide to semiconductor grade silicon.

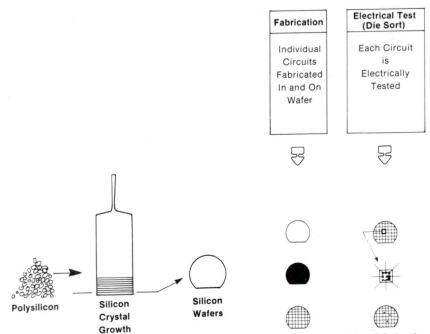

Fabrication	Electrical Test (Die Sort)
Individual Circuits Fabricated In and On Wafer	Each Circuit is Electrically Tested

Polysilicon   Silicon Crystal Growth   Silicon Wafers

Figure 1.12   Crystal growth and wafer preparation.

Figure 1.13   Wafer fabrication (and electrical test).

integrated circuits are actually formed on the wafer surface. Up to several thousand identical devices are formed on each wafer, although two to three hundred is a more common number. The area on the wafer occupied by the discrete device or integrated circuit is called a *chip* or *die*. The manufacturing process is also called *fabrication*, fab, chip fabrication, or *microchip fabrication*.

Chips are fabricated by both merchant and captive producers. Merchant suppliers manufacture just chips and sell them on the open market. Captive suppliers are firms whose final product is a computer, communications system, etc., and who produce chips in-house for their own products. Some firms produce chips for in-house use and also sell on the open market, and others produce specialty chips in-house and buy others on the open market. During the 1980s the trend has been to a greater percentage of chips being fabricated in captive fabrication areas. IBM's captive fabrication areas produce the largest number of chips in the United States and may be the world's largest chip producer.

Following wafer fabrication, the wafers go to stage four (Fig. 1.14), the *packaging* operation (also known as *assembly*). The first step after

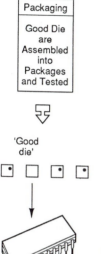

Figure 1.14   Packaging stage.

wafer fabrication is the testing of the chips on the wafer and separation of the wafer into the individual chips.

Following the separation process comes the actual packaging. A protective chip package is necessary to protect the chip from contamination and abuse and to provide a durable and substantial electrical lead system to allow connection of the chip onto a printed circuit board or directly into an electronic product. Packaging takes place in a different department of the semiconductor producer and quite often in an offshore plant. While the vast majority of chips are packaged in individual enclosures, some are mounted directly in hybrid circuits or mounted onto a printed circuit board.

Hybrid circuits consist of a ceramic substrate on which standard electrical and solid-state devices are mounted. The devices are connected by conductive or resistive paths on the surface of the substrate. These paths are formed by silk-screening inks containing a proper filler or from thin films evaporated onto the surface and patterned by photolithography techniques. The term *hybrid* refers to the mix of solid-state and conventional electrical devices present in the same circuit. Hybrid circuitry technology was conceived by RCA in 1958.

Hybrid circuits offer the advantages of structural rigidity and low leakage between devices due to the insulating property of the ceramic. On the down side, they have a much lower density than integrated circuits and have a higher cost.

Semiconductor Industry Growth

Process and product trends

Since 1947, the semiconductor industry has seen the continuous development of new and improved processes. These process improvements have, in turn, led to the more highly integrated and reliable circuits that have, in their turn, fueled the continuing electronics revolution. These process improvements fall into two broad categories: process and structure. Process improvements are those that allow the fabrication of the devices and circuits in ever higher density, quantity, and reliability. The structure improvements are the invention of new device designs that allow greater circuit performance, power control, and reliability.

The advancement of chip density from the SSI level to ULSI chips has been accompanied by two trends, increasing chip size and the reduction of the size of the individual circuit components. Discrete and SSI chips average about 100 mils (0.1 in) on a side. ULSI chips are in the 500 mils (0.5 in) per side range.

Integrated-circuit chip size has grown from the higher number of devices crowded into the circuits. However, these higher densities would not be possible without a commensurate decrease in the size of the individual devices (Fig. 1.15). This decrease has been brought about by dramatic increases in the imaging process, known as photolithography.

The measurement of this progress is the *circuit image size*, also known as line width, critical dimension, or feature size. This measurement refers to the smallest or most critical dimension present on the chip surface. Since the introduction of SSI circuits with feature sizes of 20 to 40 μm, the trend has been steadily downward to the present-day submicron (less than 1 μm) sizes.

An analogy used to explain these trends is the layout of a neighbor-

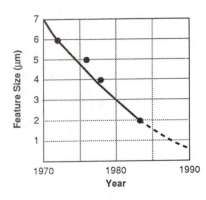

Figure 1.15 IC minimum feature size trends.

hood of single-family homes. The density of the neighborhood is a function of the house size, lot size, and the width of the streets. Increasing the density can be accomplished by building more houses, resulting in a larger neighborhood. Or we can put the houses closer together by reducing the lot and street size. Another possibility is to reduce the size of the individual houses. However, at some point the boulevards cannot be reduced anymore in size or they won't be wide enough for autos, and, at some point, the houses cannot be reduced in size and still function as dwelling units.

The same types of options take place in chip design. The individual parts of the devices have been reduced in size (houses) and moved closer together (lot and street size). The results are increasing chip size even as the chip density is growing faster.

There are several benefits to the reduction of the feature size and its attendant increase in circuit density. At the circuit performance level, there is an increase in circuit speed. With less distance to travel and with the individual devices occupying less space, information can be put into and gotten out of the chip in less time. Anyone who has waited for their personal computer to perform a simple operation can appreciate the effect of faster performance in a huge mainframe computer with its many thousands of integrated circuits. These same density improvements result in a chip or circuit that requires less power to operate. The small power station required to run the ENIAC has given way to powerful lap-top computers that run on a set of batteries.

Perhaps the most significant effect of these process and product improvements is the cost of the chips. Figures 1.16 and 1.17 show the year-by-year drop in transistor and memory chips. These chip prices have constantly declined even as the performance of the chips has increased. These two factors, increased performance at lower cost, have

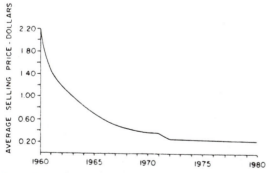

Figure 1.16 Average selling price of discrete transistors. (*From Semiconductor Industry Assoc., 1980–1981 Yearbook and Directory.*)

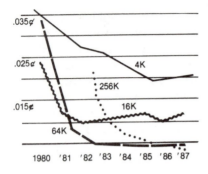

Figure 1.17   Price of chips per bit of memory.

driven the switch from conventional electronics to solid-state electronics. An example is the automobile. A 1980s auto carries some $500 worth of on-board computing power. That same amount of power would have cost some $3000 in a 1970 auto. Considering that in 1970 the total cost of a car was about $3000, this is a remarkable achievement. Even more impressive is the personal computer. Today, for a moderate price a desk-top computer can deliver the same power as an IBM mainframe manufactured in 1970.

The availability of cheap chips has led to their inclusion in many different products. Figure 1.18 is a list of products using solid-state chips from 1950. Figure 1.19 shows the distribution of chip use in major market segments in 1984. This mix has stayed relatively constant throughout the 1980s.

To offset the decline in chip prices, while delivering a more sophisticated circuit, the semiconductor industry has had to continually make dramatic improvements in processing technology, including vendor-supplied materials and equipment. The fabrication processes have become more numerous, with the use of automation becoming an absolute requirement.

Overall the semiconductor industry has experienced worldwide continual growth. From its birth in the 1950s it is now approaching worldwide sales of $100 billion a year. The millions of chips are supplied by factories located throughout the world.

The industry developed in the high-technology centers of Route 128 around Boston, Silicon Valley in northern California, and at Texas Instruments in Dallas, Texas. These areas maintained their dominance until the 1970s when the process technology moved from laboratory-type processes to equipment-based ones. Once the processes were transferred into pieces of equipment, the stage was set for dissemination throughout the world. Along with the process improvements came a more detailed understanding of the physics of solid-state devices, also allowing the mastering of the technology by student engi-

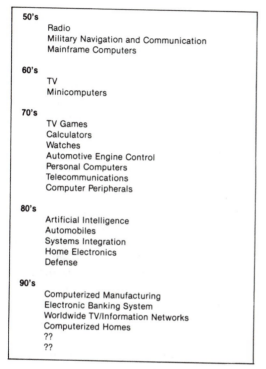

50's
    Radio
    Military Navigation and Communication
    Mainframe Computers

60's
    TV
    Minicomputers

70's
    TV Games
    Calculators
    Watches
    Automotive Engine Control
    Personal Computers
    Telecommunications
    Computer Peripherals

80's
    Artificial Intelligence
    Automobiles
    Systems Integration
    Home Electronics
    Defense

90's
    Computerized Manufacturing
    Electronic Banking System
    Worldwide TV/Information Networks
    Computerized Homes
    ??
    ??

Figure 1.18   Semiconductor uses since 1950.

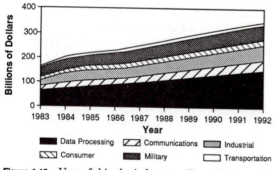

Figure 1.19   Uses of chips by industry. *(From Dataquest.)*

neers worldwide. A third factor driving the spread of global semiconductor manufacturing is the many cross-licensing and joint ventures between American firms and foreign suppliers that also has the effect of distributing the technology.

The 1980s saw the rise of Japanese semiconductor producers, followed closely by the "Four Tigers" of Hong Kong, Taiwan, Singapore, and Korea. It is reasonable to expect that by the turn of the century semiconductor technology will be a part of every nation's manufacturing base.

Four Decades of Progress

The history of the semiconductor industry is one of continual developments and advances. Each new level of process expertise and material refinement has made possible newer and more powerful circuits. The first four decades of the semiconductor era have seen a maturing industry emerging to world dominance.

The development decade (1951–1960)

While the tremendous advantages of solid-state electronics were recognized early, the advances possible from miniaturization were not realized until two decades later. During the 1950s engineers set to work and defined many of the basic processes and materials still used today.

The Bell transistor came to be known as the point-contact transistor because the surface contacts were made by springing two thin wires onto the semiconductor surface (Fig. 1.3). This system was unreliable and quickly gave way to the alloy-junction transistor (Fig. 1.20). In this process, the contacts were alloyed to the surface. Another early technique was the grown junction, where the required semiconductor junction was grown into the semiconductor crystal during the crystal-growing process (Fig. 1.21). These advances all came out of Bell Labs.

These transistors were of the bipolar type (see Chap. 14), which became the dominant transistor construction technique well into the 1970s. The term *bipolar* refers to the presence of both negative and positive currents in the transistor during operation (Fig. 1.22). The other major method of building a solid-state transistor is the FET (field-effect transistor). (William Shockley published the operational basics of a field-effect transistor in 1951.) These transistors operate with only one type of current and are also called unipolar devices. The FET came to the marketplace in volume in a structure known as the *metal oxide semiconductor* (MOS) transistor.

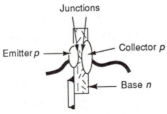

Junctions

Emitter *p* — — Collector *p*

— Base *n*

Figure 1.20 Alloy junction transistor.

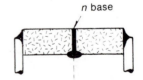

n base

Figure 1.21 Grown junction transistor.

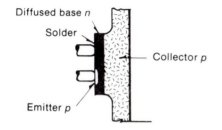

Diffused base *n*

Solder

Collector *p*

Emitter *p*

Figure   1.22   Mesa   transistor
with diffused emitter and base.

William Shockley and Bell Labs get much of the credit for the
spread of semiconductor technology. Shockley left Bell Labs in 1955
and formed Shockley Laboratories in Palo Alto, California. While his
company did not survive, it established semiconductor manufacturing
on the West Coast and provided the start for what eventually became
known as Silicon Valley. Bell Labs helped the fledgling industry with
the decision to license its semiconductor discoveries to a host of companies.

All of the early semiconductor devices were made utilizing the material germanium. Texas Instruments changed that trend with the introduction of the first silicon transistor in 1954. The issue over which
material would be dominant was settled in 1956 and 1957 by two more
developments from Bell Labs: diffused junctions and oxide masking.
Diffused junctions were an important part of the early mesa transistor
brought to the marketplace (using silicon) by Texas Instruments. The
mesa transistor was named for its mesa-like shape that allowed electrical contacts to the subsurface N-type and P-type regions.

It was the development of oxide masking that ushered in the
"silicon age." Silicon dioxide (SiO_2) grows uniformly on silicon and,
having a similar index of expansion, allows high-temperature processing without warping. Silicon dioxide is a dielectric material which allows it to function on the silicon surface as an insulator. Additionally,
SiO_2 is an effective block to the dopants that form the N-type and P-
type regions in silicon.

The net effect of these advances was planar technology (introduced,
as mentioned, by Fairchild Camera in 1960; see Fig. 1.23). With the
aforementioned techniques it was possible to form (diffusion) and protect (silicon dioxide) junctions during and after the manufacturing
process. Also, the development of oxide masking allowed all of the
transistor formation to take place from the top surface of the wafer,
that is, in one *plane*. It was this process that set the stage for Robert
Noyce's development of thin film wiring.

Bell Labs conceived of forming transistors in a high-purity layer of
semiconducting material deposited on top of the wafer. Called an
epitaxial layer, this discovery allowed higher-speed devices and pro-

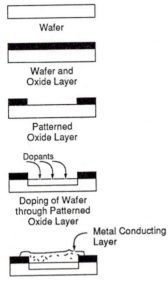

Wafer

Wafer and
Oxide Layer

Patterned
Oxide Layer

Dopants

Doping of Wafer
through Patterned
Oxide Layer

Metal Conducting
Layer

Figure 1.23 Basics of silicon planar processing.

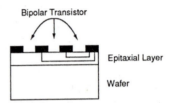

Bipolar Transistor

Epitaxial Layer

Wafer

Figure 1.24 Double diffused bipolar transistor formed in epitaxial layer.

vided a scheme for the closer packing of components in a bipolar circuit (Fig. 1.24).

The 1950s was indeed the golden age of semiconductor development. During this incredibly short time almost all of the basic processes and materials were discovered. The decade opened with the knowledge of and capability to manufacture small volumes of crude devices in germanium and ended with the first integrated circuit and the material silicon firmly established as the semiconductor of the future.

The processing decade (1961–1970)

The 1960s was the decade of the process engineer and start-up companies. The chip-price erosion trend of the industry had been well established in the 1950s. In the 1960s, many new players came into the field and established ones solidified their positions. The field got crowded, which only served to create more downward price pressure. It fell to the process engineers to develop high-yield volume processes to crank out the needed chips.

During this period, they got help from a growing number of vendors. Previously, much of the equipment was made in-house or modified for semiconductor use. Likewise, most of the chemicals were purchased at industrial-grade levels and "cleaned up" in-house. By the 1960s the number of fabrication areas had grown sufficiently, and processes were approaching a level of commonalty that attracted semiconductor specialty suppliers.

Kodak Corporation was in the game early with its line of Kodak Photo Resist (KPR), Kodak Metal Etch Resist (KMER), and Kodak Thin Film Resist (KTFR)—*negative photoresists*. The Shipley Company introduced its line of *positive resists*, and Waycoat developed its line of negative photoresists, the first based on a synthetic polymer. On the equipment front, BTU and Thermco sold diffusion and oxidation tube furnaces, and K&S and Kasper supplied contact aligners critical for the photomasking operation. Three traditional chemical giants, Merck, Monsanto, and Dow Corning helped fuel the growth by becoming suppliers of polished silicon wafers.

Many of the key company players of the 1970s and 1980s threw their hats in the ring during the 1960s. Robert Noyce left Fairchild to found Intel (with Andrew Grove and Gordon Moore). Charles Sporck also left Fairchild to lead National Semiconductor to be an industry leader. Signetics became the first company dedicated exclusively to the fabrication of integrated circuits. However, price dropping had a purging effect, and companies gave up fabrication operations. It was rare when a fabrication area was closed down for good. Usually a buyer with a new circuit idea was waiting in the wings to start production in the closed facility.

Price dropping was accelerated by the development of a plastic package for silicon devices in 1963. Also in that year, RCA announced the development of the insulated field-effect transistor (IFET), which paved the way for the MOS industry. RCA also pioneered the first complementary MOS (CMOS) circuits.

During this period, the name of the marketing game was to design a new chip and ride the sales curve up before the competition could bring out a competing design. At the same time, in the fabrication areas, the process engineers labored to keep up with higher yield and shrinking die-size and volume requirements.

In the 1970s, the semiconductor business moved from a laboratory-type of small batch processors to those of higher volume. The volume-related dependency of manufacturing costs and corporate profit was established and the industry was ready for the third stage, production.

The production decade (1971–1980)

At the start of the 1970s the industry was manufacturing primarily integrated circuits at the medium-scale integration (MSI) level. The move to profitable, high-yield large-scale integration (LSI) devices was being somewhat hampered by mask-caused defects and the damage inflicted on the wafers by the contact aligners. In fact, all of the existing processes in one way or another represented a barrier to the volume production of higher-level circuits.

The mask and aligner defect problem was solved with the development of the first practical projection aligner by the Perkin and Elmer Company. The decade also saw the improvement of clean-room construction and operation and the introduction of ion implantation machines and of electron beam machines for high-quality mask generation. Continuous spin and bake and develop and bake systems came into universal use and the mask steppers used in mask making began to show up in fabrication areas for wafer imaging.

The equipment during this decade changed from manual to automatic operation with on-board, solid-state controllers and sequencers. Button-to-button operation became the guiding principle of equipment suppliers. The move from operator control to automatic control of the processes increased both wafer throughput and uniformity. Reliance on equipment-based processes drove that segment of the industry to new heights, to a level of some $10 billion a year in sales.

The automation decade (1981–1990)

All of the marketplace pressures and trends in semiconductor processing led to higher levels of automation in the fabrication process. The work done in the 1970s resulted in automatic machines that, for the most part, would load the wafers, process them through predetermined steps, and return them to transfer carriers with the push of a button.

These advances are needed for the production of advanced chips. A characteristic of the industry is a wide distribution of chip sophistication. While the leading-edge chips have feature sizes at the submicron level, a surprising number of "older" circuits still have significant market life. These older circuits do not require the sophistication and process control of submicron, ULSI circuits, and their lower profit margins cannot support the major expense of fabrication area upgrades. The 1980s opened with a wide variety of processing modes, beginning with the manual processes of the early 1970s to the most sophisticated, almost operatorless free, fabrication areas.

With most of the individual processes individually automated, the focus in the 1980s has been on elimination of operators from the fabrication areas and automation of material delivery and movement. The elimination of operators is required by the fact that humans are one of the major sources of contamination in the process. Even elaborate gowning procedures cannot control particulate contamination generated by people moving about the wafers. (These issues are examined in more detail in Chap. 4.) The automation of the individual processes has challenged the industry to develop methods of transferring the wafers between the various machines, an aspect that is the focus of the automation decade, with the fabled "lights out" fabrication area being the goal.

Most advanced fabrication areas have some sort of computer-integrated manufacturing (CIM) system in place. These systems control the movement of the thousands of wafers through the processes and keep track of the accounting and production control aspects of the business. Computer-aided manufacturing (CAM), the control of machine parameters from a central computer, is also making inroads into wafer-fabrication facilities.

The Japanese semiconductor industries' decisions to make chips required in vast quantities, such as the DRAMs, dynamic random access memory, has led them to take some leads in the area of automation. Many industry forecasters feel that automation is the key to dominance in the worldwide semiconductor industry.

After the control of human-generated contamination and material delivery, automation will focus on closed-loop control of the processes. Currently the majority of the machines process the wafers by predetermined processes. Some of the steps have feedback loops, such as the photomasking alignment and exposure machines. Light integrators in the machines monitor and control the light sources until the required amount of energy has struck the wafer. The control circuits take into account minor changes in the light systems to deliver a constant result.

This type of closed-loop control is not generally available for other critical processes. The reasons are varied. In some cases, such as oxidation, it is difficult to design a detector that can operate in an extremely high-temperature, oxidizing atmosphere to measure a growing silicon dioxide film. In other processes, such as ion implantation, it is very difficult to measure the accumulation of implanted ions under the wafer surface as the process is occurring.

Despite these problems, the truly automated fabrication area will have to include closed-loop processing, along with the other automation techniques to achieve the dream of a "lights out" fabrication area.

Key Concepts and Terms

Capacitor	Chip size
Diode	Discrete device
Feature size	Four stages of semiconductor
Hybrid circuits	Integrated circuit
Integration scale and circuit density	Planar processing
Price erosion	Resistor manufacturing

Semiconducting material   Solid-state devices

Transistor   Wafer

Review Questions

1. List the four types of discrete devices.
2. Describe the advantages of solid-state devices over vacuum tubes.
3. An MSI circuit has more components than a ULSI circuit (**true or false**).
4. Describe the difference between a hybrid and an integrated circuit.
5. State the stage of processing that uses wafers.
6. State the stage of processing that uses chips.
7. List two other terms used to describe chips.
8. Describe what is meant by the term "feature size."
9. Name two other terms used for "feature size."
10. Describe the functions of a semiconductor package.

2

Semiconductor Materials and Process Chemicals

Overview

The unique functions of semiconductor devices and circuits are the direct result of the electrical and physical properties of semiconducting materials. In the first part of this chapter, the basics of atoms, electrical classification of solids, and intrinsic and doped semiconductors are explained. In the second part, the basics of chemistry as they apply to process chemicals are detailed.

Objectives

Upon completion of this chapter, you should be able to:

1. Identify the parts of an atom.
2. Name the two unique properties of a doped semiconductor.
3. List at least three semiconducting materials.
4. Explain the advantages and disadvantages of gallium arsenide as compared to silicon.
5. Explain the difference in composition and electrical functioning of N-type and P-type semiconducting materials.
6. Describe the properties of resistivity and resistance.
7. Identify the differences between acids, alkalies, and solvents.
8. List the four states of nature.
9. Give the definition of an atom, molecule, and ion.
10. Explain four or more basic chemical-handling safety rules.

Atomic Structure

The Bohr atom

The understanding of semiconductor materials requires a basic knowledge of atomic structure.

Atoms are the building block of the physical universe. Everything in the universe is made from only 96 stable materials (and 12 unstable ones) known as *elements*. Each element has a *different* atomic structure. The different structures give rise to the different properties of the elements.

The concept of an atom is arrived at by a mental process of division. It goes like this. Consider a small piece of gold, which is one of the stable elements. Gold has a list of well-known properties (good conductor, malleable, corrosion resistant). Now start dividing that piece of gold into smaller and smaller pieces. At some point you will have a final small piece that still retains the properties of gold. This final small piece is the atom of gold.

Further subdivision of the atom will result in three subatomic particles. They are protons, neutrons, and electrons. Each of these subatomic particles has its own properties. It is only in a particular combination and structure that they form the element gold. The structure of the atom most used to understand physical, chemical, and electrical differences between different elements was first proposed by the famous physicist, Niels Bohr (Fig. 2.1).

The Bohr atom model has the positively charged protons and neutral neutrons located together in the nucleus of the atom. The negatively charged electrons move in defined orbits about the nucleus, similar to the movement of the planets about the sun.

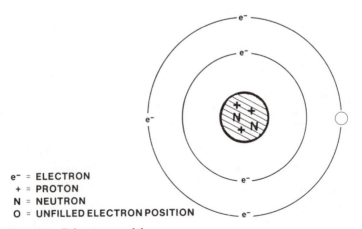

e⁻ = ELECTRON
+ = PROTON
N = NEUTRON
O = UNFILLED ELECTRON POSITION

Figure 2.1   Bohr atom model.

Each orbit has a maximum number of positions available for electrons. In some atoms not all of the positions are filled, leaving a "hole" in the structure. When a particular electron orbit is filled to the maximum, additional electrons must go into the next outer orbit.

The periodic table of the elements

The elements differ from each other in the number of electrons, protons, and neutrons in their atoms. Fortunately, nature combined the subatomic particles in an orderly fashion. An examination of some of the rules governing atomic structure is helpful in understanding the properties of semiconducting materials and process chemicals. Atoms (and therefore the elements) range from the simplest, hydrogen, to the most complicated one, lawrencium.

Hydrogen consists of only one proton in the nucleus and only one electron. This arrangement illustrates the first two rules of atomic structure.

1. In each atom of an element there is an equal amount of protons and electrons.

2. Each atom of an element contains a specific number of protons and different elements have a different number of protons. For example, a hydrogen atom has one proton in its nucleus while an oxygen atom has eight.

These facts lead to the assignment of a number to each of the elements. Known as the *atomic number*, it is equal to the number of protons (and therefore electrons) in the atom. The basic reference to the elements is the *periodic table* (Fig. 2.2). The periodic table has a box for each of the elements (identified by two letters). The atomic number is in the upper left-hand corner of the box. Thus calcium (Ca) has the atomic number 20, and we know immediately that calcium has 20 protons in its nucleus and 20 electrons in its orbital system. (Neutrons are electrically neutral particles that, along with the protons, make up the *mass* of the nucleus.)

Figure 2.3 shows the atomic structure of elements 1, hydrogen, 3, lithium, and 11, sodium. When constructing the diagrams, several rules were observed in the placement of the electrons in their proper orbits. The rule is that each orbit n (n = orbit number starting from the nucleus) can hold $2n^2$ electrons. Solution of the math for orbit 1 dictates that the first electron orbit can hold only two electrons. This rule forces the third electron of lithium into the second ring. The rule limits the number of electrons in the second ring to a maximum of 8 and that of the third ring to 16. So, for example, when constructing

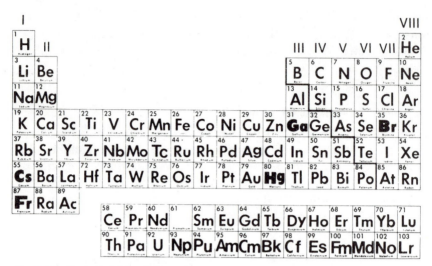

Figure 2.2   Periodic table of elements.

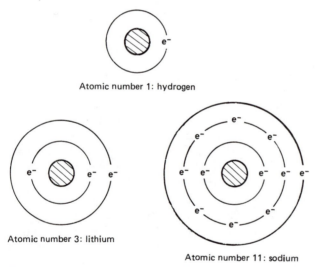

Atomic number 1: hydrogen

Atomic number 3: lithium

Atomic number 11: sodium

Figure 2.3   Atomic structures of hydrogen, lithium, and sodium.

the diagram of the sodium atom, with 11 protons and electrons, the
first two orbits take up 10 electrons, leaving the eleventh in the third
ring.

These three different atoms, hydrogen, lithium, and sodium, have a
commonality. Each has an outer ring with only one electron in it. This
illustrates another observable fact of elements.

3. Elements with the same number of outer-orbit electrons have similar properties. This rule is reflected in the periodic table. Note that hydrogen, lithium, and sodium appear in the table in a vertical column labeled I. This column number represents the number of electrons in the outer ring, and all the elements in each column share similar properties. It is no accident that the three best electrical conductors (copper, silver, and gold) all appear in the same column, Ib (Fig. 2.4).

Two more rules of atomic structure are relevant to the understanding of semiconductors.

4. Elements are stable with a filled outer ring or with eight electrons in the outer ring.

5. Atoms seek to combine with other atoms to create the stable condition of full orbits or eight electrons in their outer ring through *sharing*.

Rules 4 and 5 influence the creation of N-type and P-type semiconductor materials as explained in the section on doped semiconductors.

Electrical Conduction

Conductors

An important property of many materials is their ability to conduct electricity or support an electrical current flow. An *electrical current* is simply a flow of electrons. Electrical conduction takes place in elements and materials where the attractive hold of the protons in the nucleus of the atom on the electrons is relatively weak. In such a material the electrons can be easily moved. This condition exists in most metals.

The property of materials that conducts electricity is measured by a factor known as *conductivity*. The higher the conductivity the better

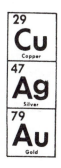

Figure 2.4 The three best electrical conductors.

the conductor. Conducting ability is also measured by the reciprocal of the conductivity, which is *resistivity*. The lower the resistivity of a material, the better the conducting ability.

$$C = 1/\rho$$

where C = conductivity and ρ = resistivity in ohms-centimeter ($\Omega \cdot cm$)

Dielectrics

At the opposite end of the conductivity scale are materials that exhibit a large force attraction between the nucleus and the orbiting electrons. The net effect in this type of material is a great deal of resistance to the movement of electrons. These materials are known as *dielectrics*. They have low conductivity and high resistivity. In electrical circuits and products, dielectric materials such as glass are used as insulators.

An electrical factor related to the degree of conductivity of a material is the *resistance* of a specific piece of the material. The resistance is a factor of the resistivity and dimensions of the material. Resistance to electrical flow is measured in ohms as illustrated in Fig. 2.5.

The formula defines the electrical resistance of a specific volume of a specific material (in this illustration the volume is a rectangular bar with dimensions D, W, and L). The relationship is analogous to density and weight, *density* being a material property and *weight* the force exerted by a specific volume of the material.

Intrinsic semiconductors

Semiconducting materials, as the name implies, are materials that have some electrical conducting ability. There are two elemental

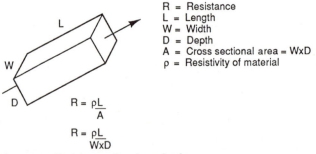

R = Resistance
L = Length
W = Width
D = Depth
A = Cross sectional area = WxD
ρ = Resistivity of material

$$R = \rho\frac{L}{A}$$

$$R = \rho\frac{L}{WxD}$$

Figure 2.5 Resistance of rectangular bar.

semiconductors, silicon and germanium, both found in column IV (Fig. 2.6) of the periodic table. In addition, there are some tens of material compounds (a *compound* is a material containing two or more chemically bound elements) that also exhibit semiconducting properties. These compounds come from elements found in columns III and V, such as gallium arsenide and gallium phosphide. Others are compounds from elements from columns II and VI. The term *intrinsic* refers to these materials in their purified state and not contaminated with impurities or other substances.

Doped semiconductors

Semiconducting materials, in their intrinsic state, are not useful in solid-state devices. However, through a process called doping, specific elements can be introduced into intrinsic semiconductor materials. Using the proper elements, intrinsic semiconductor materials display two unique properties that *are* the basis of solid-state electronics. The two properties are

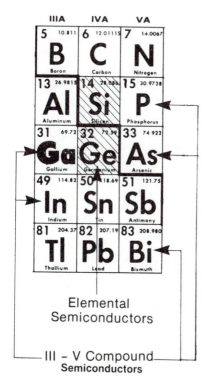

Elemental
Semiconductors

III – V Compound
Semiconductors

Figure 2.6 Semiconductor materials.

1. Precise resistivity control through doping
2. Electron and hole conduction

Resistivity of doped semiconductors

Metals have a conductivity range limited to 10^4 to 10^6 $\Omega \cdot$ cm The implications of this range are illustrated by an examination of the resistor represented in Fig. 2.5. Given a specific metal with a specific resistivity, the only way to change the resistance of a given volume is to change the dimensions. If, on the other hand, we had a material in which we could change the resistivity, we would have more degrees of freedom in the design of the resistor.

Semiconductors are such a material. Their resistivity can be extended over the range of 10^{-3} to 10^3 $\Omega \cdot$ cm by the addition of dopant atoms. Semiconducting materials can be doped into a useful resistivity range by elements that make the material either electron-rich (N type) or hole-rich (P type).

Figure 2.7 shows the relationship of the doping level to the resistivity of silicon. The x axis is labeled as the *carrier concentration* because the electrons or holes in the material are called *carriers*. Note that there are two curves: N type and P type. That is due to the different amounts of energy required to get either an electron or a hole to move and thus carry a current in the material. As the curves indicate, it takes less of a concentration of N-type dopants than P-type dopants to create a given resistivity in silicon. Another way to express this phenomenon is that it takes less energy to move an electron than to move a hole.

In terms of percentage, it takes only from 0.000001 to 0.1% of a dopant to bring a semiconductor material into a useful resistivity range. This property of semiconductors allows the creation of regions of very precise resistivity values in the material.

Electron and hole conduction

Another limit of a metal conductor is that it conducts electricity only through the movement of electrons. Metals are permanently N type. Semiconductors can be made either N type or P type by doping with specific dopant elements. N- and P-type semiconductors can conduct electricity by either electrons or holes. Before examining the conduction mechanism, it is instructive to examine the creation of free (or extra) electrons or holes in a semiconductor solid.

To illustrate the free electron (N type), consider a piece of silicon (Si) doped with a very small amount of arsenic (As) as shown in Fig. 2.8. Assuming even sharing, each of the arsenic atoms would be sur-

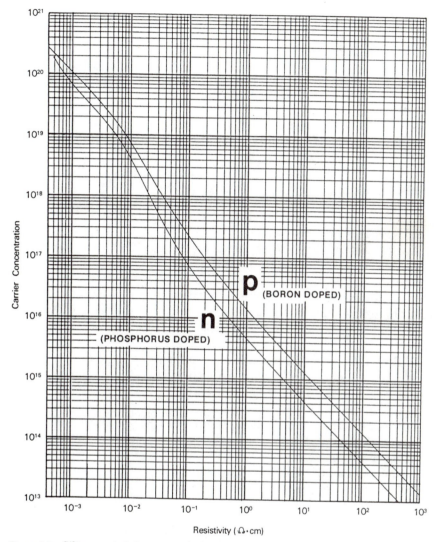

Figure 2.7 Silicon resistivity versus doping (carrier) concentration. *(After Thurber et al., Natl. Bur. Standards Spec. Publ. 400–64, May 1981, tables 10 and 14.)*

rounded by silicon atoms. Applying the rule from page 25 that atoms attempt to stabilize by having eight electrons in their outer ring, the As atom is shown sharing four electrons from its neighboring Si atom. However, arsenic is from column V, which means it has five electrons in its outer ring. The net result is that four of them pair up with electrons from the silicon atoms, leaving one left over. This one electron is available for electrical conduction.

Considering that a crystal of silicon has millions of atoms per cubic

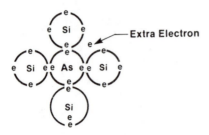

Figure 2.8 N-type doping of silicon with arsenic.

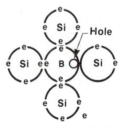

Figure 2.9 P-type doping of silicon with boron.

centimeter, there are lots of electrons available to conduct an electrical current. For silicon, the elements arsenic, phosphorus, and antimony create N-type conditions.

An understanding of P-type material is approached in the same manner (Fig. 2.9). The difference is that only boron, from column III of the periodic table, is used to make silicon P type. When mixed into the silicon, it, too, borrows electrons from silicon atoms. However, having only three outer electrons, there is a place in the outer ring that is not filled by an electron. It is this unfilled position that is defined as a *hole*.

Within a doped semiconductor material, there is a great deal of activity. Holes and electrons are constantly being created. The electrons are attracted to the unfilled holes, in turn, leaving an unfilled position that creates another hole.

How the electrons contribute to the conduction of electricity is illustrated in Fig. 2.10. When a voltage is applied across a piece of conducting or semiconducting material (such as from a battery), the electrons move in a direction set up by the polarity of the voltage source. In other words, they move toward the positive pole of the source.

In P-type material (Fig. 2.11), an electron will move toward the positive pole by jumping into a hole along the direction of route t_1. Of course, when it leaves its position, it leaves a new hole. As it proceeds toward the positive pole, it creates a succession of holes. The impression to someone measuring this process with a current meter is that the material is supporting a positive current, when actually it is a negative current moving in the opposite direction. This phenomenon is called *hole flow* and is unique to semiconducting material.

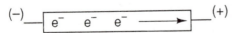

Figure 2.10 Electron conduction in N-type semiconductor material.

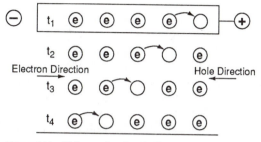

Figure 2.11 Hole conduction in P-type semiconductor material.

Classification	Electrons	Examples	Conductivity
1. Conductor	Free to Move	Gold Copper Silver	$10^4 - 10^6$ /ohm-cm
2. Insulator (Dielectric)	Bound	Glass Plastic	$10^{-22} - 10^{-10}$ /ohm-cm
3. Semiconductor a. Intrinsic	Some Available	Germanium Silicon III-IV	$10^{-9} - 10^3$ /ohm-cm
b. Doped	Controlled Amount Available	N-type Semiconductor P-type Semiconductor	

Figure 2.12 Electrical classification of materials.

The dopants that create a P-type conductivity in a semiconductor material are called *acceptors*. Dopants that create N-type conditions are called *donors*. An easy way to keep these terms straight is that acceptor has a *p* and donor is spelled with an *n*.

The electrical characteristics of conductors, insulators, and semiconductors are summarized in Fig. 2.12. The particular characteristics of doped semiconductors are summarized in Fig. 2.13.

	N-TYPE	P-TYPE
1. Conduction	Electrons	Holes
2. Polarity	Negative	Positive
3. Dopant Term	Donor	Acceptor
4. Doping Elements in Silicon	Arsenic Phosphorus Antimony	Boron

Figure 2.13 Characteristics of doped semiconductors.

Carrier mobility

It was mentioned earlier that it takes "less energy" to move an electron than to move a hole through a piece of semiconducting material. In a circuit we are interested in both the energy required to move these carriers (holes and electrons) and the speed at which they move. The speed of movement is called the *carrier mobility*, with holes having a lower mobility than electrons. This factor is an important consideration in selecting a particular semiconducting material for a circuit.

Semiconducting Production Materials

Germanium and silicon

Germanium and silicon are the two elemental semiconductors. The first transistor was made with germanium, as were the initial devices of the solid-state era. However, germanium presents problems in processing and in device performance. Its 937°C melting point limits high-temperature processing. More importantly, its lack of a naturally occurring oxide leaves the surface prone to electrical leakage.

The development of silicon–silicon dioxide planar processing solved the leakage problem of integrated circuits, flattened the surface profile of the circuits, and allowed higher-temperature processing due to its 1415°C melting point. Consequently, silicon represents over 90 percent of the wafers processed worldwide.

Compound semiconducting materials

There are many semiconducting compounds formed from elements listed in columns III and V and II and V of the periodic table. Of these compounds, the ones most used in commercial semiconductor devices are gallium arsenide (GaAs) and gallium arsenide phosphide (GaAsP). These compounds have special properties.[1] Unlike silicon or germanium, diodes made from GaAs and GaAsP give off visible and laser light when activated with an electrical current. These compounds make the light-emitting diodes (LEDs) used in so many electronic panel displays.

Another important property of gallium arsenide is its high carrier mobility. This property allows a gallium arsenide device to react to high-frequency microwaves and effectively switch them into electrical currents in communications systems. In a silicon device, the carriers move more slowly. This means that when stimulated by a microwave, a silicon device cannot react to a wave fast enough before the next wave impinges on the device.

This same property, carrier mobility, is the basis for the excitement over gallium arsenide transistors and integrated circuits. Devices of GaAs operate two to three times faster than comparable silicon devices and are the devices of choice for superfast computers and real-time control circuits such as airplane controls.

GaAs also offers two other significant advantages over silicon. First is its natural resistance to radiation-caused leakages. This phenomenon is common to all semiconductors. In the presence of radiation, such as found in space, holes and electrons are formed. These electrons and holes give rise to unwanted currents that can cause the device or circuit to malfunction or cease functioning. Devices that can perform in a radiation environment are known as *radiation hardened*. GaAs is naturally radiation hardened.

Also, GaAs is semi-insulating. In an integrated circuit this property minimizes leakage between adjacent devices, allowing a higher packing density, which, in turn, results in a faster circuit since the holes and electrons travel shorter distances. In silicon circuits, special isolating structures must be built into the surface to control surface leakage. These structures take up valuable space and reduce the density of the circuit.

Despite all the advantages, GaAs is not expected to replace silicon as the mainstream semiconducting material. The reasons reside in the trade-offs between performance and processing difficulty. While GaAs circuits are very fast, the majority of electronic products do not require their level of speed. On the performance side, GaAs, like germanium, does not possess a natural oxide. To compensate for the lack of a naturally grown oxide, layers of dielectrics must be deposited on the GaAs, which leads to longer processing and lower yields. Also, half of the atoms in GaAs are arsenic, an element very dangerous to human beings. Unfortunately, the arsenic evaporates from the compound at normal process temperatures, requiring the addition of suppression layers (caps) or pressurized process chambers. These steps lengthen the processing and add to its cost.

This same phenomenon (arsenic evaporation) goes on during the crystal-growing stage, resulting in nonuniform crystals and wafers. The nonuniformity produces wafers that are very prone to breakage during fabrication processing. Most silicon fabrication equipment cannot be used to process GaAs wafers without intolerably high breakage levels. Despite the problems, gallium arsenide is an important semiconducting material, will continue to increase in use, and will probably have a major influence on computer performance in the future.

The answer to the crystal-growing problem is a liquid-encapsulated,

	Ge	Si	GaAs	SiO$_2$
Atomic Weight	72.6	28.09	144.63	60.08
Atoms/cm^3 or Molecules	4.42×10^{22}	5.00×10^{22}	2.21×10^{22}	2.3×10^{22}
Crystal Structure	Diamond	Diamond	Zinc-Blends	Amorphous
Atoms/Unit Cell	8	8	8	—
Density	5.32	2.33	5.65	2.27
Energy Gap	0.67	1.11	1.40	8 (approx.)
Dielectric Constant	16.3	11.7	12.0	3.9
Melting Point (°C)	937°	1415°	1238°	1700° (approx.)
Breakdown Field (V/)	8 (approx.)	30 (approx.)	35 (approx.)	600 (approx.)
Linear Coefficient of Thermal Expansion $\dfrac{\Delta L}{LT} \quad \dfrac{1}{C}$	5.8×10^{-6}	2.5×10^{-6}	5.9×10^{-6}	0.5×10^{-6}

Figure 2.14   Physical properties of semiconductor materials.

crystal-growing process that is more expensive and not capable of growing crystals as large as the simpler silicon process (see Chap. 3). A comparison of the major semiconducting production materials and silicon dioxide is presented in Fig. 2.14.

Process Chemicals

It should be fairly obvious that a lot of processing is required to change raw semiconducting materials into useful devices. The majority of these processes use chemicals. In fact, microchip fabrication is primarily a chemical process, or more correctly, a series of chemical processes.

Great quantities of acids, bases, solvents, and water are consumed by a semiconductor plant. For example, the U.S. semiconductor industry uses 15 percent of all domestically produced nitrogen. When the costs of producing a chip are added up, process chemicals account for up to 40 percent of all manufacturing costs. Part of this cost is due to the extremely high purities and special formulations required of the chemicals to allow precise and clean processing. (The cleanliness requirements for semiconductor process chemicals are explored in Chap. 4. Specific chemicals and their properties are detailed in the process chapters.)

Molecules, Compounds, and Mixtures

Earlier, the basic structure of matter was explained by use of the Bohr atomic model. This model was used to explain the structural differences of the elements that make up all the materials in the physical universe. But it is obvious that the universe contains more than 103 (the number of elements) types of matter. However, it is true that the great number of materials found in nature, and made by man, are all based on those 103 elements.

The basic unit of a compound material is the molecule. The basic unit of water, for example, that still retains the properties of water is a molecule composed of two hydrogen atoms and one oxygen atom (see Figure 2.15). The multiplicity of materials comes about from the ability of atoms to bond together to form molecules.

It is inconvenient to draw diagrams as in Fig. 2.15 every time we want to designate a molecule. The more common practice is to write the molecular formula. For water, it is the familiar H_2O. This formula tells us exactly the elements and their number in the material. Chemists use the more precise term *compound* in describing different combinations of elements. Thus H_2O (water), NaCl (sodium chloride or salt), H_2O_2 (hydrogen peroxide), and As_2O_3 (arsine) are all different compounds comprised of aggregates of individual molecules.

Some elements combine into diatomic molecules. A diatomic molecule is one composed of two atoms of the same element. The familiar process gases, oxygen, nitrogen, and hydrogen, in their natural state, are all composed of diatomic molecules. Thus their formulas are O_2, N_2, and H_2.

Materials also come in two other forms, mixtures and solutions. *Mixtures* are composed of two or more substances, but the substances retain their individual properties. A mixture of salt and pepper is the classic example.

Solutions are mixtures of a solid dissolved in a liquid or two or more liquids. In a liquid, the compounds are interspersed, with the liquid

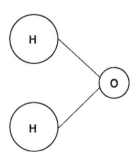

Figure 2.15   Diagram of water molecule.

having unique properties. However, the substances in a solution do not form into a new molecule. Saltwater is an example of a solution. It can be separated into its individual parts: salt and water.

Ions

The term ion or ionic is used often in connection with semiconductor processing. This term refers to any atom or molecule that exists in a material with an unbalanced charge. An ion is designated by the chemical symbol of the element or molecule followed by a superscripted positive or negative sign (Na^+, Cl^-). For example, one of the serious contamination problems is mobile ionic contamination such as sodium (Na^+). The problem comes from the positive charge carried by the sodium when it gets into the semiconductor material or device. In some processes, such as the ion implantation process, it is necessary to create an ion, such as boron (B^+), to accomplish the process.

States of Matter

Solids, liquids, gases

Matter is found in four different states. They are solids, liquids, gases, and plasma (Fig. 2.16).

- Solids are defined as having a definite shape and volume, which is retained under normal conditions of temperature and pressure.
- Liquids have definite volume but a variable volume. A liter of water will take the shape of any container it is put in.
- Gases have neither definite shape nor volume. They too will take the shape of any container, but unlike liquids, they will expand or can be compressed to entirely fill the container.

The state of a particular material has a lot to do with its pressure and temperature. Temperature is a measure of the total energy incorporated in the material. We know that water can exist in all three states (ice, liquid water, steam or water vapor) simply by changing the

- Solid
- Liquid
- Gas
- Plasma

Figure 2.16   Four states of nature

temperature and/or pressure. The influence of pressure is more complicated and beyond the scope of this text.

Plasma state

The fourth state of nature is plasma. A star is an example of a plasma state. It certainly does not meet the definitions of a solid, liquid, or gas. A plasma state is defined as a high-energy collection of ionized atoms or molecules. Plasma states can be induced in process gases by the application of high-energy, radio-frequency (RF) fields. They are used in semiconductor technology to cause chemical reactions in gas mixtures. One of their advantages is that the energy can be delivered at a lower temperature than in conventional thermal systems, such as convection heating in ovens.

Properties of Matter

All materials can be differentiated by their chemical compositions and the properties that arise from those compositions. In this section, several key properties required to understand and work with process chemicals are defined.

Temperature

The temperature of a chemical exerts great influence on its reactions with other chemicals, whether in an oxidation tube or in a plasma etcher. Additionally, safe use of some chemicals requires knowledge and control of their temperatures. Three temperature scales are used to express the temperature of a material. They are the Fahrenheit, centigrade (or Celsius), and Kelvin scales (Fig. 2.17).

The Fahrenheit scale was developed by Gabriel Fahrenheit, a German physicist, using a water and salt solution. He assigned to the solution's freezing temperature the value of zero degrees Fahrenheit (°F). Unfortunately the freezing temperature of pure water is more useful, and we have ended up with the Fahrenheit scale having a water freezing point at 32°F and a boiling point of 212°F, with 180 degrees between the two points.

The Celsius, or centigrade, scale is more popular in scientific endeavors. It more sensibly sets the freezing point of water at 0°C and boiling at 100°C. Note that there are exactly 100 centigrade degrees between the two points. This means that a one-degree change in temperature as measured on the centigrade scale requires more energy than a one-degree change on the Fahrenheit scale.

The third temperature scale is the Kelvin scale. It uses the same

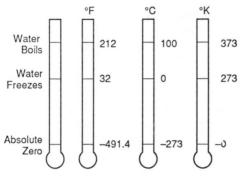

Figure 2.17   Temperature scales.

scale factor as the centigrade scale but is based on absolute zero. Absolute zero is the theoretical temperature at which all atomic motion would cease. This value corresponds to −273°C. On the Kelvin scale water freezes at 273° K and boils at 373° K.

Density, specific gravity, and vapor density

An important property of matter is density. When we say that something is *dense*, we refer to its mass or weight per unit volume. A cork has a lower density than an equal volume of iron. Density is expressed as the weight, in grams, per cubic centimeter of the material. Water is the standard, with 1 cm^3 (at 4°C) weighing 1 g. The densities of other substances are expressed as a ratio of their density to that of a comparable volume of water. Silicon has a density of 2.3. Therefore, a piece of silicon 1 cm^3 in volume will weigh 2.3 g.

Specific gravity is a term used to reference the density of liquids and gases. Again it is a number which is the ratio of the density of the substance to that of water. Gasoline has a specific gravity of 0.75, which means it is 75 percent as dense as water.

Vapor density is a density measurement of gases under certain conditions of temperature and pressure. The reference is air, with 1 cm^3 having an assigned density of 1 g. Hydrogen has a vapor density of 0.60, which makes it 60 percent of the density of a similar volume of air. A container of hydrogen, with the same mass as a container of air, will weigh 60 percent less than a similar container of air.

Pressure and vacuum

Another important aspect of matter is pressure. Pressure, as a property, is usually used in connection with liquids and gases. It is defined as the force per unit area exerted against the surface of the container.

It is the gas pressure in a cylinder that forces the gas out into a process chamber. All process machines using gases must have gauges to measure and control the pressure.

Pressures are expressed in pounds per square inch of area (psia), in atmospheres, or in torrs. An atmosphere (atm) is the pressure exerted by the atmosphere surrounding the earth at a specific temperature. Thus a high-pressure oxidation system operated at 5 atm contains a pressure 5 times that of the atmosphere.

One atmosphere of air has a pressure of 14.7 psia. Pressures inside gas tanks are measured in psig or pounds per square inch gauge. This means that the gauge reading is absolute; it does not include the pressure of the outside atmosphere.

Vacuum is also a term and condition encountered in semiconducting processing. Vacuum is actually a condition of low pressure. Generally pressures below standard atmospheric pressures are referred to as vacuum. But a vacuum condition is measured in units of pressure.

Low pressures tend to be expressed in torrs. The unit is named after the Italian scientist, Torricelli, who made many of the early discoveries in the field of gases and their properties. A torr is defined as the equivalent of one millimeter of mercury in a pressure-measuring device known as a *manometer*.

Imagine the effect on the column of mercury in the manometer in Fig. 2.18a of increasing the pressure above atmospheric pressure. As the pressure goes up, it pushes down the mercury in the dish and raises the mercury in the column. Now imagine what happens as air is extracted from the system (Fig. 2.18b) below atmospheric pressure, creating a vacuum. As long as there are any gas molecules or atoms in the manometer, some small pressure will be exerted on the mercury in the dish, and the mercury in the column will rise some small but finite amount. The amount of the rise as measured in millimeters (mm) is relative to the pressure, or in this case the vacuum.

Vacuum systems for evaporation, sputtering, and ion implantation are operated at vacuums (pressures) of 10^{-6} to 10^{-9} torrs. Trans-

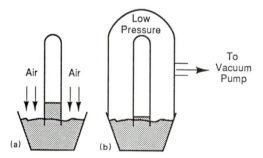

Figure 2.18 Pressure vacuum measurement.

lated into a vacuum system containing a simple manometer, this means that the column of mercury would rise only 0.000000001 (1 × 10^{-9}) to 0.000001 (1 × 10^{-6}) mm, a *very* small length! In actual practice, a mercury manometer cannot measure these extremely low pressures. Other, more sensitive gauges are used.

Acids, Alkalies, and Solvents

Acids and alkalies

Semiconductor processing requires large amounts of liquid chemicals to etch, clean, and rinse the wafers and packages. For example, a medium-size photomasking area may use up to 30,000 gallons of sulfuric acid per year. These chemicals are identified by chemists into three major classifications.[2]

- Acids
- Alkalies
- Solvents

Acids and alkalies differ from each other and each reacts chemically with other materials because of the presence of specific ions in the liquid. Acids contain *hydrogen ions* while alkalies (also called bases) contain *hydroxyl ions*. An examination of the water molecule explains the differences.

The chemical formula for water normally is written as H_2O. It can also be written in the form HOH. When separated into its parts, we find that water is made up of a positively charged hydrogen ion (H^+) and a negatively charged hydroxyl ion (OH^-). When water is mixed with other elements, either the hydrogen or hydroxyl ion combines with other substances (Fig. 2.19). Liquids that contain the hydrogen ion are called *acids*. Liquids that contain the hydroxyl ion are called *alkalies* or *bases*. Acids and bases are commonly found in the home;

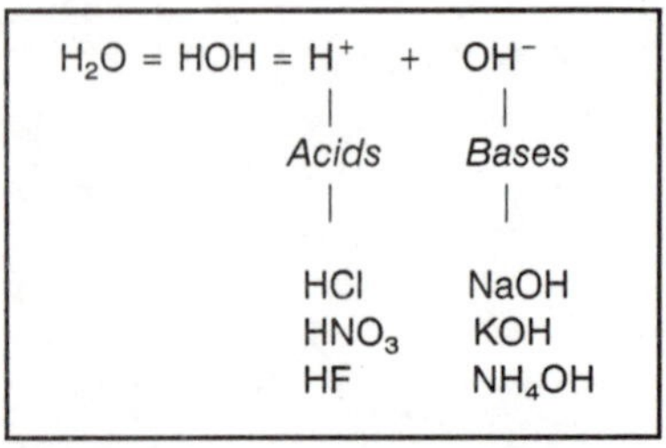

```
H₂O = HOH = H⁺   +   OH⁻
                |        |
             Acids    Bases
                |        |
              HCl      NaOH
              HNO₃     KOH
              HF       NH₄OH
```

Figure 2.19   Acid and base solutions.

lemon juice and vinegar are acids, and ammonia and baking soda in a solution with water are bases.

Acids are further divided into two categories, organic and inorganic. Organic acids are those that contain hydrocarbons while inorganic do not. Sulfonic acid is an organic acid, and hydrofluoric acid (HF) is an inorganic acid.

The strength and reactivity of acids and bases are measured by the pH scale (Fig. 2.20). This scale ranges from 0 to 14, with 7 being a neutral point. Water is neutral, neither an acid nor a base; therefore, it has a pH of 7. Strong acids, such as sulfuric acid (H_2SO_4), will have low pHs of 0 to 3. Strong bases, such as sodium hydroxide (NaOH), have high pH values greater than 7. Both acids and bases are reactive with skin and other chemicals and should be stored and handled with all the prescribed safety precautions.

Solvents

Solvents are liquids that do not ionize; they are neutral on the pH scale. Water is a solvent; in fact, it is the solvent with the greatest ability to dissolve other substances. Alcohol and acetone are commonly used solvents in the wafer-fabrication process. Most of the solvents in fabrication processing are volatile, flammable, or combustible. It is important to use them in properly exhausted stations and observe prescribed precautions in their storage and use.

The material safety data sheet

For every chemical brought into a manufacturing site, the supplier must provide a form called the material safety data sheet (MSDS).

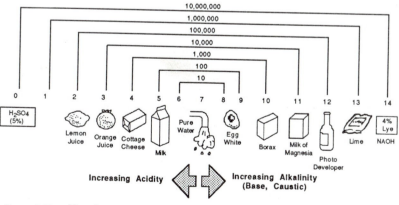

Figure 2.20   pH scale.

This form contains information on the chemical, its storage, related health concerns and first aid, and use information about the chemical. Under current regulations, the MSDSs must be filed on the site and be available to employees.

Key Concepts and Terms

Acid, alkalies, solvents

Bohr atom model

Carrier mobility

Conductivity and resistivity

Conductor

Dielectric

Dopant

Doped semiconductor

Electron, proton, neutron

Element

Gallium

Hole

Intrinsic semiconductor

Ion

MSDS

Molecule and compound

N type and P type

Periodic table of the elements

pH scale

Pressure and vacuum

Silicon and germanium

Solid, liquid, gas

Temperature scales

III-V compounds

Review Questions

1. Describe the electrical difference between a conductor, a dielectric, and a semiconductor.
2. Why are doped semiconductors required for solid-state devices?
3. Which has a higher resistivity, a metal or an intrinsic semiconductor?
4. Give two reasons why silicon is the most common semiconducting material.
5. Name one element that will make silicon N type.
6. A P-type semiconductor exhibits a _____ (negative or positive) current.
7. Indicate which classification (acid, base, solvent) is described below.
   a. Contains OH^- ions
   b. Contains H^+ ions
   c. Is neutral on the pH scale

8. Acids have a pH between _____ and _____.
9. Bases have a pH between _____ and _____.
10. What is the pH of water?

References

1. R. E. Williams, *Gallium Arsenide Processing Techniques*, Artech House, Inc., Dedham, Mass., 1984.
2. P. Van Zant, *Safety First Manual*, Semiconductor Services, Redwood City, Calif., 1984, Module 4.

Crystal Growth and Wafer Preparation

Overview

In this chapter, the conversion of the intrinsic semiconducting materials into raw wafers and the steps required to produce fabrication-grade polished wafers are explained.

Objectives

Upon completion of this chapter you should be able to:

1. Explain the difference between crystalline and noncrystalline materials.
2. Describe the conditions that characterize a polycrystalline and a single-crystal material.
3. Draw a diagram of the two major wafer crystal orientations used in semiconductor processing.
4. Explain the Czochralski, the float-zone, and the liquid crystal-encapsulated Czochralski methods of crystal growing.
5. Draw a flow diagram of the wafer-preparation process.
6. Explain the use and meaning of the flats ground on wafers.
7. Describe the benefits in the wafer-fabrication process that come from edge-rounded wafers.
8. Describe the benefits in the wafer-fabrication process that come from flat and damage-free wafer surfaces.

Semiconductor Materials Preparation

The first stage of semiconductor manufacturing is the extraction and purification of the raw material(s) from the earth. The material must

$$2SiHCl_3 \text{ (gas)} + 3H_2 \text{ (gas)} \rightarrow 2Si \text{ (solid)} + 6HCl \text{ (gas)}$$

Figure 3.1   Hydrogen reduction of trichlorosilane.

meet electronic or semiconductor specifications, which generally means that impurities have been reduced to no more than several parts per million (ppm).

Actual purification involves a chemical reaction. For silicon, the starting point is the conversion of the ore to a silicon-bearing gas such as silicon tetrachloride or trichlorosilane. The gases are reacted with hydrogen (Fig. 3.1) to produce the semiconductor-grade silicon. The silicon produced has a crystal structure, which is a necessary material requirement for making semiconducting devices.

Crystalline Materials

Different materials differ from one another in many ways. One way is the organization of the atoms in the material. In some materials, such as silicon and germanium, the atoms occupy very definite positions relative to each other. These positions are repeated throughout the material. Such materials are called *crystals*. Materials without a definite arrangement of their atoms are called *amorphous*. Plastics are examples of amorphous materials.

Unit Cells

There are actually two levels of atomic organization possible for crystalline materials. First is the organization of the individual atoms. The atoms in a crystal arrange themselves at specific points in a structure known as a *unit cell*. The unit cell is the first level of organization in a crystal. The unit cell structure is found everywhere in the material.

Another term used to reference crystal structures is *lattice*. A crystalline material is said to have a specific lattice structure, and the atoms are located at specific points in the lattice structure.

The number of atoms, relative positions, and binding energies between the atoms in the unit cell give rise to many of the characteristics of the material. Each crystalline material has a unique unit cell. Silicon atoms have 16 atoms arranged into a diamond structure (Fig. 3.2). GaAs crystals have 18 atoms in a unit cell configuration called a *zinc blende* (Fig. 3.3).

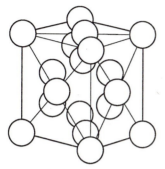

Figure 3.2   Unit cell of silicon.

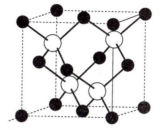

Figure 3.3   GaAs crystal structure.

Polycrystals and Single Crystals

The second level of organization within a crystal is related to the organization of the unit cells. In intrinsic semiconductors, the unit cells are *not* in a regular arrangement to each other. The situation is similar to a disorderly pile of sugar cubes, with each cube representing a unit cell. A material with such an arrangement has a polycrystal structure. The second level of organization occurs when the unit cells (sugar cubes) are all neatly and regularly arranged relative to each of the others (Fig. 3.4). Materials thus arranged have a single- (or mono-) crystal structure.

Single-crystal materials have more uniform and predictable properties than polycrystalline materials. The stringent demands of semiconductor device and circuit performance require the uniform properties of single-crystal structures.

Crystal Orientation

Besides the requirement of a single-crystal structure for a wafer, there is the requirement of a specific crystal orientation. This concept can be

Polycrystalline

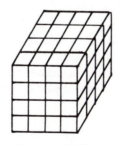

Single crystalline

Figure 3.4   Poly- and single-crystal structures.

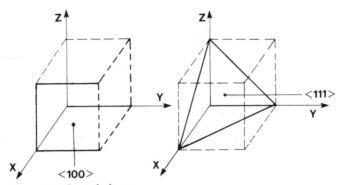

Figure 3.5   Crystal planes.

visualized by considering different planes of a unit cell. When a large single crystal is grown, these planes exist within the crystal (Fig. 3.5).

Wafers can be cut from the crystal at several different angles, with each angle representing a different plane. Each of these slices through the crystal would also cut through each of the unit cells, exposing different planes within the cell. Each plane is unique, differing in atom count and binding energies between the atoms. These differences result in different chemical, electrical, and physical properties associated with each plane. A specific crystal plane must be specified for the wafers.

The planes are identified by a series of three numbers known as *Miller indices*. Calculation of Miller indices is an exercise in geometry beyond the scope of this book. However, the reader should know that each set of Miller indices identifies the location of a specific plane in the crystal. Figure 3.5 shows two simple cubic unit cells nestled into the origin of an *XYZ* coordinate system.

The two most common orientations used for silicon wafers are the ⟨100⟩ and the ⟨111⟩ planes. The plane designation is verbalized as the one-oh-oh plane and the one-one-one plane. The brackets indicate that the three numbers are Miller indices.

⟨100⟩ oriented wafers are used for fabricating metal oxide semiconductor (MOS) devices and circuits, whereas ⟨111⟩ oriented wafers are used for bipolar devices and circuits. GaAs wafers are also cut along the ⟨100⟩ planes of the crystal. In actual practice, the large crystals are grown along the specific crystal planes, as explained in the following section on crystal growth.

Note that the ⟨100⟩ plane in Fig. 3.6 has a square shape, while the ⟨111⟩ plane is triangular in shape. These orientations are revealed when wafers are broken as shown in Fig. 3.6. The ⟨100⟩ wafers break

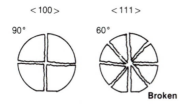

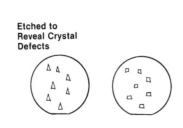

Figure 3.6   Wafer orientation in-dicators.

into quarters, or with right-angle (90°) breaks. The ⟨111⟩ wafers break into triangular pieces.

Crystal Growth

Semiconductor wafers are cut from large ingots of the semiconducting material. These ingots, called crystals, are grown from chunks of the intrinsic material that have a polycrystal structure and are undoped. The process of converting the polycrystal chunks to a large crystal of single-crystal structure, of the correct orientation, and containing the proper amount of dopant is called *crystal growing*. Three different methods are used to grow crystals, the Czochralski, the liquid crystal-encapsulated Czochralski, and the float-zone.

Czochralski (CZ) method

The majority of silicon crystals are grown by the CZ method (Fig. 3.7). The equipment consists of a quartz crucible heated by surrounding coils that carry radio-frequency (RF) waves. The crucible is loaded with polycrystalline chunks of the semiconductor material and small amounts of dopant. The dopant material is selected to create either an N-type or a P-type crystal. First the polycrystalline chunks and dopants are heated to the liquid state by the heating coils (Fig. 3.8). Next a seed crystal is positioned to just touch the surface of the liquid material (called the *melt*). The seed is a small crystal that has the same orientation as the desired final larger crystal. Seeds can be pro-

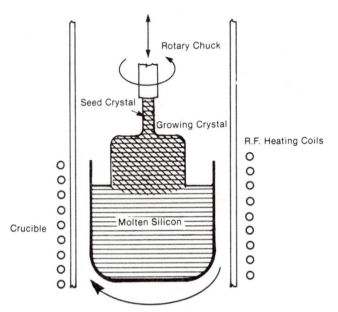

Figure 3.7 Czochralski crystal-growing system.

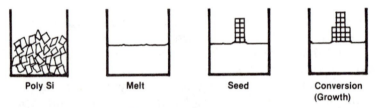

Figure 3.8 Crystal growth from a seed.

duced by chemical vapor techniques. In practice they are pieces of previously grown crystals and are reused.

Crystal growth starts as the seed is slowly raised above the melt. The surface tension between the seed and the melt causes a thin film of the melt to adhere to the seed and cool. During the cooling, the atoms in the melted semiconductor material orient themselves to the crystal structure of the seed. The net effect is that the crystal orientation of the seed is imparted into the growing crystal. The dopant atoms in the melt become incorporated into the growing crystal, creating an N-type or a P-type crystal.

To achieve doping uniformity, crystal perfection, and diameter control, the seed and crucible are rotated in opposite directions during the entire crystal-growing process. Process control requires a complicated feedback system, integrating the parameters of rotational speed, pull speeds, and melt temperature. The CZ method is capable of producing crystals several feet in length and with diameters up to 12 or more inches.

Liquid crystal-encapsulated Czochralski (LEC) method

LEC crystal growing[1] is used for the growing of gallium arsenide crystals. It is essentially the same as the standard CZ process but with a major modification for gallium arsenide. The modification is required because of the evaporative property of the arsenic in the melt. At the crystal-growing temperature, the gallium and arsenic react and the arsenic can evaporate, resulting in a nonuniform crystal.

Two solutions to the problem are available. One is to pressurize the crystal-growing chamber to suppress the evaporation of the arsenic. The other is the LEC process (Fig. 3.9). LEC uses a layer of boron trioxide (B_2O_3) floating on top of the melt to suppress the arsenic evaporation. In this method, a pressure of about 1 atm is required in the chamber.

Float-zone method

Float-zone crystal growth is one of the several processes explained in this text that were developed early in the history of the technology and are still used for special needs. A drawback to the CZ method is the inclusion of oxygen from the crucible into the crystal. For some devices, higher levels of oxygen are intolerable. For these special cases, the crystal might be grown by the float-zone technique, which produces a lower oxygen-content crystal.

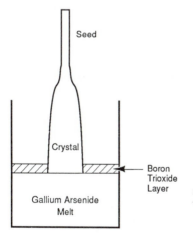

Figure 3.9 LEC system of crystal growth.

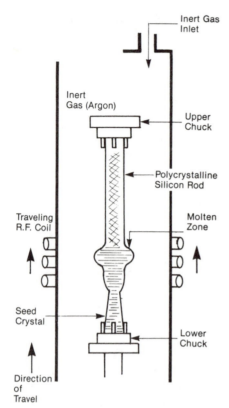

Figure 3.10 Float-zone crystal-growing system.

Float-zone crystal growth (Fig. 3.10) requires a bar of the polysilicon and dopants that has been cast in a mold. The seed is fused to one end of the bar and the assemblage placed in the crystal grower. Conversion of the bar to a single-crystal orientation starts when an RF coil heats the interface region of the bar and seed. The coil is then moved along the axis of the bar, heating it in small sections to the liquid point. Within the molten region, the atoms in the bars line up to the atoms of the seed. Thus the entire bar is converted to a single crystal with the orientation of the starting seed.

Generally, float-zone crystal growing cannot produce the large diameters capable with the CZ process, has an upper limit on resistivity, and produces a lower-quality crystal (higher dislocation density). The two methods are compared in Fig. 3.11.

Crystal and Wafer Quality

Semiconductor devices require a high degree of crystal perfection. But, even with the most sophisticated techniques, a perfect crystal is

PARAMETER	CZ	FLOAT ZONE
Large Crystal	Yes	Difficult
Cost	Lower	
Dislocations	$0 - 10^4/cm^2$	$10^3 - 10^5/cm^2$
Resistivity	Up to 100 ohm-cm	2000 ohm-cm Max.
Radial Resistivity	5 - 10%	5 - 10%
Oxygen Content	$10^{16} - 10^{18}$ atoms/cm^3	0 - Very Low

Figure 3.11 Comparison of CZ and float crystal-growing methods.

unobtainable. The imperfections, called *crystal defects*, result in device process problems by causing uneven silicon dioxide film growth, poor epitaxial film deposition, and uneven doping layers in the wafer. In finished devices, the crystal defects cause unwanted current leakages and may prevent the devices from operating at required voltages. There are three major categories of crystal defects:

1. Point defects

2. Dislocations

3. Growth defects

Point defects

Point defects come in two varieties. One comes about when contaminants in the crystal become jammed into the crystal structure, causing strain. The second is known as a *vacancy*. In this situation, an atom is missing from a location in the structure (Fig. 3.12). Vacancies are a natural phenomenon that occur in every crystal. Unfortunately vacancies occur whenever a crystal or wafer is heated and cooled, as in the fabrication process. The minimization of vacancies is one of the driving forces behind the desire for low-temperature processing.

Dislocations

Dislocations are a misplacement of the unit cells in a single crystal. They occur during the growth process. Dislocations can be imagined

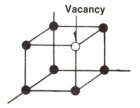

Figure 3.12 Vacancy crystal defect.

Figure 3.13 Crystal slip.

as a pile of sugar cubes with one of the cubes slightly out of alignment with the others.

Dislocations occur from growth conditions and lattice strain in the crystal. They also occur in wafers from physical abuse during the fabrication process. A chip or abrasion of the wafer edge serves as a lattice strain site that can generate a line of dislocations that progresses into the wafer interior with each subsequent high-temperature processing of the wafer. Wafer dislocations are revealed by a special etch of the surface. A typical wafer has a density of 200 to 1,000 dislocations per square centimeter.

Etched dislocations appear on the surface of the wafer in shapes indicative of their crystal orientation. ⟨111⟩ wafers etch into triangular dislocations, and ⟨100⟩ wafers show "squarish" etch pits (Fig. 3.6).

Growth defects

During crystal growth, certain conditions can result in structural defects. One is *slip*, which refers to the slippage of the crystal along crystal planes (Fig. 3.13). Another problem is *twinning*. This is a situation where the crystal grows in two different directions from the same interface. Both of these defects are cause for rejection of the crystal.

Wafer Preparation

End cropping

After removal from the crystal grower, the crystal goes through a series of steps that result in the finished wafer. First is the cropping off of the crystal ends. The crystal comes out of the grower with tapered ends that are removed with a saw.

Diameter grinding

Despite the efforts made to control the crystal diameter during growth, there is a variation over the length of the crystal (Fig. 3.14). This variation is unacceptable for the final wafers. In the fabrication process, the wafers will be stored and handled in a variety of holders

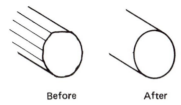

Before   After

Figure 3.14   Crystal diameter grinding.

that require a consistent specific diameter. Poor diameter control can result in warped and broken wafers.

Diameter grinding is a mechanical operation performed in a centerless grinder. This machine grinds the crystal to the correct diameter without the necessity of clamping it into a lathe-type grinder with a fixed center point.

Crystal orientation, conductivity, and resistivity check

Before the crystal is submitted to the actual wafering steps, it is necessary to determine if it meets orientation and resistivity specifications. The crystal orientation (Fig. 3.15) is determined by either x-ray diffraction or collimated light refraction. In both methods, an end of the crystal is etched or polished to remove saw damage. Next, the crystal is mounted in the refraction apparatus and the x rays or collimated light are reflected off the crystal surface onto a photographic plate (x rays) or screen (collimated light). The pattern formed on the plate or screen is indicative of the crystal plane (orientation) of the grown crystal. The pattern shown in Fig. 3.15 is representative of a ⟨100⟩ orientation.

Most crystals are grown several degrees off the major ⟨111⟩ or ⟨100⟩ plane. This off-orientation is required for several reasons, particularly in ion implantation processing. (The reasons are covered in Chap. 11.)

During the orientation determination, the crystal sits in a glue on a cutting block. The crystal is set in the glue in the proper position to

Light Source

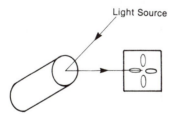

Figure 3.15   Crystal orientation determination.

ensure that, when the wafers are cut from the crystal, they are cut at the correct orientation.

Since each crystal is doped, an important electrical check is conductivity type (N or P) to ensure that the right dopant type was used. A hot-point probe connected to a polarity meter is used to generate holes or electrons (depending on the type) in the crystal. The conductivity type is displayed on the meter.

The amount of dopant put into the crystal is determined by a resistivity measurement using a four-point probe. (See Chap. 13 for a description of this measurement technique.) The curves in Fig. 3.16 show the relationship between resistivity and doping concentration for N-type and P-type silicon.

The resistivity is checked along the axis of the crystal due to dopant variation during the growing process. This variation results in wafers that fall into several resistivity specification ranges. Later in the process, the wafers will be grouped by resistivity range to meet customer specifications.

Flat grinding

Once the crystal is oriented on the cutting block, a flat is ground along the axis (Fig. 3.17). This flat will show up on each of the wafers and is called the *major flat*. The position of the flat is along one of the major crystal planes, as determined by the orientation check.

In the fabrication process, the flat functions as a visual reference to the orientation of the wafer. It is used to place the first pattern mask on the wafer so that the orientation of the chips is always to a major crystal plane.

On most crystals there is a second, smaller, secondary flat ground on the edge. The location of the secondary flat to the major flat is a code that tells the wafer-fabrication department both the orientation and the conductivity type of the wafer. The code is shown in Fig. 3.18.

Wafer slicing

The wafers are sliced from the crystal with the use of diamond-coated, inside-diameter saws (Fig. 3.19). These saws are thin circular sheets of steel with a hole cut out of the center. The inside of the hole is the cutting edge and is coated with diamonds.

For use, the saw is tightened onto a circular frame much like a musical drumhead and is secured around the circumference of the drum. This type of saw gives the rigidity required to produce thin wafers with flat and parallel surfaces. A typical 5-in-diameter wafer will be about 22 mil in diameter. An inside-diameter saw has rigidity but

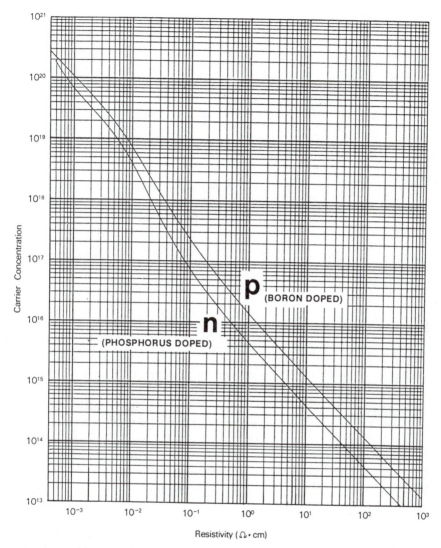

Figure 3.16 Silicon carrier concentration versus resistivity. *(After Thurber et al., Natl. Bur. Standards Spec. Publ. 400–64, May 1981, tables 10 and 14.)*

without being very thick. These factors reduce the *kerf* (cutting width) size, which, in turn, prevents sizable amounts of the crystal from being wasted by the slicing process.

Rough polish

The surface of a semiconductor wafer has to be free of irregularities and saw damage and be absolutely flat. The first requirements come

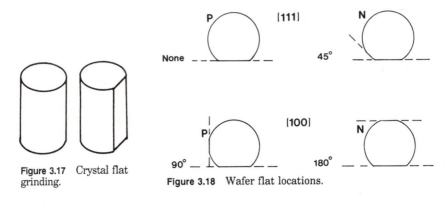

Figure 3.18 Wafer flat locations.

Figure 3.17 Crystal flat grinding.

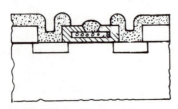

Figure 3.19 Inside diameter saw wafer slicing.

from the very small dimensions of the surface and subsurface layers making up the device. They have dimensions in the 0.5- to 2-μm range. To get an idea of the relative dimensions of a semiconductor device, imagine that the cross section in Fig. 3.20 is as tall as a house wall, about 8 ft. On that scale, the working layers on the top of the wafer would all exist in only the top 2 in of the wall.

The flatness requirement is an absolute requirement for small-dimension patterning (Chap. 9). Several of the advanced patterning processes project the required pattern image onto the wafer surface. If the surface is unflat, the projected image will be distorted just as a slide image will be out of focus on a nonflat screen.

The flatting and polishing process proceeds in two steps, rough polish and chemical-mechanical polishing (Fig. 3.21). Rough polishing is

Figure 3.20 Cross section of MOS transistor.

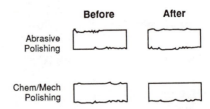

Figure 3.21 Abrasive and chemical-mechanical surface polishing.

a conventional, abrasive, slurry lapping process, but fine-tuned to semiconductor requirements. A primary purpose of the rough polish is to remove the surface damage leftover from the wafer-slicing process.

Chemical-mechanical polishing

The final polishing step is a combination of chemical etching and mechanical buffing. The wafers are mounted on rotating holders and lowered onto a rotating surface that is flooded with a mild etchant solution. The etchant grows a thin layer on the wafer that is almost simultaneously removed by the buffing action. The net effect is a very controlled polishing process capable of incredible flatness. If a semiconductor wafer surface was extended to 10,000 ft (the length of an airport runway), it would vary about ±2 in over its entire length.

Chemical-mechanical polishing is one of the techniques developed by the industry that has allowed its advancement to larger wafers and chips. Several variations of the process have been patented.

Backside processing. In most cases, only the front side of the wafer goes through the extensive chemical-mechanical polishing. The backs may be left rough or be etched to a bright appearance. For some device use, the backs may receive a special process to induce crystal damage, called *backside damage*. Backside damage causes the growth of dislocations that radiate up into the wafer. These dislocations can act as a trap of mobile ionic contamination introduced into the wafer during the fabrication process. The trapping phenomenon is called *gettering*.

Edge grinding. Edge grinding is a mechanical process that leaves the wafer with a rounded edge (Fig. 3.22). This rounding minimizes edge chipping and damage during fabrication that can result in wafer breakage or serve as the nucleus for dislocation lines.

Wafer evaluation

Before packing, the wafers (or samples) are checked for a number of parameters as specified by the customer. Figure 3.23 illustrates a typ-

Before   After

Figure 3.22 Wafer edge grinding.

The complete specification for this product includes all general requirements of specification M1.1

Dimension and Tolerance Requirements

Property	Min.	Max	Units
DIAMETER	99.0	101.0	mm
	3.898	3.976	in
THICKNESS, CENTER POINT	600	650	μm
	0.0237	0.0255	in
PRIMARY FLAT LENGTH	30.0	35.0	mm
	1.181	1.377	in
SECONDARY FLAT LENGTH	16.0	20.0	mm
	0.629	0.787	in
BOW		60	μm
		0.0023	in
TOTAL THICKNESS VARIATION		50	μm
		0.0019	in

For referee purposes, metric(SI) units apply. Conversion to U.S. Customary equivalents was done following the max-min convention in which the minimum values are rounded up and the maximum values are rounded down to ensure that the equivalent range is always inside the referee range. This is appropriate for quality control measurements to ensure that product shipped is within specification. If the equivalent unit system is used for incoming inspection measurements, the rightmost digit of the minimum values should be reduced by 1 and the rightmost digit of the maximum values should be increased by 1 to avoid rejection of material that is within the speciifcation when measured by the referee system of units. NOTE – The significance of the rightmost digit may vary depending on the quantity being measured and the precision of the test prcedure. Refer to the relevant test method for precision data which can be used to construct appropriate guard bands.

Orientation and Flat Location Requirements

Property	Requirement
PRIMARY FLAT ORIENTATION[A]	{110} ±1°
SECONDARY FLAT ORIENTATION	
(111) p-type	No secondary flat
(100) p-type	(90° ± 5°) from primary flat
(111) n-type	(45° ± 5°) from primary flat
(100) n-type	(180° ± 5°) from primary flat
SURFACE ORIENTATION	
On-orientation for (111) and (100)	± 1°
Off-orientation [111] toward nearest	2.5° ± 0.5°
<110> on a plane parallel to primary flat.	4.0° ± 0.5°
ORTHOGONAL MISORIENTATION[B]	± 5° max.

[A] For (111) slices $(1\bar{1}0)$, $(01\bar{1})$ and $(\bar{1}01)$ planes are equivalent, allowable planes. For (100) slices, the allowable equivalent {110} planes are $(01\bar{1})$, (011), $(0\bar{1}1)$, $(0\bar{1}\bar{1})$.

[B] The contribution of 5° of orthogonal misorientation to the total off-orientation angle will be less than 0.5°.

Figure 3.23 Semi wafer specifications. [Semi Standard M1.1. STD.6 for 100-mm Polished Monocrystalline Silicon Slices (625-μm Thickness).] (© *Semiconductor Equipment and Materials Institute 1979, 1983.*)

ical wafer specification for a semiconductor wafer. Of primary concern are surface problems such as particulates, stain, and haze. These problems are detected with the use of high-intensity lights or automated inspection machines.

Oxidation

Silicon wafers may be oxidized before shipment to the customer. The silicon dioxide layer serves to protect the wafer surface from scratches and contamination during shipping. Most companies start the wafer-fabrication process with an oxidation step, and buying the wafers with an oxide layer saves a manufacturing step. (Oxidation processes are explained in Chap. 7.)

Packaging

While much effort goes into producing a high-quality and clean wafer, the quality can be lost during shipment to the customer, or worse, from the packaging method itself. Therefore, there is a very stringent requirement for clean and protective packaging. The packaging materials are of nonstatic, nonparticle-generating materials, and the equipment and operators are grounded to drain off static charges that attract small particles to the wafers. Wafer packaging takes place in clean rooms.

Key Concepts and Terms

Backside damage	Point defects
Chemical-mechanical	Polycrystalline
Crystal	$\langle 100 \rangle$ plane
Crystal dislocations	$\langle 111 \rangle$ plane
Crystal growing	Seed crystal
Crystal orientation	Single crystal
Czochralski	Slip
Edge rounding	Unit cell
Float zone	Vacancy
Growth defects	Wafer flat code
Melt	Wafer slicing

Review Questions

1. In a polycrystal structure, the atoms are not arranged (true/false).

2. In a single-crystal structure, the unit cells are not arranged (true/false).

3. Draw a cubic unit cell and identify the $\langle 100 \rangle$ plane.

4. $\langle 111 \rangle$ oriented wafers are used for _____ (bipolar, MOS) devices.

5. What is the orientation of a semiconductor crystal if the seed has a $\langle 100 \rangle$ orientation?

6. Draw a diagram of a CZ crystal grower and identify all the major parts.

7. During crystal growth, the molten material is changed from single-crystal structure to a polycrystal structure (true/false).

8. Why are wafers edge rounded?

9. Draw a flow diagram of the wafer-preparation process.

10. Give two reasons why semiconductor wafers require a flat surface.

References

1. R. E. Williams, *Gallium Arsenide Processing Techniques*, Artech House, Inc., Bedham, Mass., 1984, p. 37.

Contamination Control

Overview

In this chapter, the effects of contamination on device processing, device performance, and device reliability are detailed along with the types and sources of contamination found in a fabrication area. The three clean-room strategies and major contamination-control procedures are explained. The specifications for clean-room chemicals and materials are presented.

Objectives

Upon completion of this chapter, you should be able to:

1. Identify the three major effects of contamination on semiconductor devices and processing.
2. List the major sources of contamination in a fabrication area.
3. Define the "class number" of a clean room.
4. List the particle density of class 100, 10, and 1 fabrication areas.
5. Describe the role of positive pressure, air showers, and adhesive mats in maintaining cleanliness levels.
6. List at least three techniques used to minimize contamination from fabrication personnel.
7. Identify the three contaminants present in "normal" water and their control in semiconductor plants.
8. Describe the differences between normal industrial chemicals and semiconductor-grade chemicals.
9. Name two problems associated with high static levels and two methods of static control.

The Problem

Semiconductor devices are very vulnerable to many types of contaminants. They fall into four major classes. They are:

1. Particles
2. Metallic ions
3. Chemicals
4. Bacteria

A major vulnerability of semiconductor devices, especially dense integrated circuits, is due to the small feature sizes and the thinness of deposited layers on the wafer surface. These dimensions are in the micron range. A micron or micrometer (μm) is very small. A centimeter contains 10,000 μm. Another way to envision a micron is that a human hair is about 100 μm in diameter (Fig. 4.1). The small physical dimensions of the devices make them very vulnerable to particulate contamination in the air, coming from workers, generated by the equipment, and present in processing chemicals (Fig. 4.2).

In Chap. 2 it was established that semiconductor devices are fabricated through our ability to create the two conductivity types in wafers, control the resistivity level through doping, and create N-P junctions in the surface of the devices. These three properties are achieved by the purposeful introduction of specific dopants into the crystal and

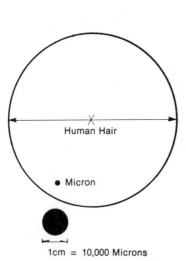

Figure 4.1 Relative size of one micron.

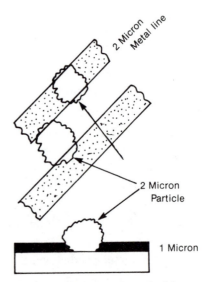

Figure 4.2 Relative size of airborne particulates to wafer dimensions.

into the wafer. These desired effects are achieved with very small amounts of the dopants. Unfortunately it only takes a small amount of certain contaminants in the wafer to change the electrical characteristics, changing device performance and reliability.

The contaminants causing these types of problems are known as mobile ionic contaminants (MICs). They are atoms of metals that exist in the material in an ionic form. Further, these metallic ions are highly mobile in semiconductor materials. This mobility means that the metallic ions can move inside the device, even after passing electrical testing and shipping, causing the device to fail. Unfortunately, the metals (Fig. 4.3) that cause these problems in silicon devices are present in most chemicals. Sodium is usually the most prevalent mobile ionic contaminant in most untreated chemicals and is the most mobile in silicon. Consequently, control of sodium is a prime goal in silicon processing. The MIC problem is most serious in MOS devices, a fact that has led some chemical suppliers to develop MOS or low-sodium-grade chemicals. These labels refer to low-mobile ionic contaminant levels.

The third major contaminant in semiconductor process areas is unwanted chemicals. Process chemicals and process water can be contaminated with trace chemicals that interfere with the wafer processing. They may result in unwanted etching of the surface, create compounds that cannot be removed from the device, or cause nonuniform processes. Chlorine is such a contaminant and is rigorously controlled in process chemicals.

Bacteria is the fourth major contaminant class. Bacteria are organisms that grow in water systems and on surfaces not cleaned regularly. Bacteria, once on the device, act as particulate contamination or may contribute unwanted metallic ions to the device surface.

IMPURITY	PARTS PER MILLION
Na	0.20
Pb	0.20
Zn	2.00
Cr	0.20
Ni	0.25
Na	0.10
Si	0.07
Ca	0.12
Al	0.05
Mo	0.03
Cu	0.02
Mn	0.015

Figure 4.3   Mobile ionic contaminants in a typical wet chemical.

The four types of contaminants affect the processing and devices in three specific performance areas. They are:

1. Device processing yield
2. Device performance
3. Device reliability

Device processing yield

Device processing in a contaminated environment can cause a multitude of problems. Contamination may change the dimensions of device parts, change the cleanliness of the surfaces, and/or cause pitted layers. Within the fabrication process are a number of quality checks and inspections specifically designed to detect contaminated wafers. High levels of contamination result in fewer wafers completing the process and higher costs.

Device performance

A more serious problem is related to small pieces of contamination that may escape the in-process quality checks. Whereas the wafer may appear to be clean, undetected particulates, unwanted chemicals, and/or high levels of mobile ionic contaminants in the wafer can change the electrical performance of the devices. This problem usually shows up at the postfabrication electrical test.

Device reliability

Device reliability is the most insidious of the contamination failures. Small amounts of metallic contaminants can get into the wafer during processing and not be detected during normal device testing. However, in the field, these contaminants can travel inside the device and end up in electrically sensitive areas, causing failure. This failure mode is a primary concern of the space and defense industries.

In the rest of this chapter, the sources, nature, and control of the types of contamination that affect semiconductor devices are identified. With the advent of LSI level circuits in the 1970s, the control of contamination became essential to the industry. Since that time a great deal of knowledge about and control of contamination has been learned. Contamination control is now a discipline of its own and is one of the critical technologies that has to be mastered to profitably produce solid-state devices.

The remarks about contamination control in this chapter are requirements for fabrication areas, mask-making areas, and areas in which semiconductor equipment and materials are manufactured.

Contamination Sources

General sources

Contamination in a clean room is defined as anything that interferes with the production of the product and/or its performance. The stringent requirements of solid-state devices define levels of cleanliness that far exceed those of almost any other industry. Literally everything that comes in contact with the product during manufacture is a potential source of contamination. The major sources are:

1. Air
2. The production facility
3. Clean-room personnel
4. Process water
5. Process chemicals
6. Process gases
7. Static charge

Each source produces specific types and levels of contamination and requires special controls to render them acceptable in the clean room.

Air

Normal air is so laden with contaminants that it must be treated before entering a clean room. A major problem is airborne particles, referred to as *particulates* or *aerosols*. Normal air contains copious amounts of small dust and particles, as illustrated in Fig. 4.4. A major problem with small particles is that they "float" and remain in the air for long periods of time. These particles are called aerosols. Air-cleanliness levels in clean rooms are identified by the particulate diameters and their density in the air.

Air quality is designated by the *class number* of the air in the area as defined in Federal Standard 209B.[1] This standard designates air quality in the two categories of particle size and density. The class number of an area is defined as the number of particles above a specified diameter in a cubic foot of air. The air in a typical city, filled with smoke, smog, and fumes, can contain up to 5 million particles per cubic foot, which is a class number of 5 million.

Figure 4.5 shows the relationship between particulate diameter and density as defined by Federal Standard 209B. The exact class number

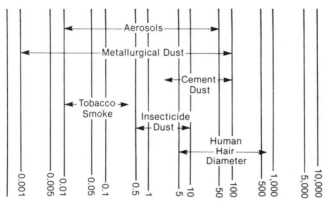

Figure 4.4 Relative size of airborne particulates (microns).

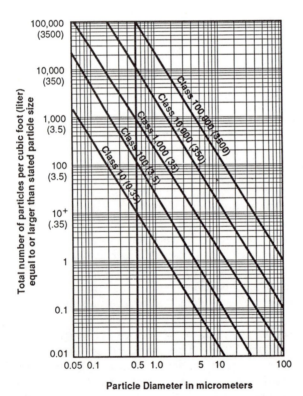

Figure 4.5 Particle size, density, and class number relationship.

Environment	Class Number	Particle Size, μm
VLSI Area	10	0.3
VLF Hood	100	0.5
Assembly Area	10,000	0.5
House Room	100,000	
Outdoors	> 500,000	

Figure 4.6 Class numbers for different environments.

and allowable particle size required for a particular processing area are determined by the feature size of the devices being produced. A rule of thumb is that the allowable particle size should not exceed one-half of the feature size. Thus a VLSI fabrication area producing circuits with a 1-μm feature size should have air with no more than 0.5-μm-diameter particles. And those particles should have a density (or class number) of 10 or less.

Figure 4.6 lists the class numbers and associated particle size for various environments. Federal Standard 209B, as of 1989, specifies cleanliness levels only down to class 100 levels. Studies are ongoing to specify class 10 and class 1 environments. The specifications used for these cleaner environments are extrapolated from existing 209B values.

Clean Air Strategies

The design of a clean room is integral to its ability to produce contamination-free wafers. A major consideration in the design is the maintenance of clean air in the process areas. Three distinct strategies are used:

1. Clean work stations

2. Tunnel design

3. Total clean room

Clean-room work station strategy

Air-filtered production areas first became a need in the space industry. It was found that even single specks of dust could cause a satellite to fail. Consequently, NASA and their contractors constructed assembly rooms that featured ceiling and wall filters. They also developed much of the basic understanding of contamination control, from operator gowning to clean materials.

The semiconductor industry adopted many of these techniques. One problem, however, was that the cleanliness levels achieved in the

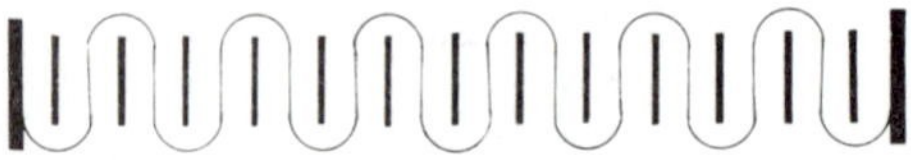

Figure 4.7   HEPA filter design.

small rooms could not be maintained when the design was translated to larger fabrication areas with more production workers. The early semiconductor industry went from the ceiling and wall filter design to a clean-room work station strateg. This strategy was based on the concept of keeping the wafers clean by processing them only in individual work stations that featured nonshedding materials and filtered air. Outside the work stations, the wafers were stored and moved in covered boxes.

The fabrication area consisted of a large room with the work stations (called *hoods*) arranged in rows so that the wafers could move sequentially through the process, never being exposed to dirty air. The filters in these clean hoods are of a type called a high-efficiency particulate attenuation (HEPA) filter. These filters are constructed of fragile fibers with many small holes and folded into the filter holder in an accordion design (Fig. 4.7). The high density of small holes and large area of the filter medium allow the passage of large volumes of air at low velocity. The low velocity contributes to the cleanliness of the hood by not causing air currents. The low velocity is also necessary for operator comfort. A typical airflow is 90 to 100 ft/min.[2] HEPA filters have a filtering efficiency of 99.99 percent plus and are the filters of choice in all three of the clean-room strategies.

Typically, a clean hood (Fig. 4.8) has a HEPA filter mounted in the top. Air is drawn from the room through a prefilter by a fan and forced through the HEPA filter. The air leaves the filter in a laminar pattern and at the work surface turns and exits the hood. A shield directs the exiting air over the wafers in the hood. The formal name for the work station is a vertical laminar flow (VLF) station. The term VLF is derived from the laminar nature of the airflow. Some work stations are designed with the HEPA filter in the back of the work surface. These stations are called horizontal laminar flow (HLF) hoods.

Both types of stations keep the wafers clean in two ways. First is the filtered air inside the hood. The second cleaning action is the slight positive pressure built up in the station. This pressure prevents airborne dirt from operators and from the aisle areas from entering the hood.

A special design of VLF hood is required for wet chemical processing stations (Fig. 4.9). These stations must be connected to an exhaust system to remove the chemical fumes. The fumes can pose both safety and contamination problems. In this design, care must be taken to

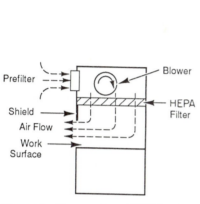

Figure 4.8 Cross section of VLF hood.

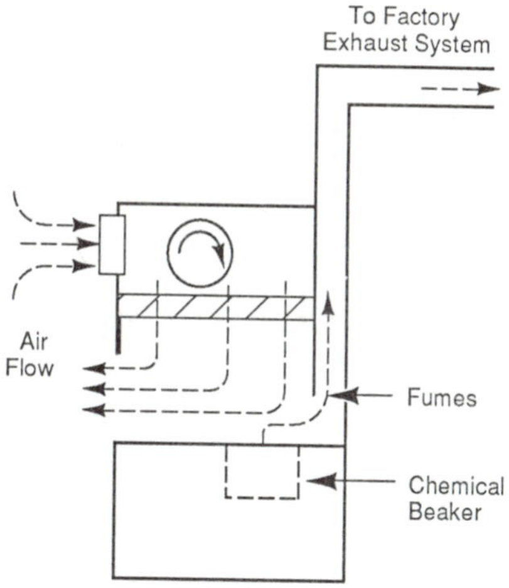

Figure 4.9 Cross section of a VLF–fume-exhaust hood.

balance the VLF air and the exhaust to maintain the required class number in the station. Cleanliness also requires that wafers be stored only in the cleaner front area of the work surface.

Tunnel concept

As more critical particulate control became necessary, it was noted that the VLF hood approach had several drawbacks. Chief among them was the vulnerability to contamination from the many personnel moving about in the room. People entering and exiting the fabrication area had the potential of contaminating all the process stations in the area.

This particular problem is solved by dividing the fabrication area into separate tunnels or bays (Fig. 4.10). Instead of the individual VLF hoods, HEPA filters are built into the ceilings and serve the same purpose. The wafers are kept clean by the filtered air from the ceiling filters and less vulnerable to personnel-generated contamination because of fewer workers in the immediate vicinity. On the downside, tunnel arrangements cost more to construct and are less versatile when the process changes.

Total clean-room strategy

Developments in clean-room design and filtering technology have allowed a return to the open fabrication area (Fig. 4.11). In the latest version, air filtering is accomplished by HEPA filters in the ceiling

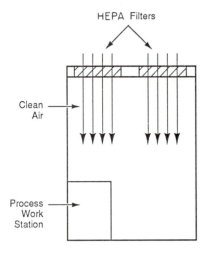

Figure 4.10   Cross section of clean-room tunnel.

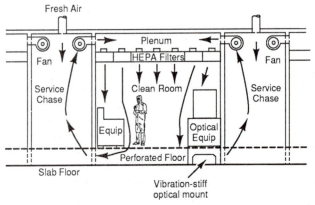

Figure 4.11   Cross section of laminar flow clean room. *(Courtesy of Semiconductor International.)*

with returns in the floor to give a continuous flow of clean air. The work stations are tabletops with perforations to allow the filtered air to pass through the tabletop uninterrupted.

Modern clean-room technology can produce fabrication areas that can perform at class 10 or better conditions. An important clean-room parameter is recovery. This parameter is the time it takes for the filters to return the area to acceptable conditions after a shift start, personnel break, or other disturbance.

Temperature, humidity, and smog

In addition to control of particulates, the air parameters of temperature, humidity, and smog must be specified and controlled in a clean room. Temperature control is necessary for operator comfort and process control. Many of the wet processes of etching and cleaning take place in non-temperature-controlled baths and rely on the room temperature for control. This control is important. A rule of thumb is that chemical reactions, such as an etch rate, change by a factor of 2 with a 10° (centigrade) change in temperature. A typical temperature range is 72°F, ± 2°F.

The relative humidity is also a critical process parameter, especially in patterning areas. In this area, thin films of a polymer are put on the wafer to act as an etch stencil. If the humidity is too high, the wafers collect moisture, preventing the polymer from sticking. The situation is the same as applying paint on a wet surface. On the other side, low humidity can foster the buildup of static charge on the wafer surface. The charge causes the wafer to attract particles out of the air. Relative humidity is controlled between 15 and 50 percent.

Smog is another airborne contaminant in a clean room. Again the problem is most critical in the patterning areas. A step in the patterning process is similar to photographic film developing, a chemical process. Ozone, a major component of smog, interferes with the development process and must be controlled. Ozone is filtered out of the air by installing carbon filters in the incoming air ducts.

Clean-Room Construction

Selection of the clean air strategy is the first step in the design of a clean room. Every clean room is a trade-off of cleanliness versus cost. A typical clean room can carry a price tag of up to $2,000 per square foot. Whatever the final design, every clean room is built upon basic principles. The overall design principle is to build a sealed room that is supplied with clean air, is built with materials that are non-contaminating, and includes systems to prevent accidental contamination from the outside or from operators.

Construction materials

The inside of a clean room is constructed entirely of materials that are nonshedding. This includes wall coverings, process station materials, and floor coverings. All piping holes are sealed, and even the light fixtures must have solid covers. Additionally, the design should minimize flat surfaces that can collect dust. Stainless-steel materials are favored for process stations and work surfaces.

Clean-room design

Both the design of a clean room and its operation must be set up to keep dirt and contamination from getting into the room from the outside. Figure 4.12 shows a layout for a typical class 10 VLSI processing area. Seven techniques are used to keep out and control dirt. They are:

1. Adhesive floor mats
2. Gowning area
3. Air pressure
4. Air showers
5. Service bays
6. Double-door pass-throughs
7. Static control

Adhesive floor mats. At the entrance to every clean room is a floor mat with an adhesive surface. The adhesive pulls off and holds dirt adhering to the bottoms of shoes. In some clean rooms, the entire floor has a surface treated to hold dirt.

Gowning area. A major part of a clean room is the gowning area or anteroom. This area is a buffer between the clean room and the plant. It quite often is supplied with filtered air from ceiling HEPA filters. In this area, the operator's clean-room apparel is stored in lockers. It is also the area where the clean-room personnel change into their clean-room garments. The management of this area varies with the degree of cleanliness required in the clean room. Quite often it is managed to the same stringent requirements of the clean room itself. Often the gowning area is divided into two sections by a bench. The operators don the garments on one side and put on their shoe coverings on the

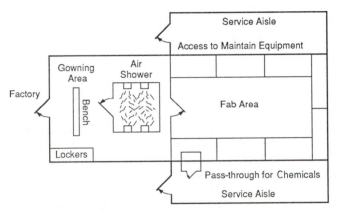

Figure 4.12   Fab area with gowning area, air showers, and service aisles.

bench. The purpose is to keep the area between the bench and the clean room at a higher cleanliness level.

A good clean-room procedure is to ensure that the doors between the factory and the clean room are never opened at the same time. This procedure ensures that the clean room is never exposed directly to the dirtier factory areas. Clean-room management also includes lists of materials and garments that can and cannot come into the gowning room. Some areas will provide hallway lockers for coats, etc.

Air pressure. A key design element is the air pressure balances between the clean room, the gowning room, and the factory. The well-designed facility will have the air of these three sections balanced such that the highest pressure is in the clean room, the second highest in the gowning area, and the lowest in the factory hallways. The higher pressure in the clean room prevents airborne particles from entering when doors are opened.

Air showers. The final design element protecting the clean room from outside contamination is the air shower located between the gowning room and the clean room. Clean-room personnel enter the air shower where high-velocity air jets blow off particles on the outside of the garments. An air shower will have an interlocking system to prevent both doors from being opened at the same time.

Service bays. A clean room is really a series of rooms (Fig. 4.12) within the factory, each contributing to the maintenance of the clean room. In the center is the processing clean room. Surrounding it is a bay area that is maintained at some designated class number that is generally higher than the clean room. In the bay are the process chemical pipes, electrical power lines, and clean-room materials. Critical-process machines are backed up to the wall dividing the clean room and the bay. This arrangement allows technicians to service the equipment from the back without entering the clean room.

Double-door pass-throughs. The bay also serves as a semi-clean area for the storage of materials and supplies. They are put into the clean room through double-door, pass-through units that protect the cleanliness of the clean room. Pass-through units may be simple double-door boxes or have a supply of positive-pressure filtered air with interlocking devices to prevent both doors from being opened at the same time. All materials and equipment brought into the clean room should be cleaned prior to entry.

Static control.   One of the consequences of moving to higher-density circuits with submicron feature sizes is the vulnerability of the devices to smaller particles of contamination. One source of particulate contamination is the static charges that build up on the wafers, the storage boxes, work surfaces, and equipment. Each of these items can carry static charges as high as 50,000 V (volts) that attract aerosols out of the air and from personnel garments. The attracted particles end up contaminating the wafers.

Static also represents a device operational problem. It occurs in devices with thin dielectric layers, as in MOS gate regions. An electric static discharge (ESD) of up to 10 A (amperes) is possible. This level of ESD can physically destroy an MOS device or circuit. ESD is a particular worry in device-packaging areas. This problem requires that sensitive devices, such as large-array memories, be handled and shipped in holders of antistatic materials.

Static is controlled by prevention and discharge techniques (Fig. 4.13). Prevention techniques include use of antistatic materials in garments and in-process storage boxes. In some areas, a topical antistatic solution may be applied to the walls to prevent the buildup of static charge. These solutions work by leaving a neutralizing residue on the surface. Generally, they are not used in critical stations because of the possible contaminating effect of the residue.

Discharge techniques include the use of ionizers and grounded static-discharge straps. Ionizers are placed just underneath the HEPA filters where they function to neutralize any charge buildup in the fil-

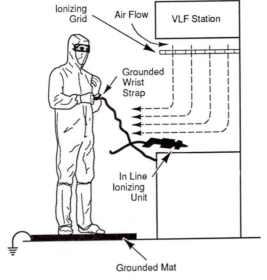

Figure 4.13   Static-charge reduction techniques.

tered air. Ionizers are also placed on nitrogen blow-off guns for the same effect. Some stations will have a portable ionizer blowing ionized air directly on the wafers being processed. Static discharge is also accomplished by grounding operators with wrist straps, having grounded mats at critical stations, and grounding work surfaces.

Personnel-Generated Contamination

Clean-room personnel are one of the biggest sources of contamination. A clean-room operator, even after showering and sitting, can give off between 100,000 and 1,000,000 particles per minute.[3] This number increases dramatically when a person is in motion. At 2 miles an hour, a human being gives off up to 5 million particles per minute. The particles come from flakes of dead hair and normal skin flaking. Additional particle sources are hair sprays, cosmetics, facial hair, and exposed clothing.

Normal clothing can add more millions of particles to the area even under a clean-room garment. In clean rooms with very high cleanliness levels, operators will be directed to wear street clothing that is made of tight-weave, nonshedding materials. Garments made of wools and cottons are to be avoided, as are ones with high collars.

A human breath also contains high levels of contaminants. Every exhale puts numerous water droplets and particles into the air. The breath of smokers carries millions of particles for a long time after a cigarette is finished. Body fluids such as saliva contain sodium, a killer to many semiconductor devices. While healthy human beings are sources of many contaminants, sick individuals are even worse. Specifically, skin rashes and respiratory infections are additional sources of contaminants. Some fabrication areas reassign personnel with certain health problems.

Given the scope of the problem, the only feasible way to render humans acceptable in a clean room is to cover them up. This is exactly what is required for access to any clean room. The style and material of clothing selected for clean-room personnel depend on the level of cleanliness required. For a typical class 10 area (Fig. 4.14), the clothing material will be nonshedding and may contain conductive fibers to draw off static charge.

Every part of the body is covered up. The head will have an inner cap that keeps the hair in place. This is covered by an outer shell that is designed to fit close to the face and has snaps or a tail for securing the headgear under the body-covering smock. Covering the face will be a mask. Masks vary from surgical types to full ski mask style designs. In some clean rooms, both an inner and outer face mask are required. The eyes, which are a major source of fluid particles, are covered by glasses (usually safety glasses) with side shields.

Figure 4.14   Worker in clean-room garments.

Body covers are oversuits that have closures for the legs, arms, and neck. Well-designed suits will have covers over the zippers and no outside pockets.

The feet are covered with shoe coverings, some with attached leggings that come up the leg. In static-sensitive areas, straps are available to drain off static charge.

Hands are covered with at least one pair of gloves. Favored are medical-type plastic gloves that permit good tactile feeling. In some areas, a second pair of gloves is required. Gloves should be pulled up over the sleeves to prevent contamination from traveling down the arm and into the clean room.

Skin flaking can be further controlled with the use of special lotions that moisten the skin. Any lotions used must be sodium- and chlorine-free.

In general, the order of gowning is from the head down. The theory is that dirt stirred up at each level is covered up by the next lower garment. Gloves are put on last. The garments and procedures needed to control contamination from clean-room workers are well known. However, the primary level of defense is the dedication and training of the operators. It is easy for an area to become lax in maintaining clean-room discipline and to suffer high levels of contamination.

Process Water

During the course of fabrication processing, a wafer will be chemically etched and cleaned many times. Each of the etching or cleaning steps

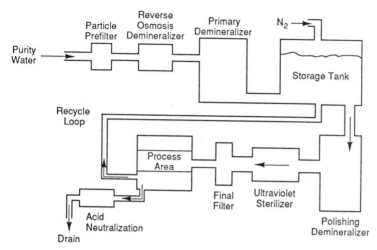

Figure 4.15 Typical deionized water system.

is followed by a water rinse. Throughout the entire process, the wafer may spend a total of several hours in water rinse systems. Given the vulnerability of semiconductor devices to contamination, it is imperative that all process water be treated (Fig. 4.15) to meet very specific cleanliness requirements.

Water from a city system contains unacceptable amounts of the following contaminants:

1. Dissolved minerals

2. Particulates

3. Bacteria

The dissolved minerals come from salts in normal water. In the water, the salts separate into ions. For example, salt (NaCl) breaks up into Na^+ and Cl^- ions. Each is a contaminant in semiconductor devices and circuits. They are removed from the water by reverse osmosis (RO) and ion-exchange systems.

The process of removing the electrically active ions changes the water from a conductive medium to a resistive one. This fact is used to improve the quality of deionized (DI) water. Deionized water has a resistivity of 18,000,000 $\Omega \cdot$ cm at 25°C. It is called 18-megohm water. Figure 4.16 shows the effect on the resistivity of water when various amounts of dissolved minerals are present.

The resistivity of all process water is monitored at many points in the fabrication area. The goal and specification is 18 megΩ in VLSI areas, although some fabrication areas will run with 15-megΩ water levels. Solid particles (particulates) are removed from the water by

Resistivity Ohms–cm 25°C	Dissolved Solids (ppm)
18,000,000	0.0277
15,000,000	0.0333
10,000,000	0.0500
1,000,000	0.500
100,000	5.00
10,000	50.00

Figure 4.16 Resistivity of water versus concentration of dissolved solids.

sand filtration, earth filtration, and/or membranes to submicron levels. Bacteria and fungi find water a favorable host. They are removed by sterilizers that use ultraviolet radiation to kill the bacteria and filters to remove them from the stream of water.

The cost of cleaning process water to acceptable levels is a major operating expense of a fabrication area. In most fabrications, the process stations are fitted with water meters that monitor the used water. If the water falls within a certain range, it is recycled in the water system for clean-up. Excessively dirty water is treated as dictated by regulations and discharged from the plant. A typical fabrication system is shown in Fig. 4.16. Water stored in the system is blanketed with nitrogen to prevent the absorption of carbon dioxide. Carbon dioxide in the water interferes with resistivity measurements, causing false readings.

Process Chemicals

The acids, bases, and solvents used to etch and clean wafers and equipment have to be of the highest purity for use in a fabrication area. The contaminants of concern are metallics, particulates, and other chemicals. Unlike water, process chemicals are purchased and used as they come into the plant. Industrial chemicals are rated by grade. They are commercial, reagent, electronic, and semiconductor grade. The first two are generally too dirty for semiconductor use. Electronic grade and semiconductor grade are cleaner chemicals, but levels of cleanliness vary from manufacturer to manufacturer.

Trade organizations such as SEMI are attempting to establish cleanliness specifications within the industry. However, most semiconductor plants purchase their chemicals to in-house generated specifications. Of primary concern are the metallic mobile ionic contaminants. These are usually limited to levels of 1 part per million (ppm) or below. Some suppliers are making available chemicals with MIC levels of only 1 part per billion (ppb). Particulate filtering levels are specified at 0.2 μm or lower.

Chemical purity is indicated by its assay number. The assay number indicates the percentage of the chemical in the container. For example, an assay of 99.9 percent on a bottle of sulfuric acid means that the bottle contains 99.9 percent sulfuric acid and 0.01 percent of other substances.

The delivery of a clean chemical to the processing area involves more than just making a clean chemical. Care must be taken to clean the inside of the containers, using containers that do not dissolve, particulate-free labels, and placing the clean bottles in bags before shipping.

In the interest of economy, some firms purchase clean process chemicals in bulk quantities. Special care must be maintained to ensure that piping and transfer vessels are cleaned regularly to prevent contamination. A particular worry with bulk chemical systems and secondary vessel filling is cross contamination. Vessels should be reserved for one type of chemical only.

Process Gases

In additional to the many wet (liquid) chemical processes, a semiconductor wafer is processed with many gases. The gases are both the air-separation gases, oxygen, nitrogen, and hydrogen, and specialty gases such as arsine and carbon tetrafluoride.

Like the wet chemicals, they have to be delivered clean to the process stations. Gas quality is measured in four categories:

1. Percentage of purity

2. Water vapor content

3. Particulates

4. Metallic ions

Extremely high purity is required for all process gases. Of particular concern are the gases used in oxidation, sputtering, plasma etch, chemical vapor deposition (CVD), reactive ion etch, ion implantation, and diffusion. All of these processes involve chemical reactions that are driven by some energy source. If the gases are contaminated with other gases, the anticipated reaction can be significantly altered or the result on the wafer changed. For example, chlorine contamination in a tank of argon used for sputtering can end up in the sputtered film with disastrous results for the device. Gas purity is specified by the assay number, with typical values ranging from 99.99 to 99.9999 percent, depending on the gas and the use in the process.

The control of water vapor is also critical. Water vapor is a gas and can enter into unwanted reactions just like other contaminating gases. In fabrication areas, processing silicon wafers with water vapor present is a particular problem. Silicon oxidizes easily wherever free oxygen or water is available. Control of unwanted water vapor is necessary to prevent accidental oxidation of silicon surfaces. Water-vapor limits are 3 to 5 ppm.

The presence of particulates and/or metallics in a gas has the same effect on the processing as in wet chemicals. Consequently, gases are filtered to the 0.2-μm level and metallic ions are controlled to parts per million or lower.

The air-separation gases are stored on the site in the liquid state. In this state they are very cold, a situation that freezes some contaminants in the bottom of the tanks. Specialty gases are purchased in high-pressure cylinders. Since many of the specialty gases are toxic or flammable, they are stored in special cabinets outside the plant.

The gases are distributed into the process areas by a network of pipes. This network must be leak-free for safety and contamination reasons. Further, the system must undergo extensive cleaning after installation to keep the gases clean on their way to the process stations.

Clean-Room Materials and Supplies

In addition to the process chemicals, it takes a host of other materials and supplies to process the wafers. Each of these must meet cleanliness requirements. Logs, forms, and notebooks used will be either of nonshedding coated paper or of a polymer plastic. Pencils are not allowed and pens are the nonretracting type.

Wafer storage boxes are made of specific non-particle-generating materials as are carts and tubing materials. Cart wheels and tools are used without greases or lubricants. In many areas, mechanics' tools and tool boxes are cleaned and left inside the clean room.

Clean-Room Maintenance

Regular maintenance of the clean room is essential. The cleaning personnel must wear the same garments as the operators. Clean-room cleaners must be carefully specified. Normal household cleaners are far too dirty for use in a clean room. Special care must be taken when using vacuum cleaners. Special clean-room vacuums with filtered exhausts are available. Many clean rooms will have built-in vacuum systems to minimize dirt generation during cleaning.

The wipe-down of process stations is done with special low-particulate wipes and sponges. The wiping procedure is critical. Wall surfaces should be wiped from top to bottom and deck surfaces from back to front. Cleaning chemicals in spray bottles should be sprayed into wipes, not onto the surfaces. This simple procedure minimizes unwanted overspray onto the wafers and equipment.

Figure 4.17 summarizes the specifications for clean-room cleanliness. Figure 4.18 shows clean-room and process sources of contamination.

SOURCE	SPECIFICATION
Air	
Particulates LSI	Class 100 @ 0.5 micron
VLSI	Class 10 @ 0.3 micron
Humidity	15–50%
Temperature	68–74°F
Photochemical Smog	2 pphm (parts per hundred million)
Deionized Water	
Resistivity	15–18 meg Ω cm
Particulate	Less than 100 particulates per centiliter after 0.5 micron filtration
Bacteria	Less than 100 colonies per ml sample after 0.5 micron filtration
Chemicals	
Metallic Impurities	Less than 1 ppm
Filtration	To 0.2 micron
Gases	
Purity	Greater than 99.9%
Water Vapor	Less than 5 ppm
Filtration	To 0.3 micron
Static Charge	Less than 50 volts

Figure 4.17   Summary of clean-room requirements.

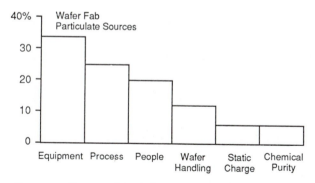

Figure 4.18   Sources of particulate contamination. This analysis, shown at SEMI Forecast by Dr. C. Rinn Cleavelin, Texas Instruments, revealed equipment-generated particles as the top enemy in 1985. *(Courtesy of Semiconductor Equipment and Materials Institute.)*

Key Concepts and Terms

Aerosols	Mobile ionic contaminants
Air shower	Particulates
Anteroom	Process gases
Bacteria	Process wet chemicals
Clean-room class number	Service bay
Clean-room design	Temperature, humidity, smog
Contamination sources	Total clean-room strategy
DI water	Tunnel strategy
Gowning requirements	Vertical laminar flow
HEPA filter	VLF work station strategy
Micron	

Review Questions

1. State the class number required for LSI processing and VLSI processing.

2. List three techniques used to keep contamination out of a clean room.

3. Draw a cross section of a "total" clean room, showing the filters and air patterns.

4. What does VLF mean?

5. What is the DI water specification?

6. Name three contaminants found in normal water.

7. What are the two ways that static is harmful in a fabrication area?

8. Explain the problem associated with high humidity in a fabrication area.

9. What is a mobile ionic contaminant and why is it unwanted in a fabrication area?

10. How does a service bay contribute to the cleanliness of a clean room?

References

1. Clean Room and Work Station Requirements, Federal Standard 209B, Sec. 1–5, Apr. 24, 1973, Office of Technical Services, Dept. of Commerce, Washington, D.C.
2. Operator Training Course, Class-10 Technologies, Inc., San Jose, Calif., 1983, p. 13.
3. *The New American Revolution*, Araclean Services, La Grange, Ill., 1984, p. 9.

Overview of
Wafer Fabrication

Overview

This chapter will present an overview of the wafer-fabrication process. The four basic planar operations performed on the wafer are explained along with the process sequences used to create the circuit components on the chip surface. Circuit design is traced from the functional diagram to the production of a photomask. Finally, both wafer and chip features and terminology are detailed.

Objectives

Upon completion of this chapter you should be able to:

1. Identify and explain the four basic wafer operations.
2. Identify the parts of a wafer.
3. Draw a flow diagram of the circuit-design process.
4. Explain the definition and use of a composite drawing and mask set.
5. Draw cross sections showing the doping sequence of basic operations.
6. Draw cross sections showing the metallization sequence of basic operations.
7. Draw cross sections showing the passivation sequence of basic operations.
8. Identify the "parts" of an integrated circuit chip.

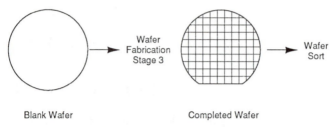

Figure 5.1   Wafer fabrication stage.

Goal of Wafer Fabrication

In Chap. 1 the four stages (materials preparation, crystal growth and wafer preparation, wafer fabrication, and packaging) were identified. The first two stages have been explored. In this chapter the fundamentals of wafer fabrication are explained.

Wafer fabrication is the series of processes used to create the semiconductor devices in and on the wafer surface. The polished wafers come into fabrication with blank surfaces and exit some six to eight weeks later with the surface covered with hundreds of chips (Fig. 5.1).

Wafer Terminology

A completed wafer is shown in Fig. 5.2. The regions of a wafer surface are:

1. Chip, die, device, circuit, microchip, or bar. All of these terms are used to identify the identical patterns covering the majority of the wafer surface.

2. Scribe lines, saw lines, streets, and avenues. These areas are small separations between chips. They become the areas used to separate the chips from each other.

3. Engineering die, test die. These chips are different from the regular device or circuit die. They contain special devices and circuit elements that can be electrically tested during the fabrication processing.

4. Edge die. The edges of the wafer contain partial die patterns. The partial die will not function. The number and area occupied by the edge die is a function of the chip size and the wafer diameter. One of the driving forces behind larger wafer diameters is to minimize the area occupied by the edge die.

5. Wafer crystal planes. The cutaway section illustrates the crystal

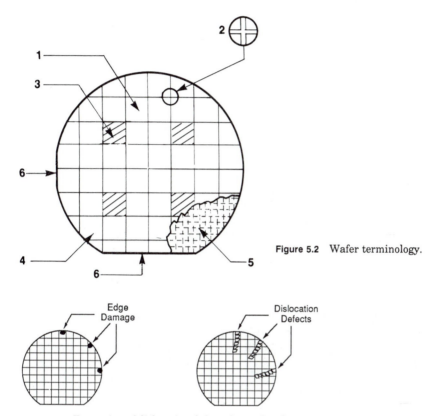

Figure 5.2   Wafer terminology.

Figure 5.3   Formation of dislocation defects from edge damage.

structure of the wafer under the circuit layers. The diagram shows that the chip edges are oriented to the wafer crystal structure.

6. Wafer flats. The depicted wafer has a major and minor flat, indicating that it is a P-type ⟨100⟩ oriented wafer.

Figure 5.3 shows the formation of damaging dislocation lines forming from wafer-edge damage sites.

Basic Wafer-Fabrication Operations

Wafer-fabrication areas around the world produce billions of chips with thousands of different functions and designs. The designs of the devices and circuits are based on a number of different transistor structures. Bipolar and MOS transistors are the major structure designs, but there are numerous variations of them (see Chap. 16). Ad-

- Layering
- Pattering
- Doping
- Heat Treatments

Figure 5.4   Basic wafer fabrication operations.

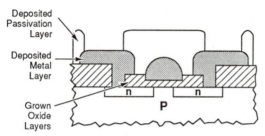

Deposited Passivation Layer

Deposited Metal Layer

Grown Oxide Layers

n   n

P

Figure 5.5   Cross section of completed metal gate MOS transistor with grown and deposited layers.

ditionally, there are usually several choices of processes or materials available to create the individual layers of any particular structure.

The result is thousands of different process flows with hundreds of process options. But within all this process diversity only four (Fig. 5.4) operations are performed on a wafer in the fabrication process. They are:

1. Layering

2. Patterning

3. Doping

4. Heat treatments

Layering

Layering is the operation used to add thin layers to the wafer surface. An examination of the MOS transistor structures in Fig. 5.5 shows a number of layers that have been added to the wafer surface. These layers are either insulators, semiconductors, or conductors. They are of different materials and are grown or deposited by a variety of techniques.

The layers are added to the surface by two major techniques, growing or deposition (Fig. 5.6). Oxidation is a technique of growing a silicon dioxide layer on a silicon wafer. Common deposition techniques are chemical vapor deposition (CVD), evaporation, and sputtering. Figure 5.7 lists common layer materials and layering techniques. (The

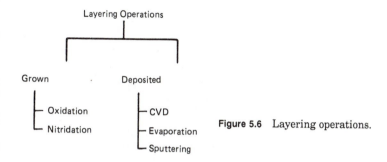

Figure 5.6   Layering operations.

Layers	Thermal Oxidation	Chemical Vapor Deposition	Evaporation	Sputtering
Insulators	Silicon Dioxide	Silicon Dioxide Silicon Nitrides		Silicon Dioxide Silicon Monoxide
Semiconductors		Epitaxial Silicon Poly Silicon		
Conductors			Aluminum Aluminum/Silicon Aluminum/Copper Nichrome Gold	Tungston Titanium Molybdenum Aluminum Aluminum/Silicon Aluminum/Copper

Figure 5.7   Table of layers, processes, and materials.

details of each are explained in the process chapters. The role of the different layers in the structures is explained in Chap. 16.)

Patterning

Patterning is the series of steps that results in the removal of selected portions of the added surface layers (Fig. 5.8). After removal, a *pattern* of the layer is left on the wafer surface. The material removed may be in the form of a hole in the layer or just a remaining island of the material.

The patterning process is known by the names photomasking, masking, photolithography, and microlithography. It is the patterning

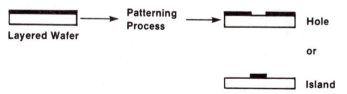

Figure 5.8   Patterning.

operation that creates the surface parts of the devices that make up a circuit. The goals of the operation are to create the *parts* in the exact dimensions (feature size) required by the circuit design and place them in their proper location on the wafer surface.

Patterning is the most critical of the four basic operations. This operation sets the critical dimensions of the devices. Errors in the patterning process can cause distorted or misplaced patterns that result in changes in the electrical functioning of the device or circuit. Another problem is defects. The patterning process requires the use of a delicate stencil material known as *photoresist*. Contamination in the patterned area can damage this layer, causing malfunctioning devices. This contamination problem is magnified by the fact that patterning operations are performed on the wafer from five to fifteen times in the course of the wafer-fabrication process.

Doping

Doping is the process that puts specific amounts of dopants in the wafer surface through openings in the surface layers (Fig. 5.9). The two techniques are thermal diffusion and ion implantation, which are detailed in Chap. 11.

Thermal diffusion is a chemical process that takes place when the wafer is heated to the vicinity of 1000°C and exposed to vapors of the proper dopant. Ion implantation is a physical process in which the dopant atoms are ionized, accelerated to a high speed, and "shot" into the wafer surface.

The purpose of the doping operation is to create either N-type or P-

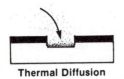

Thermal Diffusion

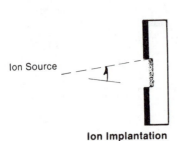

Ion Source

Figure 5.9   Doping.

Ion Implantation

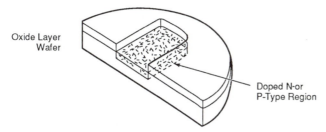

Figure 5.10   Formation of doped N- or P-type region in wafer surface.

type pockets in the wafer surface (Fig. 5.10). These pockets form the N-P junctions required for operation of the transistors, diodes, capacitors, and resistors of the circuit.

Heat Treatments

Heat treatments are the operations in which the wafer is simply heated and cooled to achieve specific results. In the heat treatment operations, no additional material is added or removed from the wafer.

An important heat treatment takes place after ion implantation. The implantation of the dopant atoms causes a disruption of the wafer crystal structure which is repaired by a heat treatment, called *anneal*, at about 1000°C. Another takes place after the conducting stripes of metal are formed on the wafer. These stripes carry the electrical current between the devices in the circuit. To ensure good electrical conduction, the metal is "alloyed" to the wafer surface by a heat treatment, which takes place at 450°C. Figure 5.11 shows a list of common process techniques grouped by their basic operation in the fabrication process.

Construction of a Semiconductor Circuit

The construction of a semiconductor circuit is a multistep process involving up to three hundred individual steps in the wafer-fabrication stage. The four basic operations are used in specific sequences to build the parts of the devices in and on the wafer surface. For illustration, let's examine the fabrication of one MOS metal gate transistor in an integrated circuit. In this section, the building of the transistor is illustrated to explain the construction sequence. (The functions of the individual parts of this type of transistor and the operation of the transistor are explained in Chap. 14.)

Circuit design

Before fabrication starts, a specific function and design for the circuit must be established. The design process starts with a determination of

Operation	Purpose	Techniques
Layering	Grow or deposit a thin layer on the wafer surface.	Oxidation 　Atmospheric Pressure 　High Pressure CVD 　Epi 　Low Pressure 　Plasma Evaporation 　Metals Sputtering 　Metals 　Insulators
Patterning	Selective removal of the top LAYER(s) on the wafer.	Resist 　Positive 　Negative Exposure 　Contact 　Proximity 　Projection 　Direct Step 　E Beam 　X Ray Etch 　Wet 　Dry 　Lift-Off 　Ion Milling 　RIE
Doping	Change conductivity type and resistivity of selected portions of the wafer surface.	Thermal Diffusion 　Open Tube 　Closed Tube 　Doped Film Ion Implantation
Heat Treatments	Heat and/or cool the wafer for various effects.	Heat 　Hot Plate 　Convection IR Cool 　Freeze plate

Figure 5.11   Summary of basic wafer fabrication operations.

the functioning of the circuit. The circuit designer will start with a block functional diagram of the circuit such as the logic diagram in Fig. 5.12. Next, the designer translates the functional diagram to a schematic diagram (Fig. 5.13). This diagram shows the number and location of the various circuit components. Each component is represented by a symbol. Accompanying the schematic diagram are the electrical parameters (voltage, current, resistance, etc.) required to make the circuit work.

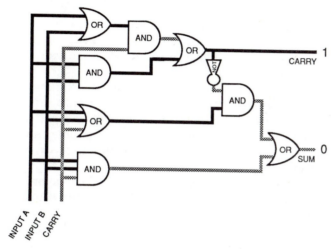

Figure 5.12 Example functional logic design of a simple circuit.

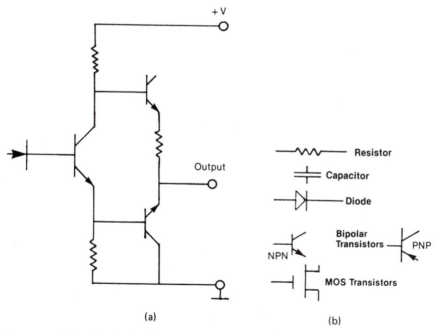

Figure 5.13 Example circuit schematic diagram with component symbols.

These first two steps are common to all electrical-circuit design. The third step, circuit layout, is unique to semiconductor circuits. Layout starts with the translation of each of the circuit components to its relative final dimensions, as they will be formed in and on the wafer surface.

Layout is a complicated process that is done on sophisticated computer-aided design (CAD) systems. The computer draws a composite picture of the circuit surface showing all of the sublayer patterns. This drawing is called a *composite* (Fig. 5.14). The composite drawing is analogous to the blueprint of a multistory office building as viewed from the top and showing all of the floors.

Since buildings and semiconductor circuits are built one layer at a time, it is necessary to separate the composite drawing into the individual layer drawings as shown in parts 1 to 7 in Fig. 5.14. After separation, each drawing is digitized. Digitizing is the tracing of the layer drawing on a computerized X-Y plotting table and storage of the drawing in the computer memory.

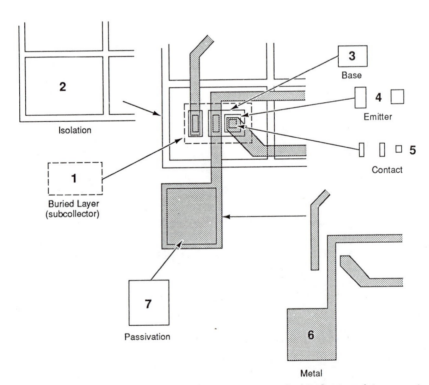

Figure 5.14 Composite and individual layer drawings of a bipolar transistor.

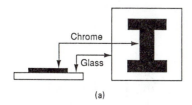

(a)

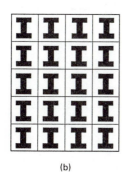

(b)

Figure 5.15 (a) Chrome on glass reticle of simple pattern; (b) photomask of same pattern.

Reticle and masks

The information from the digitizing step is used to produce a reticle. A reticle is a "hard copy" of the individual drawing recreated in a thin layer of chrome deposited on a glass or quartz plate (Fig. 5.15a). The reticle may be used directly in the patterning process or may be used to produce a photomask. A photomask is also a glass plate with a thin chrome layer on the surface. After production, it is covered with many copies of the circuit pattern (Fig. 5.15b). It is used to pattern a whole wafer surface in one pattern transfer. (The details of the reticle and mask-making processes are given in Chap. 11.)

The process sequence of the pattern from the functional diagram to the wafer surface is shown in Fig. 5.16. Reticles and masks are produced in a separate department or are purchased from outside vendors. They supply the fabrication area with a separate set of reticles or masks for each type of circuit.

Example fabrication process

The actual construction of the circuit proceeds in steps starting with a polished wafer. The cross section sequence in Fig. 5.17 shows the step-by-step formation of an MOS metal-gate transistor structure formed in a silicon wafer.

Step 1: Layering. The building starts with an oxidation of the wafer

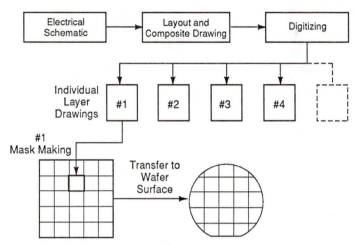

Figure 5.16 Pattern creation sequence.

surface to protect it and serve as a doping barrier. This silicon dioxide layer is called the *field oxide*. Once the wafer is oxidized, it goes to the patterning area for the first mask application.

Step 2: Patterning. The patterning process leaves two holes in the field oxide that define the source and drain areas of the transistor.

Step 3: Doping. Next, the wafer goes to a doping process where an N-type dopant is put through the openings to create two N-type pockets in the wafer surface.

Step 4: Patterning. In step 4, another patterning operation is used to remove the field oxide in the gate region between the source and drain.

Step 5: Layering. The exposed silicon in the gate region is reoxidized with a thin layer known as the gate oxide. At the same time, an oxide grows in the source and drain holes.

Step 6: Patterning. In this patterning step, two holes are patterned in the reoxidized source and drain regions. These holes are called *contact holes* and will allow direct contact of the metallization layer with the source and drain regions.

Step 7: Layering. Next in the sequence is the deposition of a conductive metal layer. This layer blankets the entire surface of the wafer.

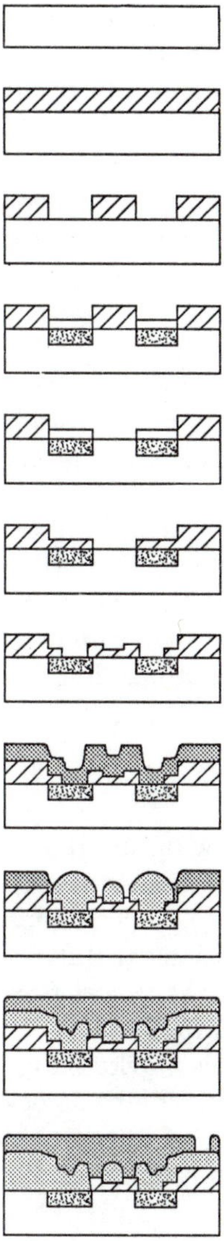

Cross Section	Step	Operation	Name/Purpose
		Starting wafer	
	1	Layering	Field Oxide
	2	Patterning	Source/drain holes
	3	Doping Layering	N-type doping and reoxidation of source/drain
	4	Patterning	Gate region is formed
	5	Layering	Gate oxide is grown
	6	Patterning	Contact holes are patterned into source/drain regions
	7	Layering	Conducting metal layer is deposited
	8	Patterning	Metal layer is patterned
	9	Heat Treatment	Metal is alloyed to layer
	10	Layering	Protective passivation layer is deposited
	11	Patterning	Passivation layer is removed over metal pads

Figure 5.17 Formation of metal gate MOS transistor.

Step 8: Patterning. After deposition, the wafer goes back to the patterning area where portions of the metallization layer are removed from the chip area and the scribe lines. The remaining portions connect all the parts of the surface components to each other in the exact pattern required by the circuit design.

Step 9: Heat Treatment (Alloy). Following the metal patterning step, the wafer goes through a heating process in a nitrogen gas atmosphere. The purpose of the step is to "alloy" the metal to the exposed source and drain regions and the gate region to ensure good electrical contact.

Step 10: Layering. The final layer of this device is a protective layer known variously as a *scratch* or *passivation layer*. Its purpose is to protect the components on the chip surface during the testing and packaging processes.

Step 11: Patterning. The last step in the sequence is a patterning process that removes the scratch protection layer over the metallization terminal pads on the periphery of the chip. This step is known as the *pad mask*.

The 11-step process illustrates how the four basic fabrication operations are used to build a particular transistor structure. The other components of the circuit are formed along with the transistors in the same sequence of steps. Other transistor types, such as bipolar and silicon gate MOS, are formed by the same basic four operations, but using different materials and in different sequences.

Fabrication process sequences

The process illustrated shows three sequences that show up in all process flows. The doping sequence (steps 1, 2, and 3) is used to create doped regions in the wafer surface. The metallization sequence (steps 6, 7, 8, and 9) is used to provide the metal interconnections of the circuit. The passivation sequence (steps 10 and 11) is used in most process flows to provide the top layer protection for the chip.

Figure 5.18 shows this process flow with the sequences marked. Figure 5.19 shows the same process flow but separated into the basic operations. This figure also illustrates the movement of the wafer through the various areas within the fabrication clean room.

Chip Terminology

Figure 5.20 is a photomicrograph of a bipolar medium-scale integration (MSI) integrated circuit. This level of integration was chosen so that the surface details could be seen. The components of higher-density circuits are so small that they cannot be distinguished on a photomicrograph of the entire chip. The chip features are:

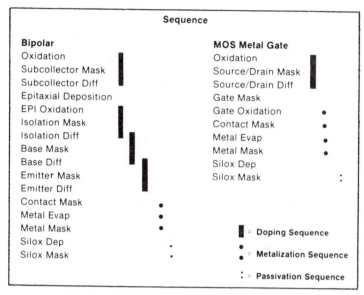

Figure 5.18 Comparison of bipolar and metal gate MOS process steps with sequences indicated.

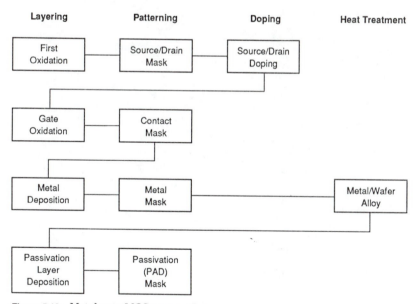

Figure 5.19 Metal gate MOS process flow.

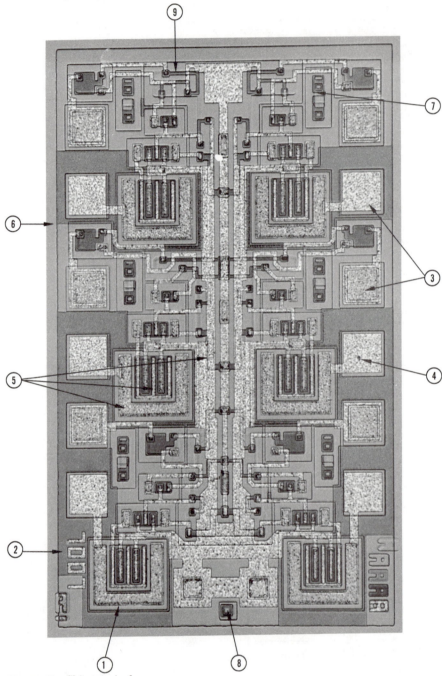

Figure 5.20 Chip terminology.

1. A bipolar transistor

2. The circuit designation number

3. Bonding pads for connecting the chip into a package

4. A piece of contamination on a bonding pad

5. Metallization lines

6. Scribe (separation) line

7. Unconnected component

8. Mask alignment marks

9. Resistor

Wafer Sort

Following the wafer-fabrication process comes a very important testing step, wafer sort. This test is the report card on the fabrication process. During the test, every individual chip is electrically tested for electrical performance and circuit functioning. Wafer sort is also known as *die sort* or *electrical sort*.

For the test, the wafer is mounted on a vacuum chuck and aligned to thin electrical probes that contact each of the bonding pads on the chip (Fig. 5.21). The probes are connected to power supplies that test the circuit and record the results. The number, sequence, and type of tests are directed by a computer program. The goal of the test is twofold. First is the evaluation of the effectiveness of the fabrication process in producing working chips. Second is the identification of working chips before they go into the packaging process. The working and nonworking chips are identified by a drop of ink on the *nonworking* chips or by a computer map of the wafer indicating the status of the chips.

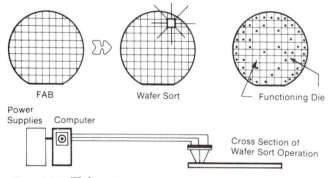

Figure 5.21 Wafer sort.

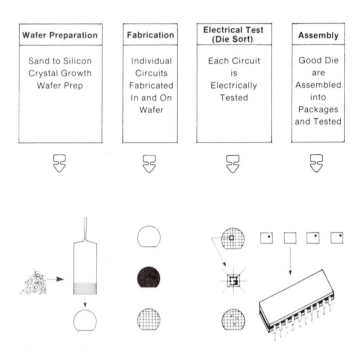

Wafer Preparation	Fabrication	Electrical Test (Die Sort)	Assembly
Sand to Silicon Crystal Growth Wafer Prep	Individual Circuits Fabricated In and On Wafer	Each Circuit is Electrically Tested	Good Die are Assembled into Packages and Tested

Figure 5.22 Integrated circuit manufacturing sequence.

Wafer sort is one of the principal yield calculations in the chip production process. The details of wafer sort and the implications of the yield are addressed in Chap. 6.

Packaging

Wafer fabrication and wafer sort complete stage three of the chip production process. Stage four is packaging where the chips are incorporated into a protective package (Fig. 5.22). The wafers are transferred to a packaging area on the same site or to a remote location. Many semiconductor producers package their chips in offshore facilities. (Chapter 18 details the packaging process.)

Key Concepts and Terms

Chip, die, device, circuit, bar
Chip terminology
Circuit-design steps
Circuit layout
Composite drawing

Engineering test die
Heat treatment operation
Layering operation
Metallization sequence
Wafer sort

Review Questions

1. Name one layering technique.
2. Which basic operations use the ion implant technique?
3. List the four basic wafer-fabrication operations.
4. Draw and label cross sections of a chip illustrating a doping sequence.
5. Describe a composite drawing.
6. Which basic operation uses a photomask?
7. What parameter is tested at wafer sort (wafer thickness, defect density, circuit function)?
8. What step in the circuit-design process uses a CAD system?
9. List the steps and basic operations in the metallization sequence.
10. What is the role of the holes created in the contact mask step?

6

Process Yields

Overview

High process yields are essential to the production of reliable chips at a profit. In this chapter, the major yield measurement points are identified along with the major influences on each of them. Typical yields at each of the measurement points and for different circuits are presented.

Objectives

Upon the completion of this chapter you should be able to:

1. Name the three major yield measurement points in the process.
2. Explain the effect of wafer diameter, die size, die density, number of edge die, and defect density on the wafer sort yield.
3. From a fabrication yield sheet, calculate the accumulative fabrication yield.
4. Be able to explain and calculate an overall process yield.
5. Explain the four major influences on fabrication yield.
6. Sketch a yield versus time curve for different process and circuit maturities.
7. Explain the relationship between high process yields and device reliability.

Yield Measurement Points

It is impossible to discuss semiconductor manufacturing without mentioning the term *yield*. To the casual observer it would seem that the industry is fixated on their production yields. That observation is accurate. It is accurate because of the demanding nature of the process and the sheer number of processes required to produce a packaged chip. These two factors result in a production process that typically ships only 20 to 40 percent of the chips it commits into the wafer-fabrication line.

These yields seem extraordinarily low to most manufacturing engineers. Yet when one considers the goal of producing millions of micron-size patterns in layers that are equally thin, at very stringent cleanliness levels, all within the confines of a $\frac{1}{2}$-in^2 chip, it is a testament to the industry that any chips at all are produced.

Another factor that helps keep yields depressed is the nonrepairable nature of most production mistakes. While defective automobile parts can be replaced, no such option is available in semiconductor manufacturing. The final fact of semiconductor processing is that scrap wafers cannot be recovered or recycled.

Added to these process factors is the volume nature of the business. High capital costs and a higher than average percentage of engineering personnel translate to a high-overhead situation. This high overhead coupled with competition that keeps downward pressure on selling prices requires that most chip producers run a high-volume, high-yield process to stay in business.

Given all of the factors the preoccupation with yield is understandable. Most suppliers of equipment and materials tout the yield improvements possible with their products. Likewise, process engineering groups have as their prime responsibility the maintenance and raising of process yields. Yield percentages are calculated at every process station. However, three major yield points are calculated and monitored to measure and control the process (Fig. 6.1).

Accumulative Wafer-Fabrication Yield

The first major yield measurement is calculated at the completion of the last wafer-fabrication process step. This yield is called by a variety of names, including fab yield, line yield, accumulative fab yield, or "cum" yield.

- Wafer Fabrication
- Wafer (Die) Sort
- Packaging

Figure 6.1 Major yield measurement points.

The cum yield is expressed as a percentage and calculated two different ways. One is simply to divide the number of wafers exiting the fabrication area by the number that were started. This simplistic calculation is seldom used because most fabrication lines produce a number of different circuits simultaneously. Keep in mind that different circuits have different feature size and density factors, and that most often the fabrication line will be running a number of different process flows to produce the various circuits. These factors translate to a fabrication line producing a variety of chips by a variety of processes, each with a different number of steps and its own degree of difficulty and cum yield.

Also, a wafer-fabrication line carries a large inventory of in-process wafers, and the cycle times stretch out to 4 to 6 weeks. These factors often result in uneven movement through the line. Thus the number of wafers exiting the process seldom relate directly to the number started. Using a simplistic counting of wafers in and wafers out would not reflect the true yield of each of the individual circuit types.

In most cases the cum yield is calculated by multiplying together the yields at each of the process steps for a particular circuit type. This chain multiplication ends in the accumulated yield for that process flow. Figure 6.2 lists an 11-step process such as the one illustrated in Chap. 5. Typical individual process step yields are listed in column 3

STEP	Wafers In	Yield*	Wafers Out	Accumulated Yield
1. Field Oxidation	1000	99.5	995	99.5
2. Source/Drain Mask	995	99.0	985	98.5
3. Source/Drain Doping	985	99.3	978	97.8
4. Gate Region Mask	978	99.0	968	96.8
5. Gate Oxidation	968	99.5	964	96.4
6. Contact Hole Mask	964	94.0	906	90.6
7. Deposit Metal Layer	906	99.2	899	89.9
8. Metal Layer Mask	899	97.5	876	87.6
9. Alloy Metal Layer	876	100	876	87.6
10. Passivation Layer Deposition	876	99.5	872	87.2
11. Passivation Layer Mask	872	98.5	859	85.9

*Yield values are typical for the particular steps.

Figure 6.2 Accumulated (wafer fab) yield calculation.

and the accumulated yield in column 5. Note that the accumulated yield equals the simple cum yield calculation for this individual circuit.

Typical wafer-fabrication yields vary from 50 to 95 percent depending on a number of factors detailed in the following text. The calculated cum yield is used for production planning and by engineering and management as a measure of the process effectiveness.

Wafer-Fabrication Yield Limiters

Wafer-fabrication yield is limited by four dominant factors. They are:

1. Number of process steps

2. Wafer breakage and warping

3. Process variation

4. Process defects

Number of process steps

Note in the calculation in Fig. 6.2 that each individual process step yield must be in the 90 percent range to produce the 85.9 percent accumulated yield. Illustrated is a fairly simple 11-step process. Since most VLSI and ULSI circuits require 50 or 60 major process steps, it is easy to appreciate the continual pressures on fabrication areas to maintain high cum yields. The first cum yield limiter is the number of process steps. The more complicated the circuit, with a high number of steps, the lower the expected cum yield.

More process steps also increase the probability that one of the other three yield limiters will affect the wafer during the process. This factor is a tyranny of numbers. For example, in order to achieve a 75 percent accumulated fabrication yield with a 50-step process, each of the individual steps would have to be 99.4 percent! A further tyranny of this type of calculation is that the cum yield can never exceed the lowest individual step yield. If one process step can only achieve a 50 percent yield, the overall cum yield can never be higher than 50 percent.

More manufacturing pressure is present in the fact that for each major process step there are a number of substeps. In the illustrated 11-step process, the first step is oxidation. That step can actually entail eight substeps, all of which represent an opportunity to break a wafer or make a mistake. (Figure 6.3 lists the substeps for a typical oxidation step.)

A figure of merit for a profitable merchant semiconductor is a min-

Substep	Number of Wafer Handlings
1. Wafers are removed from carrier and placed in cleaning boat.	2
2. Wafers are cleaned, rinsed, and dried.	1
3. Wafers are removed from cleaning boat, inspected, and placed on oxidation boat.	2
4. Boat is removed from furnace.	0
5. Wafers are removed from boat and placed back in carrier.	1
6. Test wafers are removed from carrier and measured.	2
TOTAL NUMBER OF HANDLINGS	8

Figure 6.3 Substeps of oxidation process.

imum of a 75 percent cum yield, with automated lines achieving up to a 90 percent (or better) yield.

Wafer breakage and warping

During the course of the fabrication process, the wafers are handled many times by a combination of manual and automatic techniques. A major concern is breaking of the relatively fragile wafers. Recall that a typical 6-in-diameter wafer is only about 25 mils (0.025 in) thick. Great care must be taken to handle the wafers gently and to adjust automatic handlers to minimize breakage.

The many heat treatments add to the susceptibility of the wafers to breaking. Since the wafers are crystalline materials, strains caused by high-temperature processing can result in breaking in subsequent processes. A requirement of automatic processing machines is that only full-diameter wafers can be processed. Therefore, any breakage, however small, is a cause for rejecting the wafer from the process.

Silicon wafers are relatively easy to handle with good practices and automatic equipment has reduced wafer breakage to a low level. Gallium arsenide wafers, however, are not that resilient and breakage is a major wafer-yield limiter. In gallium arsenide fabrication lines, where the circuits and devices are very high performance and command a high selling price, partial wafers may be processed, especially through the manual processes.

Along with minimizing breakage, the wafer surfaces must remain flat throughout the processing. This is especially true on fabrication lines that use projection patterning techniques. This technique projects the pattern onto the wafer surface. If the surface is warped or wavy, the projected image will become distorted and have out-of-specification image dimensions. Warping comes about from rapid heating and/or cooling of the wafers in tube furnaces. (The solution for this problem is addressed in Chap. 7.)

Process variation

As the wafer comes through the fabrication process it receives a number of doping, layering, and patterning processes, each of which must meet incredibly stringent physical and cleanliness requirements. On a daily basis, even the most sophisticated processes will vary, inducing some variation in the result on the wafer surface.

Throughout the process there are a number of inspections and tests designed to detect unwanted variations as well as frequent calibration of the equipment parameters to process specifications. Some of these tests are performed by production personnel and some by quality control organizations. All of the tests and process specifications allow for some variation. Therefore, at the end of the process, each wafer will vary from ideal specifications; that is, each wafer experiences process variations. Occasionally individual wafers or complete batches will fall outside established specs and are scrapped. One of the major benefits of automated processing is the minimization of these process variations.

Process defects

Process defects are defined as isolated regions (or spots) of contamination or irregularities on the wafer surface. These defects are often called *spot defects*. Within a circuit only *one* very small defect can render the circuit inoperable. Unfortunately, these small isolated defects are not always detectable within the fabrication process. They become evident at wafer sort as rejected chips.

The major sources of these defects are the various liquids, gases, room air, personnel, process machines, and water used in the fabrication area. Particulates and other small contaminants become lodged in or on the wafer surface. Many of these defects occur in the patterning process. Recall that the patterning process requires using a thin, fragile layer of photoresist to protect the wafer surface during the etch steps. Any holes or tears in the photoresist layer from particulates will end up as tiny etched holes in the wafer surface layer. These holes are called pinholes and are a major concern of photomasking engineers. Consequently the wafers are inspected often, usually after each major step, for contamination. Wafers that exceed the established allowable density are rejected.

Wafer-Sort Yield Factors

After fabrication, the wafers go to the wafer sort tester. During the test, each chip will be tested electrically for device specifications and functionality. Up to several hundred individual electrical tests may be performed on each circuit. While these tests measure the electrical performance of the device(s), they indirectly are measuring the precision and cleanliness of the fabrication processes. Because of natural process variations and undetected contamination, the wafer may have passed all the in-process checks and still have many chips that do not function.

Since wafer sort is a comprehensive test, many factors influence the yield. They are:

1. Wafer diameter
2. Die size (area)
3. Number of processing steps
4. Circuit density
5. Defect density
6. Crystal defect density
7. Process cycle time

Wafer diameter and edge die

The semiconductor industry went to round wafers with the introduction of silicon. The first wafers were less than 1 in in diameter. Since that time there has been a regular progression to larger-diameter wafers, with 150 mm (6 in) being the standard of VLSI fabrication lines in the late 1980s with 200-mm wafers being developed for production use in the 1990s.

The move to larger-diameter wafers has been driven by production efficiency and wafer-sort yield improvements. Production efficiency is easily understood when one considers that the increased cost of processing a larger-diameter wafer is incremental while the number of available whole chips on the wafer can increase substantially as illustrated in Fig. 6.4.

The effect of increasing the wafer diameter also has positive effects on the wafer-sort yield. The increase comes from several factors. The first has to do with the relative number of whole and edge die. Figure

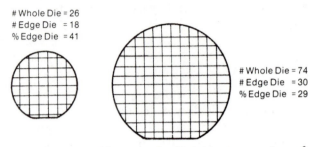

Whole Die = 26
Edge Die = 18
% Edge Die = 41

Whole Die = 74
Edge Die = 30
% Edge Die = 29

Figure 6.4 Effect of larger wafer diameter on percentage of partial die.

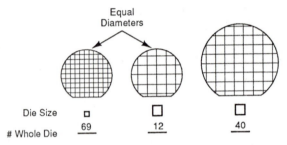

Equal
Diameters

Die Size   ▫   □   □

Whole Die   69   12   40

Figure 6.5 Effect of processing larger die on large wafers.

6.5 shows two wafers of the same diameter but with different die sizes. Note that the smaller-diameter wafer has a very large proportion of its surface covered with partial die, die that cannot function. The larger-diameter wafer with its greater number and percentage of whole die will, if all other factors are equal, have a higher wafer-sort yield.

Wafer diameter and die size

Another driving force for larger-diameter wafers is the trend to larger die sizes. As shown in Fig. 6.5, increasing the die size *without* increasing the wafer diameter also results in a wafer surface with a smaller percentage of whole die.

Maintaining a decent wafer-sort yield as the die size increases requires increasing the wafer diameter. Figure 6.6 lists the number of various size chips that will fit on different size wafers. The bottom line is that larger-diameter wafers are more cost-effective.

Wafer diameter and crystal defects

In Chap. 3 the concept of a crystal dislocation was introduced. A crystal dislocation is a point defect *in the wafer* that comes from a local

Die Size, mils	Wafer Diameter		
	3.0 in	100 mm	125 mm
100 × 100	638	1164	1788
150 × 150	276	496	780
200 × 200	144	276	432
250 × 250	88	164	260
300 × 300	60	112	180
350 × 350	42	80	129

Figure 6.6 Die size versus number of die on a wafer.

discontinuity of the crystal structure. Dislocations exist throughout the crystal structure and, like contamination and process defect density, affect the wafer-sort yield.

Dislocations also are generated during the fabrication process. They generate (or nucleate) at sites where there are chips and abrasions of the edge of the wafer. These chips and abrasions come from poor handling techniques and automatic handling equipment. The abraded area causes a crystal dislocation. Unfortunately the dislocation is propagated into the center of the wafer (Fig. 6.7) during subsequent heat treatments, such as oxidations and diffusions. The length of the dislocation line into the interior of the wafer is a function of the thermal history of the wafer. Consequently, wafers receiving more process steps and/or more heating steps will have more and longer dislocation lines and a larger area of the outside of the wafer will have lower yield. One obvious solution to the problem is larger-diameter wafers.

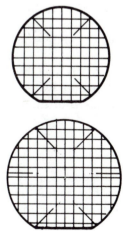

Figure 6.7 Effect of dislocations on wafer-sort yield for different wafer diameters.

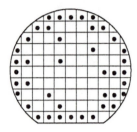

Figure 6.8 Typical location of functioning die after wafer sort.

• Nonfunctioning Die

This solution leaves a larger number of unaffected die in the center of the wafer.

Wafer diameter and process variations

The process variations discussed above under "Wafer Breakage and Warping" also limit the wafer-sort yield. In the fabrication area, process variations are detected by sampling inspection and measurement techniques. The nature of sample inspections is that not all of the variations and defects are detected so that wafers are passed on with some number of problems. These problems show up at wafer sort as failed devices.

Process variations occur at a higher rate around the edge of the wafer. For example, feature size uniformity is dependent on the uniformity of the exposing light used in the patterning process. The nature of the light systems is such that the center will be of higher uniformity than the outside edges. And in the high-temperature processes performed in tube furnaces, there is always some temperature nonuniformity across the wafers. The change in temperature results in uniformity differences on the wafer. The third contributor to this wafer edge phenomenon is contamination and physical abuse of the wafer layers that emanates from handling and touching the wafers on their edges.

All of these problems result in a lower wafer-sort yield around the edge of the wafer, as illustrated in Fig. 6.8. Larger-diameter wafers help to maintain wafer-sort yields by having a larger area of unaffected die in the center of the wafer.

Die area and defect density

The die size also affects wafer sort yield relative to the defect density on the wafer surface. The relationship is illustrated in Figure 6.9. In Fig. 6.9a, a wafer is shown with five defects and no die pattern. This situation demonstrates the fact of a background defect density that is

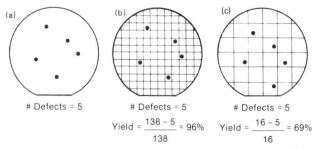

Figure 6.9 Effect of defects of sort yield for different die sizes.

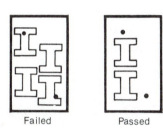

Failed   Passed

Figure 6.10 Relationship of defect location to die density.

independent of process mistakes and gross defects from contaminated chemicals and/or failure of the clean-room procedures. The wafers in Fig. 6.9*b* and *c* illustrate the effect of this background defect density on the wafer-sort yield for two different die sizes. The larger the die size for a given defect density, the lower the yield.

Circuit density and defect density

The defects on the wafer surface cause die failures by causing a malfunction of some part of the die. Some of the defects are located in nonsensitive parts of the die and do not cause a failure. However, the trend is to higher levels of circuit integration which came about because of smaller feature size and a higher density of die components. The result of these trends is a higher probability that any given defect will be in an active part of the circuit, thus lowering the wafer-sort yield as illustrated in Fig. 6.10.

Wafer-sort yield formulas

The ability to understand and predict wafer-sort yields with some accuracy is essential to the operation of a profitable and reliable chip supplier. Over the years, a number of models have been developed that relate process and size parameters to the wafer-sort yield. The three model formulas are shown in Fig. 6.11.

Exponential $Y = \dfrac{1}{e^{AD}}$

Seeds $Y = \dfrac{1}{(1 + DA)^n} \left(\dfrac{r - \sqrt{A}}{\sqrt{A}} \right)^2$

Murphy $Y - \left(\dfrac{1 - e^{-DA}}{DA} \right)^2$

where Y = percent functioning die
  A = die area
  D = defect density
  n = number of masking steps
  r = radius of wafer

Figure 6.11 Wafer sort yield models.

The exponential relationship (Fig. 6.11a) is the simplest and one of the first yield models[1] developed. It is applicable for wafers that contain over 300 die and MSI circuits of lower densities. The requirement of many die for use of this formula is because the model does not have a factor for edge die as does the Seeds model. The exponential model is a valuable learning tool because it clearly shows the primary relationships between die area, defect density, and wafer sort yield. To understand the relationship in this mathematical model remember that e is a constant with a value of 2.718. The exponential model shows that any increase in die area or defect density will increase the value of the divisor on the right-hand side of the equation and drive the wafer sort yield down.

The Seeds model adds two additional parameters to the job of predicting wafer sort yield, the number of process steps (n) and the wafer radius–die size relationship $r(r$ and $a)$. In most yield models, the factor for processing steps (n) is actually the number of patterning steps. Experience has proved that the patterning steps contribute the greatest number of point defects and therefore have a direct bearing on sort yield.

Perhaps the most widely used yield model is the Murphy relationship (Fig. 6.11). While not including a factor for the number of processing steps, it has proved a good predictor for LSI-, VLSI-, and ULSI-level circuits. Figure 6.12 shows a graphical solution of the Murphy model, where the relationships of sort yield, defect density, and die size are clearly shown.

The defect density used in all the models is not the same as a defect density determined by optical inspection of the wafer surface. The de-

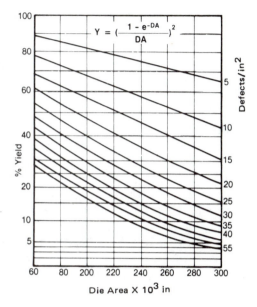

$$Y = \left(\frac{1 - e^{-DA}}{DA} \right)^2$$

Figure 6.12 Murphy's yield model, showing die yield as a function of die size defect density.

fect density that shows up in the yield models is all-inclusive; it includes contaminants and surface and crystal defects. Further, it predicts only the defects that destroy die: the "killer defects." Defects that fall in noncritical areas of the chip are not part of the models nor are situations where two or more defects fall in the same sensitive area.

It is also important to keep in mind that the yield numbers predicted by the formulas are those expected from a process that is basically under control. In reality the wafer sort yield will vary from wafer to wafer because of the normal process variations in the fabrication process. A typical wafer-sort yield plot is shown in Fig. 6.13. Note that

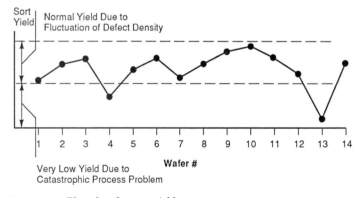

Figure 6.13   Plot of wafer sort yields.

wafer 13 falls far below the normal range of sort yields. In a situation like this, the process engineer would look for some catastrophic process failure such as an out-of-spec layer thickness or a doping layer that is too deep or too shallow.

Assembly and Final Test Yields

After wafer sort, the wafers go to the assembly (packaging) process. There they are cut into die and packaged into a protective enclosure. During this series of steps, there are a number of visual inspections and quality checks of the assembly process.

At the conclusion of the packaging process, the packaged die go through a series of physical, environmental, and electrical tests, known collectively as the *final test*. (The details of the processes, inspections, and final tests are described in Chap. 18.) After the final tests, the third major yield is calculated, which is the ratio of die passing final test compared with the number of die that passed the wafer sort test.

Overall Process Yields

The overall process yield is the mathematical product of the three major yield points (Fig. 6.14). This number, expressed as a percent, gives the percentage of shipped die as compared with the number of whole die on the starting wafer. It is an inclusive measurement of the success of the entire process.

Overall yields vary with several major factors. In Fig. 6.15 is a list of typical process yields and their calculated overall yield. In the first two columns are major process factors that influence the individual and overall yields.

First is the integration level of the particular circuit. The more highly integrated the circuit, the lower the expected yield in all categories. Column 2 lists the maturity of the manufacturing process. Process yields almost always follow an *S* curve pattern (Fig. 6.16) through the lifetime of the product in manufacturing. In the beginning the yield rises rather slowly as the initial bugs are worked out of the process. This is followed by a period when the yields rise rapidly, eventually leveling off as the limits imposed by the process maturity die size, integration level, circuit density, and defect density.

As the table in Fig. 6.15 shows, overall yields can vary from as little

---

Fab Yield (%) × Sort Yield (%) × Test Yield × = Overall Yield (%)

Figure 6.14   Overall yield calculations.

Product	Level of Integration	Position in Product Cycle	Fab Yield %	Sort Yield %	Assembly and Final Test Yield %	Overall Yield %
IC	VLSI	Introduction	60	2	85	1
IC	LSI	Mature	80	50	95	38
IC	MSI	Mature	85	60	97	49
Discrete	—	Mature	95	80	98	74

Figure 6.15 Typical yields for various semiconductor products.

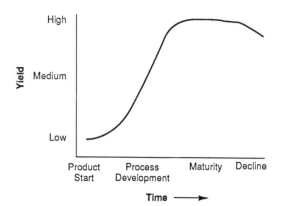

Figure 6.16 Yield changes with process maturity levels.

as 1 percent to as high as 84 percent depending on the particular factors connected with the circuit. Throughout the lifetime of the product in manufacturing the yields start low and (hopefully) increase. Semiconductor producers consider their yield performance very proprietary since profit and production control are a direct function of the process yields.

An examination of the yield values in the table reveals that wafer sort is the lowest of the three yield points. This fact illustrates why yield-improvement programs are directed at the many factors that influence the wafer sort yield. Improvement of wafer sort yields has the biggest impact on overall line performance.

Key Concepts and Terms

Assembly and final test yields
Defect density
Murphy's yield formula
Overall yield factors
Three major yield points

Typical overall yields
Wafer-fabrication yield factors
Wafer-sort yield factors
Wafer-sort yield formulas

Review Questions

1. Name the three yield measurement points in the manufacturing process.
2. Of the three yield points, which is typically the lowest?
3. Indicate if the wafer sort yield would go up or down with the following changes:
    a. Change to larger-diameter wafers
    b. Change to smaller die size
    c. More process steps
    d. Decrease in defect density
    e. Increase in die density
4. If 1000 wafers are started into the process and 800 are passed to wafer sort, what is the accumulated fabrication yield?
5. Calculate the overall yield for a process that has an 82 percent fabrication yield, a 47 percent wafer sort yield, and a 92 percent assembly and final test yield.
6. Which situation would you expect to have the highest overall yield, a 22-step LSI process or a 33-step VLSI process?

Reference

1. S. M. Sze, *VLSI Technology*, McGraw-Hill Publishing Company, New York, 1983, p. 605.

7

Oxidation

Overview

The ability of a silicon surface to form a silicon dioxide passivation layer is one of the key factors in silicon technology. In this chapter, the uses, formation, and processes of silicon dioxide growth are explained. Detailed is the all-important tube furnace, which is a mainstay of oxidation, diffusion, heat treatment, and chemical vapor deposition processes.

Objectives

Upon completion of this chapter, you should be able to:

1. List the three principal uses of a silicon dioxide layer in silicon devices.

2. Describe the mechanism of thermal oxidation.

3. Sketch and identify the principal sections of a tube furnace.

4. List the two oxidants used in thermal oxidation.

5. Sketch diagrams of the three methods of wet oxidation.

6. Draw a flow diagram of a typical oxidation process.

7. Explain the relationship of process time, pressure, and temperature on the thickness of a thermally grown silicon dioxide layer.

8. Describe the principles and uses of rapid thermal, high-pressure, and anodic oxidation.

Silicon Dioxide Layer Uses

When a silicon surface is exposed to oxygen, it is converted to silicon dioxide, whose molecule is composed of one silicon atom and two oxygen atoms (SiO_2). Although silicon is a semiconducting material, silicon dioxide is a dielectric material. This combination, a dielectric formed on a semiconductor, along with other properties of silicon dioxide, makes it one of the most commonly used layers in silicon devices. Silicon dioxide layers find use in devices to passify the silicon surface, to act as doping barriers and surface dielectrics, and to serve as dielectric parts of devices.

Surface passivation

In Chap. 4, the extreme sensitivity of semiconductor devices to contamination was examined. While a major focus of a semiconductor facility is the control and elimination of contamination, the techniques are not always 100 percent effective. Silicon dioxide layers play an important role in protecting semiconductor devices from contamination.

Silicon dioxide performs this role in two ways. First is the physical protection of the surface and underlying devices. Silicon dioxide layers formed on the surface are very dense (nonporous) and very hard. Silicon dioxide is the material that we know as glass, a hard and durable material. Thus a silicon dioxide layer (Fig. 7.1) acts as a contamination barrier by preventing dirt in the processing environment from getting to the sensitive wafer surface. The hardness of the layer protects the wafer surface from scratches and abuse endured by the wafer in the fabrication processes.

The second way silicon dioxide protects devices is chemical in nature. Regardless of the cleanliness of the processing environment, some electrically active contaminants (mobile ionic contaminants) end up on the wafer surface. Fortunately, during the thermal oxidation process these contaminants are drawn up into the silicon dioxide film where they are less harmful to the devices. In the early days of MOS device processing, it was common to oxidize the wafers and then remove the oxide before further processing to rid the surface of unwanted mobile ionic contamination.

Doping barrier

In Chap. 5 doping was identified as one of the four basic fabrication operations. Doping requires creating holes in a surface layer through which specific dopants are introduced into the exposed wafer surface.

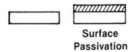

Surface
Passivation

Figure 7.1 Surface passivation with silicon dioxide layers.

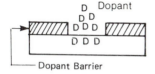

Figure 7.2 Dopant barrier with silicon dioxide layer.

In silicon technology, the surface layer is most often thermally grown silicon dioxide (Fig. 7.2). Silicon dioxide serves this purpose well for both diffusion and ion implant doping processes. All of the dopants used in silicon technology have a very slow rate of movement in silicon dioxide as compared to silicon. While the dopants penetrate to the required depth in the exposed silicon, they only penetrate a short way into the silicon dioxide surface. It only takes a relatively thin silicon dioxide layer to block the dopants from reaching the silicon surface.

Another factor favoring the use of silicon dioxide as a dopant barrier is that it possesses a thermal expansion coefficient similar to that of silicon. In diffusion doping, the wafer is heated and exposed to dopant vapors at a high temperature. The silicon dioxide expands and contracts at close to the same rate as silicon, which means that the wafer will not warp during the heating and cooling.

In the ion implant technique, dopant atoms are sprayed across the wafer surface. Those hitting the exposed silicon penetrate the wafer. The dopant atoms hitting the silicon dioxide layer are slowed down and blocked from reaching the wafer surface.

Surface dielectric

Silicon dioxide is classified as a dielectric. This means that under normal circumstances it does not conduct electricity. When dielectrics are used in electrical circuits or devices they are referred to as *insulators*. Acting as an insulator is an important role of silicon dioxide layers. Figure 7.3 shows a cross section of a wafer with a conductive layer of metal on top of a layer of silicon dioxide. The oxide prevents shorting of the metal layer to the underlying metal. In this capacity the oxide must be continuous, that is, having no holes or voids. The oxide must also be thick enough to prevent charges induced by electrical currents flowing in the metal layers from inducing unwanted charges in the

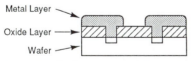

Figure 7.3 Dielectric use of silicon dioxide layer.

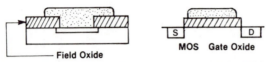

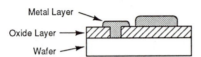

Figure 7.4 Silicon dioxide as field oxide and in MOS gate.

Figure 7.5 Silicon dioxide layer in solid state capacitor.

wafer surface. An oxide layer that covers most of the wafer surface is called a *field oxide*.

Device dielectric

In MOS technology, thin layers of silicon dioxide are grown in the gate regions (Fig. 7.4) of the transistors. The oxide functions as a dielectric whose thickness is chosen specifically to allow induction of a charge in the gate region. Within an MOS transistor, the gate is the part that controls the operation of the device. The dominance of MOS technology for very large-scale integrated (VLSI) circuits has made the formation of gate regions a prime focus of process development and concern. (The operation of MOS transistors is detailed in Chap. 16.) Thermally-grown oxides are also used as the dielectric layer in capacitors formed between the silicon wafer and a surface conduction layer (Fig. 7.5).

Thermal Oxidation Mechanism

The silicon dioxide layers used in silicon-based devices vary in thickness. Figure 7.6 lists the thickness ranges for the major uses. Most of

Silicon Dioxide Thickness, Å	Application
60–100	Tunneling Gates
150–500	Gate Oxides, Capacitor Dielectrics
200–500	LOCOS Pad Oxide
2,000–5,000	Masking Oxides, Surface Passivation
3,000–10,000	Field Oxides

Figure 7.6 Silicon dioxide layer thicknesses. *(After Wolf and Tauber.[1])*

Si (solid) + O_2 (gas) → SiO_2 (solid)

Figure 7.7 Reaction of silicon and oxygen to form silicon dioxide.

the silicon dioxide layers are grown on the silicon surface by thermal oxidation. Three methods are used: atmospheric growth, rapid thermal oxidation, and high-pressure oxidation.

For all three methods, the basic mechanism of oxide growth is the same. It is a simple chemical reaction as shown in Fig. 7.7. This reaction takes place even at room temperature. However, an elevated temperature is required to achieve quality oxides in reasonable process times for practical use in circuits and devices. Oxidation temperatures are between 900 and 1200°C.

Although the formula shows the reaction of silicon with oxygen, it does not illustrate the growth mechanism of the oxide. To understand the growth mechanism, consider a wafer placed in a heated chamber and exposed to oxygen gas (Fig. 7.8). The actual growth of the oxide takes place in two stages, a linear stage and a parabolic stage. The linear stage starts immediately when the surface of the silicon wafer is exposed to oxygen. This stage is called linear because the oxide grows in equal amounts for each unit of time. After approximately 500 angstroms (Å) of oxide is grown, a limit is imposed on the growth rate. [An angstrom is one ten-thousandth of a micron (μm); in other words there are 10,000 Å in 1 μm.]

The growth rate limit is a result of the separation of the silicon surface from the oxygen vapors by the oxide layer grown in the initial stage. For oxide growth to continue, either the silicon in the wafer must migrate to the vapor or the oxygen in the vapor must migrate to the wafer surface. In the thermal growth of silicon dioxide, the oxygen migrates (the technical term is *diffuses*) through the existing oxide layer to the silicon wafer surface. The silicon in the silicon dioxide

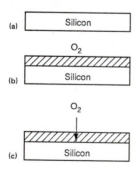

Figure 7.8 Formation mechanism of S_iO_2. (*a*) Start; (*b*) initial linear stage; (*c*) parabolic stage.

$$R = \frac{X^2}{t}$$

where R = silicon dioxide growth rate
X = oxide thickness
t = oxidation time

Figure 7.9 Parabolic relationship of SiO_2 growth parameters.

layer is provided by the wafer itself. Thus the layer of silicon dioxide grows *into* the silicon surface.

With each succeeding new growth layer, the diffusing oxygen must move further to reach the wafer. The effect is a slowing of the oxide growth rate with time. When graphed, the mathematical relationship of the oxide thickness, growth rate, and time takes the shape of a parabola, and silicon thermal oxidation is said to be a *parabolic growth relationship*. Another term used for this second stage of growth is a *transport-limited reaction*, which means that the growth rate is limited by the transportation of the oxygen through the oxide layer already grown. The formula in Fig. 7.9 expresses the fundamental parabolic relationship for oxide layers above approximately 1200 Å. A more detailed explanation of the growth mechanism of silicon dioxide is presented in *Silicon Processing* by Wolf and Tauber.[1]

The major implication of this parabolic relationship is that thicker oxides require much more time to grow than thinner oxides. For example, growth of a 2000-Å (0.20-μm) film at 1200°C in dry oxygen requires 6 minutes (Fig. 7.10).[2] To double the oxide thickness to 4000 Å requires some 220 minutes, over 36 times as long. This longer oxidation time presents a problem for semiconductor processing. When pure dry oxygen is used as the oxidizing gas, the growth of thick oxide layers requires even longer oxidation times, especially at the lower temperatures. Generally, process engineers want to have the shortest process times possible as is consistent with quality control. The 220 minutes in the example given is excessive, since only one oxidation would be possible in one shift of operation.

One way to achieve faster oxidations is to use water vapor (H_2O) instead of oxygen as the oxidizing gas (oxidant). The growth of silicon dioxide in water vapor proceeds by the reaction shown in Fig. 7.11. In the vapor state the water is in the form $H-OH^-$. It is composed of one atom of hydrogen (H) and a molecule of oxygen and hydrogen with a negative charge (OH^-). This molecule is called the *hydroxyl ion*. The hydroxyl ion diffuses through the oxide layers already on the wafer

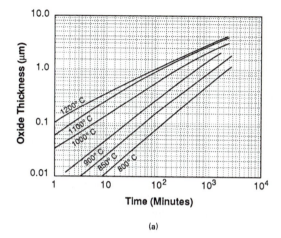

(a)

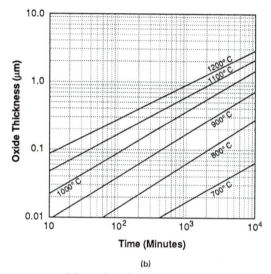

(b)

Figure 7.10 Silicon dioxide versus time for different temperatures and orientation in (a) dry oxidation and (b) steam.

faster than oxygen. The net effect is a faster oxidation of the silicon, as shown in the growth curves in Fig. 7.10.[2]

Water vapor at the oxidation temperatures is in the form of steam, and the process is called either *steam oxidation, wet oxidation*, or *pyrogenic steam*. The term wet oxidation comes from the time when liquid water was the primary water vapor source. An oxygen-only oxidation process is called *dry oxidation*. If oxygen only is used, it must

$$Si \text{ (solid)} + H_2O \text{ (gas)} \rightarrow SiO_2 + 2H_2 \text{ (gas)}$$

Figure 7.11   Chemical reaction of silicon and water vapor.

be free of any water vapor (dry) or the oxide growth would be that of water vapor.

Notice in the reaction of water vapor and silicon that there are two hydrogen molecules ($2H_2$) on the right side of the equation. Initially these hydrogen molecules are trapped in the solid silicon dioxide layer, making the layer less dense than an oxide grown in dry oxygen. However, after a heating of the oxide in an inert atmosphere, such as nitrogen (see "Oxidation Processes," p. 154), the two oxides are identical in structure and properties.

Wafer orientation

The orientation of the wafer has an effect on the oxidation growth rate. <111> planes have more silicon atoms than <100> planes. The larger number of atoms allows for a faster oxide growth on <111>-oriented wafers than for <100>-oriented wafers. Figure 7.12 shows the growth rates for the two orientations in steam.

Wafer dopant redistribution

The silicon surface being oxidized always has dopants. A production silicon wafer starts into the line doped as either an N type or P type. Later on in the process, wafers have dopant(s) in the surface from dif-

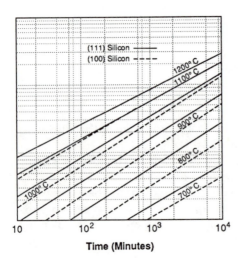

Time (Minutes)

Figure 7.12   Oxidation of <111> and <100> silicon in steam.

fusion or ion implant operations. The dopant elements used and their concentration both have effects on the oxidation growth rate. For example, oxides grown over a highly doped phosphorus layer are less dense than those grown over the other silicon dopants. These phosphorus-doped oxides also etch faster and present an etching challenge in the patterning operation due to resist lifting and rapid undercutting.

Another effect on oxidation growth rate is the distribution of the dopant atoms in the silicon after the oxidation is completed.[3] Recall that during thermal oxidation the oxide layer grows *down* into the wafer. A question is "What happens to the dopant atoms that were in the layer of silicon converted to silicon dioxide?" The answer depends on the conductivity type of the dopant. The N-type dopants of phosphorus, arsenic, and antimony have a higher solubility in silicon than in silicon dioxide. When the advancing oxide layer reaches them, they move down into the wafer. The silicon–silicon dioxide interface acts like a snowplow pushing ahead an ever-greater pile of snow. The effect is that there is a higher concentration (called *pile-up*) of N-type dopants at the silicon dioxide–silicon interface than was originally in the wafer.

When the dopant is the P-type boron, the opposite effect happens. The boron is drawn up into the silicon dioxide layer, causing the silicon at the interface to be depleted of the original boron atoms. Both of these effects, pile-up and depletion, have significant impact on electrical performance of devices. The exact effects of pile-up and depletion on the dopant concentration profile are illustrated in Chap. 17.

Thermal Oxidation Methods

The oxide formation reaction formulas include a triangle under the reaction direction arrows. These triangles indicate that the reaction requires energy to proceed. In silicon technology that energy is usually supplied by heating the wafers and is called *thermal oxidation*. Silicon dioxide layers are grown either at atmospheric pressure or at high pressure. There are two atmospheric techniques, tube furnaces and rapid thermal systems (Fig. 7.13).

Horizontal tube furnaces

The mainstay equipment for the thermal oxidation of silicon is the horizontal tube furnace. Horizontal tube furnaces have been used in the industry since the early 1960s for oxidation, diffusion, heat treating, and various deposition processes. They were first developed for diffusion processes in germanium technology and to this day are often

Thermal Oxidation		
• Atmospheric Pressure	Tube furnace	Dry Oxygen Wet Oxygen Bubbler Flash System Dry Oxidation
	Rapid Thermal	Dry Oxygen
• High Pressure	Tube Furnace	Dry or Wet Oxygen
Chemical Oxidation		
• Anodic Oxidation	Electrolytic Cell	Chemical

Figure 7.13 Oxidation methods.

called simply *diffusion furnaces*. The more correct, generic, term is a *tube furnace*.

A cross section of a basic single horizontal tube furnace is shown in Fig. 7.14. It consists of a long ceramic tube made of mulite, with three to five coils of copper tubing on the inside surface. Each of the coiled tubes defines a zone and is connected to a separate power supply operated by a proportional band controller. Inside the furnace tube is a quartz reaction tube that serves as the reaction chamber for the oxi-

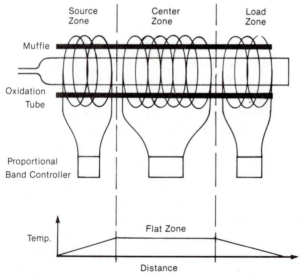

Figure 7.14 Cross section of three-zone tube furnace.

dation (or other processes). The reaction tube may itself be inside a ceramic liner called a *muffle*.

Thermocouples are positioned against the quartz tube and send temperature information to the proportional band controllers. The controllers proportion power to the coils, which in turn heat the reaction tube by radiation and conduction. These controllers are very sophisticated and can control temperatures in the center zone (flat zone) to plus or minus 0.5°. For a process that operates at 1000°C, this variation is only plus or minus 0.05 percent. For the oxidation, the wafers are placed on a holder and positioned in the flat zone. The oxidant gas is passed into the tube where the oxidation takes place. The details and the options used in actual practice are discussed in the following sections.

Horizontal tube furnace systems

A production tube furnace is an integrated system of various sections (Fig. 7.15). They are

1. Reaction chamber(s)

2. Temperature control system

3. Furnace section

4. Source cabinet

5. Wafer cleaning station

6. Wafer load station

7. Process automation

Reaction chamber. The basic tube furnace operation was previously described. An important part of the system is the reaction chamber.

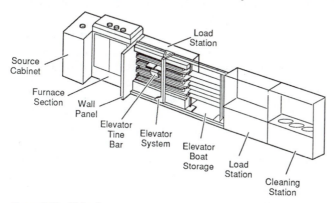

Figure 7.15 Tube furnace.

This part protects the wafers from outside contamination and serves as a heat sink to even out the temperature inside the tube. Tube furnace reaction tubes are round with a gas inlet end and a wafer load end. The gas inlet end, also known as the source end, tapers down to a ground fitting to provide a leak-free connection to the gas sources. The load end also has a ground fitting to receive either an end cap that keeps dirt out of the furnace or a wafer transfer unit (see "Wafer Loading," p. 138).

The traditional reaction chamber material is high-purity quartz. Quartz is a highly purified glass favored for its inherent stability at high temperatures and its cleanliness. Drawbacks to quartz are its fragility and its tendency to let sodium ions pass from the heating coils into the reaction chamber. Another drawback is the tendency of quartz to break up and sag at temperatures above 1200°C. The breakup is called *devitrification* and results in small flakes of the quartz tube surface falling onto the wafers. Sagging impedes the easy placement of the wafer holders in and out of the tube.

Quartz tubes require periodic cleaning. One method is in a tank of hydrofluoric (HF) acid or a solution of HF and water. This cleaning process takes place out of the fabrication area, necessitating the removal and cooling of the tube. The cooling and heating hasten the devitrification process. The HF cleans by removing a thin layer of the quartz. This continual etching eventually weakens the tube wall, limiting its lifetime.

Some firms do an in situ tube cleaning with a portable plasma generator. The plasma generator is positioned inside the tube where it activates etching gases passed into the tube. The gases etch away contaminants. A second in situ cleaning method uses etching gases in the tube that are activated by a plasma field created in the tube. This type of cleaning is more applicable to tube furnaces used for chemical vapor deposition (CVD) processes where the buildup of reaction-created particles is greater than in oxidation processes.

An alternative material to quartz process tubes and wafer holders is silicon carbide. Silicon carbide is structurally stronger and does not break down with repeated heating and cooling. This resistance to temperature cycling also makes the material a better metallic ion barrier over a longer period of time. During processes where oxygen is in the tube, the inside surface grows a thin layer of silicon dioxide. When this oxide is removed in HF, there is no attack of the tube material, which also contributes to the extended lifetime. The widespread use of silicon carbide tubes and wafer holders has been slowed by their high cost and weight.

Temperature control system.   The temperature control system consists of thermocouples touching the reaction tube that are connected to proportional band controllers that feed the power to the heating coils. Proportional band controllers maintain even temperatures in the tube by feeding in or turning off the current to the coils in proportion to the deviation of the tube temperature from the set point. The closer the tube is to the set-point temperature, the smaller the amount of power that is fed to the coils. This system allows fast recovery of the tube to a cold load without overshoot. *Overshoot* is the raising of the tube temperature too high above the desired process temperature as a result of applying too much power to the tube when the wafers are first loaded into it (Fig. 7.16).

Setup of the temperature control system usually is by long thermocouples sheathed in quartz tubes. The thermocouples are placed inside the tube and moved in increments through the furnace zones and the temperature is recorded. Adjustments are made to the controllers until the desired temperatures in the processing section of the tube (flat zone) are achieved.

Advanced systems have thermocouples positioned against the outside of the tube wall. These feed a microprocessor that, in turn, gives the information to the controllers. This system is called *autoprofiling*.[4] Good process control requires that the temperature profile inside the furnace be checked periodically with thermocouples inserted in the tube and the temperature measured by independent recorders.

The processing of larger-diameter wafers has brought with it the concern of wafer warping. Wafers that are heated or cooled rapidly will warp to the point of being useless. The degree of warping increases with higher process temperatures, that is, those above 1150°C.

Two methods are employed to minimize warping of wafers in tube furnaces. One is called *ramping* (or temperature ramping). Ramping is the procedure of maintaining the furnace at a temperature several hundred degrees below the process temperature. The wafers are slowly inserted into the furnace at this lower temperature and, after a short stabilization period, the controllers automatically take the furnace up to the process temperature. At the end of the process cycles, the furnace is cooled to the lower temperature before the wafers are removed. During the ramping process the controllers must maintain the temperature control in the flat zone.

Tube furnaces are maintained at close to the process temperature 24 hours a day due to the devitrification of quartzware and the length of time it takes to stabilize the flat zone. In the interest of economy, some fabrication areas will keep the furnaces at the lower temperature. This is called an *idle* condition.

The second antiwarping procedure is the slow loading of the wafer boat into the tube. At loading rates of about 1 in/min, warping is minimized. For large-diameter wafers and large batch sizes, both methods are used.

Another requirement of the heating system is a fast recovery time after the wafers are loaded in the tube. A full load of 200 six-inch-diameter wafers can drop the tube temperature as much as 50°C.[5] The heating system works to bring the flat zone to temperature as fast as possible without introducing warping conditions. Figure 7.16 illustrates a typical temperature–time recovery curve for a five-zone tube furnace.

Furnace section.   A production-level tube furnace will contain three or four tubes (reaction chambers) and a separate temperature control system for each of the tubes. The tubes are arranged vertically above each other in a stack. They are referred to as three- or four-stack furnaces or a furnace bank. The load end of the tube sticks into an exhaust chamber of the furnace enclosure. The exhaust draws away the spent and heated gases as they exit the tube. This section of the furnace is called a *scavenger* after its role in *scavenging* the gases from the tubes. The scavenger is connected to the facility's exhaust system, which contains a scrubber to remove toxic gases from the withdrawn gases.

The tubes within a bank may be all used for the same purpose, such as oxidations, or be designated for different operations, such as oxidation, diffusion, alloy, or CVD. The exact designation depends on the process requirements and the volume of wafers being processed. However, each of the tubes is dedicated to a particular use. To change the use requires a change of tubeware and contamination monitoring to ensure that contaminants from a previous use are not present in the individual tube section.

Source cabinet.   Each individual tube process requires a number of gases to accomplish the desired chemical reaction. In the case of oxi-

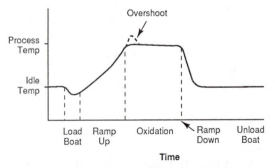

Figure 7.16 Temperature levels during oxidation.

dation, the gas oxidants of oxygen or water vapor have been detailed. In addition, an oxidation process will use a nitrogen gas flow. In fact, all of the tube processes require a source of nitrogen. The nitrogen is used during the loading and unloading stage of the process. When a tube is in the idle condition, nitrogen is kept constantly flowing through the tube. The flow serves to keep dirt out of the system and maintain the preestablished flat zone.

Each of the tube processes requires that the gases be delivered to the tube in a specified sequence, at a specified pressure, at a specific flow rate, and for a specific time. The equipment used to regulate the gas sources is located in a cabinet attached to the furnace section of the system and is known as the *source cabinet*. The cabinet has a shelf at the level of each tube. The tube is positioned so that the inlet fitting extends into the source cabinet. It is connected to a gas flow controller that contains the pressure gauges, flow meters, filters, and timers. In its simplest version, the gas flow controller consists of manually operated valves and timers. In production systems, the sequencing and timing of the various gases into the tube is controlled by a microprocessor that operates solenoids. The piping material used in gas flow controllers is stainless steel to maintain high levels of cleanliness and to minimize chemical reactions between the gas and tube material.

Often the gas flow controller is called a *jungle*. This term came about when the gas controllers were built in-house and had the look of a "jungle" of tubing and valves. Gases are supplied to the gas flow controller either by tubes from the liquid gas supplies in the pad section of the facility, or by smaller lecture bottles of gas located in the source cabinet. Gas saturated with various chemicals is supplied to the tube by bubblers. These are units that contain liquids mixed with the desired chemicals. A gas is *bubbled* through the container, becoming saturated with the vapors of the liquid in the small tank before passing into the tube.

Oxidant sources

Dry oxygen.   When oxygen is used as the oxidant, it is supplied from the facility source or from tanks of compressed oxygen located in or near the source cabinet. It is imperative that the gas be dry, that is, not contaminated with water vapor. The presence of water vapor in the oxygen would increase the oxidation rate and cause the oxide layer to be out of specification.

Water vapor sources.   There are several methods used to supply water vapor (steam) into the oxidation tube. The choice of method depends on the level of thickness and cleanliness control required of the oxide layer in the device.

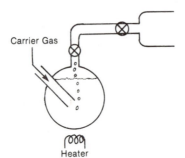

Figure 7.17 Bubbler water vapor source.

Bubblers. The traditional method of creating a steam vapor in the tube is by bubbling a carrier gas through a glass or quartz flask of heated deionized water (Fig. 7.17). The water is heated close to the boiling point (98 to 99°C), which creates a water vapor in the space above the liquid. As the carrier gas is bubbled through the water and passes through the vapor, it becomes saturated with water. Under the influence of the elevated temperature inside the tube, the water vapor becomes steam and causes the oxidation of the silicon surfaces.

Drawbacks to bubbler systems are changes in the amount of water vapor going into the tube as the water level in the bubbler changes and fluctuations in the water temperature. With bubblers there is always concern about contamination of the tube and oxide layer from dirty water or dirty flasks. This contamination potential is heightened by the need to open the system periodically to replenish the water.

Flash systems. A *flash system* is also a glass or quartz system (Fig. 7.18). The system is connected to a continuous source of deionized water through a narrow tube designed to allow small drops of the water into the flask. The drops fall onto a heated quartz surface maintained at a high enough temperature to "flash" the drops into steam. A carrier gas also fed into the system takes the steam created into the oxidation tube. Flash systems suffer the same control problems as bubblers but are somewhat cleaner in that they do not have to be replenished with water.

Dry oxidation (dryox). Bubblers and flash systems are adequate for thick and noncritical oxide layers, but the issues of strict control and absolute cleanliness were brought home to the semiconductor industry with the introduction of MOS devices. The heart of an MOS transistor is the gate structure, and the controlling layer in the gate is a thin, thermally grown oxide. Early MOS process engineers found the liquid-

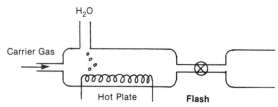

Figure 7.18 Flash system water vapor source.

water-steam systems inadequate for growing gate oxides. This need led to the development of the dry oxidation (or dry steam) process (Fig. 7.19).

In the dry oxidation system, gaseous oxygen and hydrogen are plumbed directly to the back of the oxidation tube. Inside the tube, the two gases mix and, under the influence of the high temperature, form steam. The result is a wet oxidation in steam. Dryox systems offer improved control and cleanliness over liquid systems. First, gases can be purchased in a very clean and dry state. Second, the amounts going into the tube can be very precisely controlled by the gas flow meters. Dryox is the preferred oxidation method for production for all advanced devices.

A drawback to dryox systems is the explosive property of hydrogen. At oxidation temperatures, hydrogen is very explosive. Precautions used to reduce the explosion potential include separate oxygen and hydrogen lines to the tube and pumping excess oxygen into the tube. The excess oxygen ensures that every hydrogen molecule (H_2) will combine with an oxygen atom to form the nonexplosive water molecule, H_2O. Other precautions used are hydrogen alarms and a hot filament in the source cabinet and in the scavenger end of the furnace to immediately burn off any free hydrogen before it can explode.

Wafer cleaning station

When the wafers come to a tube furnace, they first go to the attached wafer-cleaning station. The need for stringent cleanliness control has been stressed throughout this text. It is especially important before the tube operations because of the heating of the wafers. Contamina-

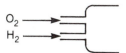

Figure 7.19 "Dryox" (dry steam) water vapor source.

tion left on the wafer surfaces can pit the surface, diffuse into the surface, or interfere with the quality of the layer being grown.

It is common practice to give the wafers a thorough chemical and physical cleaning before each tube operation. The cleaning takes place in a VLF "wet" bench that has built-in tanks for cleaning chemicals, units for rinsing the wafers in deionized water, and drying units. Wafers usually arrive in plastic carriers and are transferred into holders made of Teflon or a Teflon derivative for the chemical cleaning. Teflon is preferred because of its chemical resistance and stability in heated chemical baths. Wafer cleaning is detailed in the section on "Oxidation Processes," p. 154.

Wafer load station

After cleaning, the wafers are passed to the in-line load station. Here the wafers are inspected for cleanliness and loaded into holders for insertion into the tube. The station is located under a ceiling-mounted HEPA filter or in a VLF hood. The loading station and the cleaning station are located next to each other in such a manner that the wafers stay in Class 10 (or Class 100) conditions during the transfer.

A quick surface inspection of the wafers is normally done with the aid of an ultraviolet light. These high-intensity light sources allow the operator to see small particles and stains not visible to the naked eye. Sometimes a microscope inspection of the surface is performed.

Manual wafer loading. Wafer holders, also called *boats* or *cassettes*, for tube operations are made of either quartz or silicon carbide. The boats come in various designs (Fig. 7.20) depending on the degree of control required of the operation and the loading density. The most productive designs have the wafers standing up in machined slots, positioned

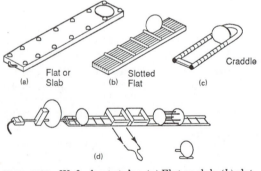

Figure 7.20   Wafer boat styles. (*a*) Flat or slab; (*b*) slotted flat; (*c*) cradle; (*d*) slotted cradle boats on paddle.

crosswise to the gas flow. In tube operations where uniformity is critical, the wafers may be loaded parallel to the gas flow.

The wafers are manually removed from the chemical cleaning boats with vacuum wands or limited-grasp tweezers (Fig. 7.21). Vacuum wands are attached to a vacuum source and are designed to allow grasping of the wafers from the backside. This arrangement minimizes damage and contamination of the sensitive front side of the wafer.

A faster boat-loading method is by dump transfer. In this method the cleaning boats and the tube boats are the same size and designed to fit together with a pin-and-slot arrangement. The operator positions the empty tube boat over the loaded cleaning boat and, holding them tightly together, rotates them until the empty boat is under the full one. During the rotation the wafers are "dumped" into the empty boat. The loaded boat is placed inside a transfer tube called an *elephant*. The elephant has a ground joint that mates with the joint on the load end of the tube. It also has a hole in the back end to admit a quartz rod that hooks into the back end of the wafer boat. The operator pushes the boat into the flat zone of the tube furnace (Fig. 7.22).

Automatic wafer loading. Production efficiency and VLSI cleanliness requirements have led to the development of a variety of automatic boat- and tube-loading mechanisms. Automatic boat loaders come in two styles: dump and pick-and-place. The dump style is an automated version of manual dump transfer. Pick-and-place machines (sometimes called *robots*) pick each of the wafers out of one cassette and

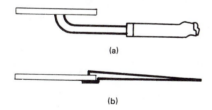

(a)

(b)

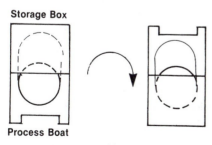

Storage Box

Process Boat

(c)

Figure 7.21 Manual wafer handling devices. (a) Vacuum pickup; (b) limited grasp tweezer; (c) flip transfer boats.

place it into an empty one. Some versions of pick-and-place machines pick up the entire load of wafers and transfer them in one operation to the empty cassette. A challenge to any wafer boat-loading system is the correct placement of test wafers within the load of device wafers as well as dopant source wafers (see "Concept of Diffusion," Chap. 11, p. 265). These wafers must be *picked* from other boats. An additional challenge is the loading of two wafers back to back in the same slot. This procedure is used to increase the productivity of the operation.

The loaded cassettes are put into the furnace by a number of techniques. One version uses an elephant but automates the push-pull action with a push rod connected to a motor. Another version has a number of cassettes placed by the operator on a platform called a *paddle*. The paddle moves the cassettes into the tube under the control of a motor (Fig. 7.22).

Boat loading can be a source of severe wafer contamination from particles scraped from the inside of the tube as the cassettes are shoved down the tube. One technique for solving this problem is to use cassettes or paddles with small wheels. The rolling wheels kick up fewer particles than does scraping. Another solution is a cantilever system, which features a rigid rod with the cassettes positioned on the end. The rod moves the cassettes down the center of the tube without ever touching its sides. Some systems leave the cassettes suspended in the center of the tube while others give the cassettes a soft landing on the tube bottom.

Automated push-pull machines are necessary to achieve the control required for slow entry and exit of the wafers to prevent warping. A typical push-pull rate is 1 in/min. A standard four-stack furnace system requires an automatic push-puller for each of the tubes. Although these systems load and unload automatically, they require an operator to place the loaded boats on the paddle.

An overriding goal of process equipment design is "hands-off" operation. That need is addressed in tube furnace systems built on the elevator design.[6] The concept employs a storage buffer of loaded tube cassettes whose tube destination is known by a computer. When it is time for the cassettes to be loaded in the tube, a robot selects the proper

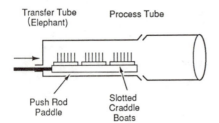

Transfer Tube
(Elephant)   Process Tube

Push Rod
Paddle

Slotted
Craddle
Boats

Figure 7.22 Transfer tube loading of wafers.

cassettes and moves to the correct tube position and loads them into the tube. The advantages of these systems are the computer production control and the need for only one loading mechanism per furnace stack. For a properly designed system, robotic activity is less contaminating to the wafers than human operators (Fig. 7.15).

Oxidation process automation

Tube furnaces have evolved from operator-controlled manual systems to various levels of automation. A modern production tube furnace is automated from wafer load to wafer unload in a one-button operation. The operator places the cassettes of cleaned wafers on the load mechanism and pushes the start button. The computer takes over and directs the tube furnace through the predetermined process, which is called a *recipe*. The times, temperatures, gas sequences, and push-pull rates are programmed into an onboard computer or a host computer. Full automation of a tube operation involves delivery of the proper wafers to the station with robots or automatic equipment taking the wafers entirely through the process and passing them on to the next process.

Vertical Tube Furnaces

Horizontal tube furnaces have enjoyed great popularity over the years. A lot of process development has taken place in the 30 years of their use. However, the needs for greater contamination control and the move to ever-increasing wafer diameters as well as the need for more productive processing has pushed horizontal tube furnaces to a limit. While contamination control has led to the development of cantilever systems, they have had to get larger and stronger as the wafers have grown in diameter and weight. Also, the larger wafer diameters require larger tube diameters, which puts additional pressures on the maintenance of extended temperature flat zones.

The larger wafer loads have caused a lengthening of the furnaces and their associated load stations. In clean-room terms, the "footprint" of the equipment is increasing. The problem is that larger footprints require larger and more expensive clean rooms.

There are also process problems associated with larger-diameter tubes. One is the problem of keeping the gas streams in a laminar flow pattern in the tube. Laminar flow is uniform with no separation of the gases into layers. It is also without the turbulence that would cause uneven reactions within the tube. These considerations have resulted in the development of vertical tube furnaces. Several designs are available and are finding their way into production processes.[7] In this configuration, the tube is held in a vertical position (Fig. 7.23) with loading taking place from the top or bottom. Tube materials and heating systems are the same as for horizontal systems.

The wafers are loaded in standard cassettes and lowered or raised into the flat zone. This action is accomplished without the particulates generated by the cassettes scraping the sides of the tubes. In this configuration, the wafers are in the most dense loading for a tube furnace. An added plus for vertical tube furnaces is the ease of rotating the wafers in the tube, which produces more uniform reactions.

Perhaps one of the most appealing aspects of vertical tube furnaces is the small footprint. The system is smaller than a conventional four-stack system. Vertical systems offer the possibility of locating the furnaces outside the clean room with only a load station door opening into the clean room. In this arrangement, the clean-room footprint of the furnace is practically zero and maintenance can take place from the service chase. Another possible arrangement of vertical furnaces is in an island configuration. The furnaces are arranged around a central robot that alternately feeds cassettes to each of the systems.

Vertical furnaces offer the promise of more uniform processing by maintaining good laminar flow. In a horizontal system, gravity tends to separate mixed gases as they flow down the tube. In a vertical system this problem is eliminated. Vertical tube furnaces offer a number of advantages over horizontal furnaces and have the potential of being

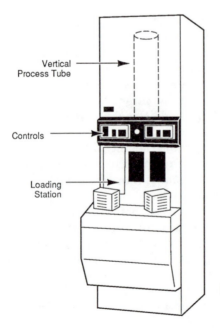

Vertical
Process Tube

Controls

Loading
Station

Figure 7.23 Vertical tube furnace.

the furnace system of choice for ultralarge-scaled intergrated (ULSI) processes.

Rapid Thermal Processing (RTP)

Each advance in the density of circuits, reduction of feature size, and increase in wafer diameter strains the applicability of traditional processes. Ion implantation has replaced thermal diffusion due to its inherent doping control. However, ion implantation requires a follow-on heating operation, called *annealing*, to cure out crystal damage induced by the implant process. The annealing step has been traditionally done in a tube furnace. Although the heating anneals out the crystal damage, it also causes the dopant atoms to spread out in the wafer, an undesirable result. This problem led to the investigation of alternate energy sources to achieve the annealing without the spreading of the dopants. The investigations led to the development of rapid thermal process (RTP) technology.

RTP technology is based on the principle of radiation heating (Fig. 7.24). Heat sources include graphite heaters and plasma arc and tungsten halogen lamps. Tungsten halogen lamps are the most popular.[8] The radiation from the heat source couples into the wafer surface and brings it up to the process temperature in seconds. The same temperature would take minutes to reach in a conventional tube furnace. Likewise, cooling takes place in seconds. With radiation heating, because of its very short heating times, the body of the wafer never comes up to temperature. For the ion implant annealing step, this means that the crystal damage is annealed while the implanted atoms stay in their original location.

By their nature, RTP systems are single-wafer processors that require automated wafer loading to gain needed production efficiencies. A problem presented by RTP systems is accurate temperature measurement and control. Temperatures are usually measured by thermocouples; however, they require contact with the heated wafer, which is impractical in a single-wafer system, and thermocouples have a response time that is longer than some RTP heating cycles. Optical

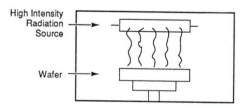

Figure 7.24   Rapid thermal processing.

pyrometers, which gauge temperatures by measuring characteristic energies given off by the heated object, are also prone to errors in an RTP system.

RTP technology is a natural choice for the growth of thin oxides used in MOS gates. The trend to smaller feature sizes on the wafer surface has brought along with it a decrease in the thickness of layers added to the wafer. Layers undergoing dramatic reduction in thickness are thermally grown gate oxides. Advanced production devices are requiring gate oxides less than 100 Å thick. Oxides this thin are sometimes hard to control in conventional tube furnaces due to the problem of quickly supplying and removing the oxygen from the system. RTP systems can offer the needed control by their ability to heat and cool the wafer temperature very rapidly. RTP systems used for oxidation, called *rapid thermal oxidation* (RTO) systems, are similar to the annealing systems but have an oxygen atmosphere instead of an inert gas.

RTP is a technology that grew out of the process needs of advanced devices and will be a dominant technology through the 1990s. Currently, there is interest in RTP for a wide range of applications, including oxidation, damage annealing, metal-wafer annealing, localized film growth, control of circuit leakage, and CVD film growth.[9]

High-Pressure Oxidation

The problem of dopant atoms spreading in a conventional tube furnace led to the development of RTP systems. Two other problems have led to the introduction of high-pressure oxidation systems. Those problems are the growth of dislocations in the bulk of the wafer and the growth of hydrogen-induced dislocations along the edge of openings in layers on the wafer surface.[10] In the first case, the dislocations cause various device performance problems. In the latter case, surface dislocations cause electrical leakage along the surface or the degradation of silicon layers grown on the wafer for bipolar circuits.

The growth of dislocations is a function of the temperature the wafer is processed at and the time it spends at that temperature. A solution to this problem is to perform thermal oxidation processes at a lower temperature. This solution by itself causes the production problem of longer oxidation times. The solution that addresses both problems is high-pressure oxidation (Fig. 7.25). These systems are configured like conventional horizontal tube furnaces but with one major exception: the tube is sealed and the oxidant is pumped into the tube at pressures of 10 to 25 atm (10 to 25 times the pressure of the atmosphere). The containment of the high pressure requires encasing the quartz tube in a stainless steel jacket to prevent it from cracking.

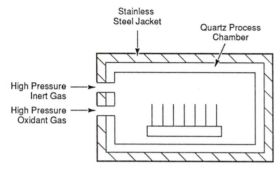

Figure 7.25   High-pressure oxidation.

At these pressures, the oxidation proceeds at a faster rate than in atmospheric systems. A rule of thumb is that a 1-atm increase in pressure allows a 30°C drop in the temperature. In a high-pressure system that increase relates to a drop of 300 to 750°C in temperature. This reduction is sufficient to minimize the growth of dislocations in and on the wafers.

Another option using high-pressure systems is to maintain the regular process temperature and reduce the time of the oxidation. Other considerations concerning high-pressure systems focus on the safe operation of the system and possible contamination from the additional pumps and piping needed to create the high pressures inside the tube.

In addition to oxidation, high-pressure systems are finding some use in CVD epitaxial depositions and for flowing glass layers on the wafer surface.[11] Both of these processes are of higher quality when performed at lower temperatures.

Wafer Preclean

The processing of the wafers through the oxidation process is divided into several distinct steps, as shown in the flow diagram in Fig. 7.26. After the wafers are logged into the station, they receive a thorough cleaning. Clean wafers are essential at all stages of the fabrication process but are especially necessary before any of the operations performed at high temperature. The cleaning techniques described are used throughout the wafer fabrication process. In the subsequent process chapters, where appropriate, the reader will be referenced back to this chapter.

Wafer cleaning could be considered a separate technology. The evolution of semiconductor processes is in many respects related directly to the ability of the industry to develop cleaning technology to keep pace with the increasing need for contamination-free wafers. Wafer surfaces can have four different types of contamination. Each repre-

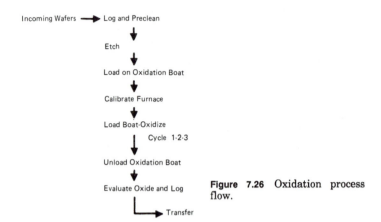

Incoming Wafers ⟶ Log and Preclean

↓

Etch

↓

Load on Oxidation Boat

↓

Calibrate Furnace

↓

Load Boat-Oxidize

| Cycle 1-2-3

↓

Unload Oxidation Boat

↓

Evaluate Oxide and Log   **Figure 7.26** Oxidation process flow.

⟶ Transfer

sents a different problem on the wafer and each is removed by different means. The four types are

1. Particulates

2. Organic residues

3. Inorganic residues

4. Unwanted oxide layers

Particulate removal

Particulates on the wafer surface vary from very large ones (50-μm size) to very tiny ones about a micron in size. The larger ones can be cleaned off by conventional chemical baths and rinses. The smaller particulates are more difficult to remove because they are often held to the surface by strong intermolecular forces. A particulate cleaning process is most often a series of steps designed to remove both the large and small particles.

The simplest particulate removal process is to blow off the wafer surface using a spray of filtered high-pressure nitrogen from a hand-held gun. The guns are located in the cleaning stations and are connected by flexible tubing to the nitrogen source. In fabrication areas where small particles are a problem, the nitrogen guns are fitted with ionizers that strip static charges from the nitrogen stream. The static neutralization of the stream prevents static charge buildup on the wafer surface, a charge that can attract more particulates to the wafer.

Nitrogen blow-off guns are effective in removing most large particles not held to the surface by molecular forces. Since the guns are hand-held, the operators must use them in a manner that does not contaminate other wafers in the station or the station itself. The operator must take care to direct the nitrogen spray away from other wafers in the station and away from the station surface to avoid blowing particles onto the wafers.

Wafer scrubbers

The stringent wafer cleanliness requirements for epitaxial growth led to the development of mechanical wafer surface scrubbers. These machines are now used wherever particulate removal is critical.

The scrubbers hold the wafer on a rotating vacuum chuck. Particulate removal takes place from the action of a rotating brush brought in near contact with the rotating wafer while a stream of deionized water is directed onto the wafer surface. Often a detergent will be added to the water to increase the cleaning effectiveness. The combination of the brush and wafer rotations creates a high-energy cleaning action at the wafer surface. In effect, the liquid is forced into the small space between the wafer surface and the brush ends where it achieves a high velocity, which aids the cleaning action. Caution must be exercised to keep the brushes and cleaning liquid lines clean to prevent secondary contamination. Also, the brush height above the wafer must be maintained to prevent scratching the wafer surface.

Scrubbers are designed as standalone units with automatic loading capabilities or are built into other pieces of equipment to clean the wafers automatically before processing.

High-pressure water cleaning

An effective method of removing statically attached particles first became a necessity for the cleaning of glass and chrome photomasks. A method developed was the high-pressure water spray, in which a small stream of water is pressurized from 2000 to 4000 psi. The stream is swept across the mask or wafer surface, dislodging both large and small particles. Often a small amount of surfactant will be added to the water stream to act as a destatic agent.

Organic residue removal

Organic residues are compounds that contain carbon, such as oils in fingerprints. These residues can be removed in solvent baths such as acetone, alcohol, or TCE. In general, solvent cleaning of wafers is avoided whenever possible due to the difficulty of drying the solvent completely off the wafer surface. Also, solvents always contain some impurities that may themselves represent a contamination source.

When solvents are used, there is a series of baths. A typical process may be one bath of TCE (or other solvent) followed by an isopropyl alcohol rinse, followed by an acetone rinse. Each of the succeeding rinses dissolves the previous solvent. Acetone is the most volatile of the solvents used and is the easiest to remove from the surface. In some processes the acetone step is followed by a rinsing with deionized water and drying. The use of acetone is discouraged because of its high flammability.

Inorganic residue removal

Inorganic residues are those that do not contain carbon. Examples are inorganic acids such as hydrochloric or hydrofluoric acid, that may be introduced from other steps in wafer processing. The inorganic residues are cleaned from the wafers in a variety of cleaning solutions described in the following section.

Chemical cleaning solutions

A wide range of cleaning processes exist in the semiconductor industry. Each fabrication area has different cleanliness needs and different experiences with different solutions. The solutions described in this section are those in common use, although there are numerous variations and different combinations of solutions from one fabrication area to another.

Liquid chemical cleaning processes are generally referred to as *wet processes* or *wet cleaning*. The cleaning takes place in glass, quartz, or polypropylene tanks fitted into the deck of the cleaning stations. Where heating of the solution is required, the tank may be sitting on a hotplate, be wrapped with heating elements, or have an immersion heater inside.

The most common chemical cleaning solution is hot sulfuric acid with an added oxidant. The sulfuric acid is an effective cleaner in the 90 to 125°C range. At these temperatures, it will remove most inorganic residues and particulates from the surface. Oxidants are added to the sulfuric acid to remove carbon residues. The chemical reaction converts the carbon to carbon dioxide, which leaves the bath as a vapor by the reaction

$$C + O_2 \rightarrow CO_2 \text{ (gas) } \uparrow$$

The oxidants normally used are hydrogen peroxide (H_2O_2), ammonium persulfate [$(NH_4)_2S_2O_8$], and nitric acid (HNO_3).

Sulfuric acid and hydrogen peroxide.   The addition of hydrogen peroxide to sulfuric acid is perhaps the most common wafer-cleaning solution in use in semiconductor plants. It is used to clean wafers at all stages of processing especially before the tube processes. It is also used as a photoresist stripper in the patterning operation. Within the industry this solution is known by a number of names, including *Carro's acid* and *piranha etch*. The latter term attests to the aggressiveness and effectiveness of the solution.

Two mixing methods are used. One is the addition of about 30% (by volume) of hydrogen peroxide to a beaker of room-temperature sulfuric acid. In this ratio an exothermic reaction takes place that quickly raises the temperature of the bath into the 110 to 130°C range. This is the temperature range for effective cleaning. As time proceeds, the reaction slows and the bath temperature falls below the effective range. At this point the bath may be recharged with additional hydrogen peroxide or discarded. Recharging the bath eventually results in a lowered cleaning rate due to the conversion of the hydrogen peroxide to water, which dilutes the sulfuric acid.

Another way to use these two chemicals is to heat the sulfuric acid to the effective cleaning temperature range and add small amounts (50 to 100 ml) of hydrogen peroxide before cleaning each batch of wafers. This method, while requiring a heating and temperature control system, maintains the bath at the proper temperature and the water created from the hydrogen peroxide evaporates out of the solution. The use of heated sulfuric acid is preferred for economic and process control reasons. It is also easier to automate this approach to mixing the two chemicals.

Sulfuric acid and ammonium persulfate. A drawback to the use of hydrogen peroxide in cleaning solutions is its reactiveness with itself and other chemicals. Hydrogen peroxide is unstable and is constantly giving off oxygen molecules. Safety demands that hydrogen peroxide bottles have vented caps. Unfortunately, the evolution of the oxygen gradually degrades the effectiveness of the chemical, especially in hot climates or heated storage rooms. Also, hydrogen peroxide is classified as an oxidizer due to its rabid reactions with many other chemicals. This property requires that special precautions be taken to keep it from mixing with other chemicals or splashing onto operators where it can burn the skin.

An oxidant that performs the same role when mixed with heated sulfuric acid but without the drawbacks mentioned is ammonium persulfate (AP),[12] a white crystalline material that contains 6.7% oxygen. The oxygen is released into the bath when it is mixed into the heated sulfuric acid. A typical process would be the addition of 40 g of AP to every 1.5 gal of sulfuric acid in the bath. The AP is added before cleaning every run of wafers. This method maintains the required level of oxidant in the bath. A drawback is the buildup of any contaminants contained in the AP as each succeeding amount is added to the bath. Being a crystal, AP does not have the deterioration problem of hydrogen peroxide or the safety problem. However, it is an operator-controlled process that is less easy to automate.

RCA clean. In the late 1960s, RCA engineers developed a two-step process to remove organic and inorganic residues from silicon wafers. The process proved to be so effective that the procedure is known simply as the *RCA clean*. The first step uses a solution of water, hydrogen peroxide, and ammonium hydroxide (H_2O, H_2O_2, NH_4OH). Solutions vary in composition from 5:1:1 to 7:2:1 and are heated to the 75 to 85°C range. The second step uses a solution of water, hydrogen peroxide, and hydrochloric acid (H_2O, H_2O_2, HCl) mixed in ratios of 6:1:1 to 8:2:1 and used at the 75 to 85°C temperature range.

There have been many adaptations and changes made to the original cleaning solutions. Whenever an RCA cleaning process is referred to, it means that hydrogen peroxide is used along with some base or acid.

Oxide removal

The ease of silicon oxidation has been mentioned. The oxidation can take place in air or in the presence of oxygen in the heated chemical cleaning baths. Often the oxide grown in the baths, while thin (100 to 200 Å), is thick enough to block the silicon surface from reacting properly during one of the other process operations. The thin surface oxide can act as an insulator, preventing good electrical contact between the silicon surface and a layer of conducting metal.

The removal of these thin oxides is a requirement in many processes. The removal chemical of choice is hydrofluoric acid (HF). Prior to an initial oxidation, when the surface is only silicon, the wafers are cleaned in a bath of full-strength HF (49%). While this step is part of the cleaning process, the action of the HF in removing the oxide is actually an etch process.

Later in the processing, when the surface is covered with previously grown oxides, thin oxides in patterned holes are etched away with a water and HF solution. These solutions vary in strength from 100:1 to 10:7 (H_2O to HF). The strength is chosen depending on the amount of oxide already on the wafer, since the water and HF solution will etch both the oxide on the silicon surface in the hole and the oxide covering the rest of the surface. A strength is chosen to ensure the removal of the oxide in the holes while not excessively thinning the other oxide layers.

Water rinsing

The chemicals used in cleaning the wafers can also be contaminants if left on the surface. Therefore every cleaning process or sequence is followed by a rinse in deionized water. The water rinse also serves to end the etching action of the oxide in the removal steps. Rinsing is done by several different methods.

Overflow rinsers. The need for atomically clean surfaces requires the complete rinsing of all chemicals from the wafer surface. This cannot be done by just dunking the wafers in a tank of water. Thorough rinsing requires a continuous supply of clean water to the wafer surface. One such method is an overflow rinser (Fig. 7.27). Physically the rinser is a box sunk into the cleaning station deck. Deionized water is brought into the bottom of the box and flows through and around the wafers, exiting over a dam into a drain system. The rinsing action of the flowing water is enhanced by a stream of nitrogen bubbles introduced into the rinser from the bottom plate. As the nitrogen *bubbles* up through the water, it aids the mixing of the chemicals on the wafer surface with the water. This type of system is often called a *bubbler*.

A rule of thumb is that adequate rinsing takes a minimum of 5 minutes with a flow rate equivalent to five times the volume of the rinser per minute (the number of turnovers per minute). If the size of the bubbler is 3 L, the flow rate should be a minimum of 15 L/min.

The rinse time is determined by measuring the resistivity of the water as it exits the rinser. The chemicals used in the cleaning processes act as charged molecules in the rinse water and their presence can be inferred from the resistivity of the water. If the water goes into the rinser at an 18-megΩ level, a reading of 15 to 18 megΩ on the exit side indicates that the wafers are cleaned and rinsed. The rinsing step is so critical that generally a minimum of two rinsers are used and the total rinsing time is set at two to five times the minimum determined by resistivity measurements. Often a water resistivity meter is mounted on the outlet to constantly measure the output resistivity. The operators are instructed to leave the wafers in the rinser until the specified resistivity is reached. Some systems will sound an alarm when the correct resistivity is reached.

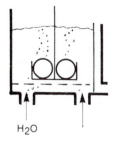

Figure 7.27   Overflow rinser.

H_2O

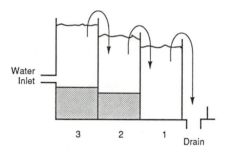

Water
Inlet

3   2   1

Drain

Figure 7.28 Three-stage cascade rinser.

Cascade rinse systems. A rinse system based on the overflow principle is the cascade system. A cascade rinse system is two or three overflow rinsers connected to each other. The design is such that the water enters only the end rinser and cascades through the downstream rinsers. The wafers start the rinse process in the downstream rinser and are moved sequentially to the rinser with the direct water supply (Fig. 7.28). When several boats of wafers are being rinsed simultaneously, this system is more efficient than others in its use of water.

Spray rinsing. Flowing water removes a water-soluble chemical from the wafer surface by a dilution mechanism. The top layer of the chemical dissolves into the water and is carried away by the flow. This action occurs over and over in a continuous manner. Rinsing is speeded up by a faster water flow rate, which can remove the dissolved chemical more quickly. Actually, it is the number of turnovers that directly influences the rinse rate. This can be understood by considering a very fast water flow rate but in a huge rinse tank. The chemical removed from the wafer surface would be evenly distributed throughout the tank and therefore some would still cling to the surface. The chemical could only be removed from the tank when enough water has flowed in and out of it to carry away the chemical.

One way to have a faster rinse rate is by using a water spray. The spray removes the chemical with a physical force from its own momentum and the many small droplets continually hitting the wafer surface have the effect of an extremely high turnover rinse. Along with more efficient rinsing, a spray rinser uses considerably less water than an overflow rinser. A problem with a spray rinser comes when the resistivity of the exiting water is measured with a resistivity monitor. Carbon dioxide from the air gets trapped in the water spray and the CO_2 molecules act as charged particles and the resistivity meter reads them as contaminants, which they aren't.

Dump rinsers. In respect to rinsing efficiency and water savings, a dump rinser (Fig. 7.29) is an attractive method. The system is like an

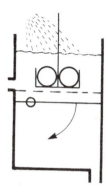

Figure 7.29   Spray/dump rinser.

overflow rinser but with a spray capability. The wafers are placed into the dry rinser and immediately sprayed with deionized water. While they are being sprayed, the cavity of the rinser is rapidly filled with water. As the water overflows the top, a trap door in the bottom swings open and the water is dumped instantly into the drain system. This fill-and-dump action is repeated several times until the wafers are entirely rinsed.

Dump rinsing is also favored because all of the rinsing takes place in one cavity, which saves equipment and space. It is also a system that can be automated, so that the operator only needs to load the wafers in (this can be done automatically) and push a button.

Spin-rinse dryers (SRDs).   After rinsing, the wafers must be dried. This is not a trivial process. Any water that remains on the surface (even atoms) has the potential of interfering with any subsequent operation. Complete drying is accomplished in a centrifugelike piece of equipment. In one version, the wafer boats are put in holders around the inside surface of a drum. In the center of the drum is a pipe with holes that is connected to a source of deionized water and hot nitrogen (Fig. 7.30).

The drying process actually starts with a rinse of the wafers as they rotate around the center pipe that sprays the water. Next, the SRD switches to a high-speed rotation as heated nitrogen comes out of the center pipe. The rotation literally throws the water off the wafer surfaces. The heated nitrogen assists in the removal of small droplets of water that may cling to the wafer.

SRDs are also built for drying single-wafer boats. The boat slips into a rotating holder in the center of a chamber. The water and nitrogen come into the chamber through its side rather than through a pipe in the center. The rinsing and drying take place as the boat spins about its own axis. This type of SRD is called an *axial dryer*. These two ma-

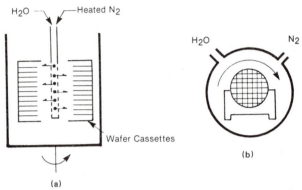

Figure 7.30 Spin rinse dryer styles. (*a*) Multiboat; (*b*) single boat axial.

chines are used for automatic wafer cleaning and etching. The required chemicals are plumbed to the machine and microprocessor-controlled valves direct the right chemicals into the chamber.

Oxidation Processes

After the wafers are precleaned and etched and loaded on the push-pull mechanism, they are ready for surface oxidation. The actual oxidation proceeds in different gas cycles (Fig. 7.31) within the furnace tube. The first gas cycle occurs as the wafers are being loaded into the tube. Since the wafers are at room temperature and precise oxide thickness is a goal of the operation, the gas metered into the tube during loading is dry nitrogen. The nitrogen is necessary to prevent any oxidation while the wafers are coming up to the required oxidation temperature.

Once the wafers are stabilized at the correct temperature, the flow gas controller switches the gas flow to the selected oxidant. For oxides greater than 1200 Å, the oxidant is usually steam from one of the sources previously discussed. For oxides less than 1200 Å, pure oxygen is usually used because of its greater process control and the cleaner, denser oxide it produces.

The thinner MOS gate oxides require very clean layers. Improvements in cleanliness and device performance are achieved when chlorine is incorporated into the oxide.[13] The chlorine tends to reduce mobile ionic charges in the oxide layer, reduce structural defects in the oxide and silicon surface, and reduce charges at the oxide-silicon interface. The chlorine comes from the inclusion of anhydrous chlorine (Cl_2), anhydrous hydrogen chloride (HCl), trichloroethylene (TCE), or trichloroethane (TCA) in the dry oxygen gas stream. The gas sources,

Cycle	Gas	Purpose
1. Nitrogen		Temperature Stabilization in an Inert Atmosphere
2. Oxygen or Water Vapor		Oxide Growth
3. Nitrogen		Stop Oxidation and Removal of Wafers in an Inert Atmosphere

Figure 7.31   Oxidation process cycles.

chlorine and hydrogen chloride, are metered into the tube along with the oxygen from separate flow meters in gas flow controller. The liquid sources, TCE and TCA, are carried into the tube as vapors from liquid bubblers. For safety and ease of delivery, TCA is the preferred source of chlorine.

The oxidation-chlorine cycle may take place in one step or be preceded or followed by a dry oxidation cycle. After the oxidation cycle, the furnace gas is switched back to dry nitrogen. The nitrogen terminates the oxidation of the silicon by diluting and removing the oxidant used. It also prevents any oxidation during the wafer exit step.

Postoxidation Evaluation

After the wafers are removed from the oxidation boats, they will receive an inspection and evaluation(s). The nature and number of the evaluations depend on the oxide layer and the precision and cleanliness required of the particular circuit being fabricated. (The details of the evaluations performed are explained in Chap. 14.)

A requirement of the oxidation process is a uniform noncontaminated layer of silicon dioxide on the wafer. As the wafers proceed through the wafer fabrication operations, there is a buildup of both thermally grown oxides and other deposited layers on the wafer surface. These other layers interfere with the determination of the quality of a particular oxide. For this reason each batch of oxidized wafers going into the tube includes a number of test wafers, with bare surfaces, placed at strategic locations on the wafer boat. Test wafers are necessary for the evaluations that are destructive or require large undisturbed areas of oxide. At the conclusion of the oxidation operation, they are used for the evaluation of the process. Some of the evaluations are performed by the oxidation operator and some are performed off-line in quality control (QC) labs.

Surface inspection

A quick check of the cleanliness of the oxide is performed by the operator as the wafers are unloaded from the oxidation boat. Each wafer is viewed under a high-intensity ultraviolet (UV) light. Surface particulates, irregularities, and stains are readily apparent in UV light.

Oxide thickness

The thickness of the oxide is of major importance. It is measured on test wafers by a number of techniques (Chap. 14). The techniques are color comparison, fringe counting, interference, ellipsometers, stylus apparatus, and scanning electron microscopes (SEMs).

Oxide and furnace cleanliness

In addition to the physical contaminants of particles and stains, the oxide should have a minimum number of mobile ionic contaminants. These are detected by the sophisticated capacitance-voltage (C/V) technique, which detects the total number of mobile ionic contaminants present in the oxide. It cannot identify the origin of those contaminants, which may come from the tubes, the gases, the wafers, or the cleaning process. Therefore, C/V evaluation is a go–no-go assessment of the wafers and serves as a check of the total furnace operation. In most fabrication lines, C/V analysis is also used to certify the cleanliness of the furnace and its associated parts. A oxide with a low-mobile ionic contamination level certifies that the entire system is clean. When the oxide fails the test, more investigation is necessary to identify the source.

A second oxide-cleanliness-related parameter is dielectric strength. This parameter is a measure of the dielectric (nonconducting) property of the oxide. It is measured by the destructive oxide rupture test.

A third cleanliness factor is the index of refraction of the oxide. Refraction is the property of a transparent substance that causes light to bend as it travels through it. The apparent versus actual location of an object on the bottom of a body of water is an example of refraction. The index of refraction of a pure oxide is 1.46. Variations from this value come about from impurities in the oxide. A constant index of refraction is the starting point for several of the interference thickness-measurement techniques. Variations in the index can lead to erroneous thickness measurements. The index of refraction is measured by interference and ellipseometry techniques (Chap. 14).

Anodic Oxidation

In addition to thermal oxidation, there is one chemical oxidation process used to grow thin oxides on silicon. The method is called *anodic*

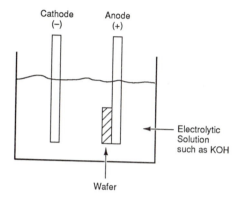

Figure 7.32   Anodic oxidation.

oxidation and is similar to chemical plating. In this application the silicon wafer is connected to a positive electrode (the anode) in a bath that also contains a negative electrode. Both are submersed in an electrolyte of KNO_3.[14] When a current is applied between the two electrodes, oxygen is created at the anode and forms silicon dioxide on the wafer surface (Fig. 7.32). The oxide formed is generally of lesser quality than a thermally grown oxide, but some renewed interest in the method has been shown for the formation of very thin gate oxides.

The principal use of anodic oxides is in profiling dopant concentrations down into the wafer surface. This is possible because in anodic oxidation the silicon from the water surface travels to the top of the oxide layer rather than oxygen traveling to the wafer surface. When an anodic oxide is etched from a silicon surface, it leaves the dopants near the surface undisturbed, unlike a thermally grown oxide. By using a four- or two-point probe to measure the resistivity of the surface and repeating the anodic oxidation and etch steps, the concentration of the dopants at different levels can be determined.

Thermal Nitridation

An important factor in the production of small high-performance MOS transistors is a thin gate oxide. However, at a thickness of approximately 100 Å silicon dioxide films tend to be of poor quality. An alternative to silicon dioxide films is a thermally grown silicon nitride film formed by the exposure of the silicon surface to ammonia (NH_3) between 950 and 1200°C.[15]

Key Concepts and Terms

Dielectric	Doping barrier
High-pressure oxidation	Oxidant sources

Oxidation

Oxidation orientation effects

Oxide

Oxide growth mechanism

Preoxidation cleaning

Silicon dioxide

Oxidation dopant effects

Oxidation process cycles

Oxide layer evaluation

Passivation layer

Rapid thermal processing

Tube furnace

Review Questions

1. Describe the role and movement of oxygen during thermal oxidation.

2. Draw a sketch of a horizontal tube furnace and identify all the sections.

3. Name the water vapor source favored for VLSI oxidation processes.

4. Why are temperature ramping techniques required?

5. What is the advantage of rapid thermal oxidation?

6. Name the two oxidants added to sulfuric acid for wafer cleaning.

7. Make a flow diagram of the oxidation steps, from incoming wafer to evaluation.

8. Name the three factors that determine the oxide thickness.

9. What advantages are offered by high-pressure oxidation?

10. List three uses of silicon dioxide layers in semiconductor devices.

References

1. S. Wolf and R. Tauber, *Silicon Processing for the VLSI Era*, Lattice Press, Sunset Beach, Calif., 1986.
2. P. Gise and R. Blanchard, *Modern Semiconductor Fabrication Technology*, Reston Books, Reston, Va., p. 43.
3. P. Gise and R. Blanchard, *Modern Semiconductor Fabrication Technology*, Reston Books, Reston, Va., p. 46.
4. J. Maliakal, D. Fisher, Jr., and A. Waugh, "Trends in automated diffusion furnace systems for large wafers," *Solid State Technology*, Dec. 1984, p. 107.
5. J. Maliakal, D. Fisher, Jr., and A. Waugh, "Trends in automated diffusion furnace systems for large wafers," *Solid State Technology*, Dec., 1984, p. 107.
6. P. Burggraaf, "'Hands-off' furnace systems," *Semiconductor International*, Sept.1987, p. 78.
7. P. Singer, "Trends in vertical diffusion furnaces," *Semiconductor International*, Apr. 1986, p. 56.
8. S. Leavitt, "RTP: On the edge of acceptance," *Semiconductor International*, Mar. 1987.

9. S. Leavitt, "RTP: On the edge of acceptance," *Semiconductor International*, Mar. 1987.
10. D. Toole and P. Crabtree, "Trends in high-pressure oxidation," *Microelectronic Manufacturing and Test*, Oct. 1988, p. 1.
11. D. Toole and P. Crabtree, "Trends in high-pressure oxidation," *Microelectronic Manufacturing and Test*, Oct. 1988, p. 8.
12. P. Van Zant, "Ammonium persulfate as a stripping and cleaning oxidant," *Semiconductor International*, Apr. 1984, p. 109.
13. Wolf and Tauber, *Silicon Processing for the VLSI Era*, p. 226.
14. P. Schmidt and W. Michel, "Anodic formation of oxide films on silicon," *Journal of the Electrochemical Society*, Apr. 1957.
15. Wolf and Tauber, *Silicon Processing for the VLSI Era*, p. 210.

8

Photolithography— Preparation to Exposure

Overview

In this chapter the all-important photolithography processes are detailed. The chapter starts with an explanation of a basic ten-step process and includes a discussion on photoresist chemistry and physical parameters. The majority of the chapter is devoted to the process steps from surface preparation to alignment and exposure. Each of the options available for the steps is explained and evaluated.

Objectives

Upon completion of this chapter, you should be able to:

1. Sketch wafer cross sections showing the basic ten-step photomasking process.

2. Explain the reaction of negative and positive photoresists to light.

3. Describe the correct resist and mask polarities required to produce holes and islands in wafer surface layers.

4. Make a list of the major process options for each of the ten basic steps.

5. Select from the list in objective 4 the processes used to pattern features in micron and submicron sizes.

6. Describe the need for, and process steps used in, double masking, multilayer resist processing, and planarization techniques.

7. Explain the use of antireflective coatings and contrast enhancement in the patterning of "small" feature sizes.

8. List the optical and nonoptical methods used for alignment and exposure.

9. Compare the equipment and advantages of each alignment and exposure method.

Introduction

Photolithography is one of the terms used to identify the basic operation of patterning. Other terms used are *photomasking, masking, oxide* or *metal removal* (OR, MR), and *microlithography*. (Recall that patterning is the process that removes specific portions of the top layer(s) on the wafer surface. See Fig. 8.1.).

Photolithography is one of the most critical operations in semiconductor processing. It is the patterning process that sets the horizontal dimensions on the various parts of the devices and circuits. The goal of the operation is twofold. First, is to create in and on the wafer surface a pattern whose dimensions are as close to the design requirements as possible. This goal is referred to as the *resolution* of the images on the wafer. The pattern dimensions are referred to as the *feature sizes* or *image sizes* of the circuit.

The second goal is the correct placement (called *alignment* or *registration*) of the circuit pattern on the wafer. The entire circuit pattern

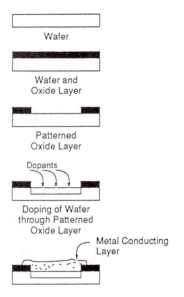

Wafer

Wafer and
Oxide Layer

Patterned
Oxide Layer

Dopants

Doping of Wafer
through Patterned
Oxide Layer

Metal Conducting
Layer

Figure 8.1 Basics of silicon planar processing.

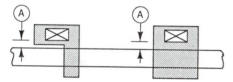

Figure 8.2 Minimum spacings (registration) for contact and metal patterns.

must be correctly placed on the wafer surface and the individual parts of the circuit must be in the correct positions relative to each other. Keep in mind that the final pattern is created from several photomasks applied to the wafer in a sequential manner. The registration requirement is similar to the correct stacking of the different floors of a building in relation to each other. It is easy to imagine the effect on the functioning of a stairwell or elevator if the floors of a building did not sit directly on top of each other. In a circuit, the effects of a misaligned mask layer can cause the entire circuit to fail.

For VLSI and ULSI work, the resolution and registration requirements are very stringent. A VLSI circuit with micron or submicron feature sizes must be formed on the wafer surface with tolerances of $\pm \frac{1}{3}$ μm. The patterns must register to each other to a specification of 0.1 μm[1] as illustrated in Fig. 8.2.

Overview of the Photomasking Process

Photolithography is a multistep pattern transfer process similar to stenciling or photography. In photolithography the required pattern is first formed in reticles or photomasks and transferred into the surface layer(s) of the wafer through the photomasking steps.

The transfer takes place in two steps. First, the pattern on the reticle or mask is transferred into a layer of photoresist (Fig. 8.3). Photoresist is a light-sensitive material similar to the coating on a regular photographic film. It changes its structure and properties when exposed to light. In the example in Fig. 8.3, the photoresist in the region exposed to the light was changed from a soluble condition to an insoluble one. Resists of this type are called *negatively acting* and the condition change is called *polymerization*. Removing the soluble portion with chemical solvents (developers) leaves a hole in the resist layer that corresponds to the opaque pattern on the reticle.

The second transfer takes place from the photoresist layer into the wafer surface layer (Fig. 8.4). The transfer occurs when the wafer etchants remove the portion of the wafer's top layer that is not covered with photoresist. The chemistry of photoresists is such that they do not dissolve in the chemical etching solutions; they are *etch-resistant*.

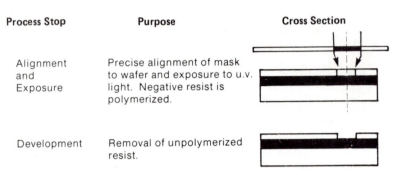

Process Stop	Purpose	Cross Section
Alignment and Exposure	Precise alignment of mask to wafer and exposure to u.v. light. Negative resist is polymerized.	
Development	Removal of unpolymerized resist.	

Figure 8.3 First pattern transfer—from mask to resist layer.

Process Step	Purpose	Cross Section
Etch	Selective removal of top surface layer	
Photoresist Removal	Clean photoresist from the wafer surface	
Final Inspection	Inspection of wafer for correctness of image transfer from photoresist to top layer	

Figure 8.4 Second image transfer.

In the example shown in Figs. 8.3 and 8.4, the result is a hole etched in the wafer layer. The hole came about because the pattern in the mask was opaque to the exposing light. A mask whose pattern exists in the opaque regions is called a *clear-field mask* (Fig. 8.5). The pattern could also be coded in the mask in the reverse, in a dark-field mask. If the same steps were followed, the result of the process would be an island of material left on the wafer surface (Fig. 8.6).

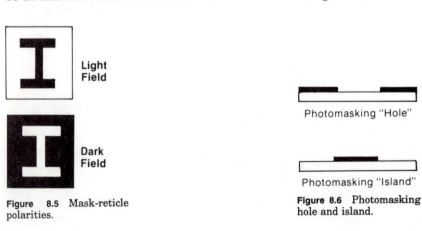

Light Field

Dark Field

Photomasking "Hole"

Photomasking "Island"

Figure 8.5 Mask-reticle polarities.

Figure 8.6 Photomasking hole and island.

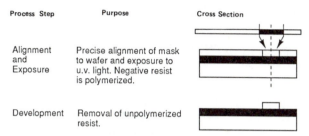

Process Step	Purpose	Cross Section
Alignment and Exposure	Precise alignment of mask to wafer and exposure to u.v. light. Negative resist is polymerized.	
Development	Removal of unpolymerized resist.	

Figure 8.7 First image transfer from a light-field mask to a positive photoresist layer.

The resist reaction to light just described is a character of negative-acting photoresists. There are also positive-acting photoresists. Within these resists, the light changes the chemical structure from relatively nonsoluble to much more soluble. The term describing this change is *photosolubilization*. Figure 8.7 shows that an island is produced when a light-field mask is used with a positive photoresist.

The result obtained from the photomasking process from different combinations of mask and resist polarities is shown in Fig. 8.8. The choice of mask and resist polarity is a function of the level of dimensional control and defect protection required to make the circuit work. These issues are discussed in the process sections of the chapter.

Ten-Step Process

Transferring the image on the reticle or mask into the wafer surface layer requires a number of steps (Fig. 8.9). There are ten basic steps with many variations. The process illustrated is shown with a light-field mask and a negative photoresist. It is broken into two sections. The first section, steps 1 to 7, shows the first transfer. Steps 8, 9, and 10 finish the image transfer into the wafer surface layer.

Understanding this basic process is facilitated by making a list of the steps and drawing in the corresponding cross section. It is also instructive to draw cross sections but to change the polarity of the mask

		Photoresist Polarity	
		Negative	Positive
MASK POLARITY	Clear Field	HOLE	ISLAND
	Dark Field	ISLAND	HOLE

Figure 8.8 Mask and photoresist polarity results.

Process Step	Purpose	Cross Section
		Top Layer
1. Surface Preparation	Clean and dry wafer surface	
		Wafer
2. Photoresist Apply	Apply a thin layer of photoresist to the wafer	Photoresist
3. Softbake	Partial evaporation of photoresist solvents to promote adhesion	
4. Alignment and Exposure	Precise alignment of mask to wafer and exposure to u.v. light. Negative resist is polymerized.	
5. Development	Removal of unpolymerized resist.	
6. Hard Bake	Final Evaporation of Solvents	
7. Develop Inspection		
8. Etch		
9. Photoresist Removal		
10. Final Inspection		

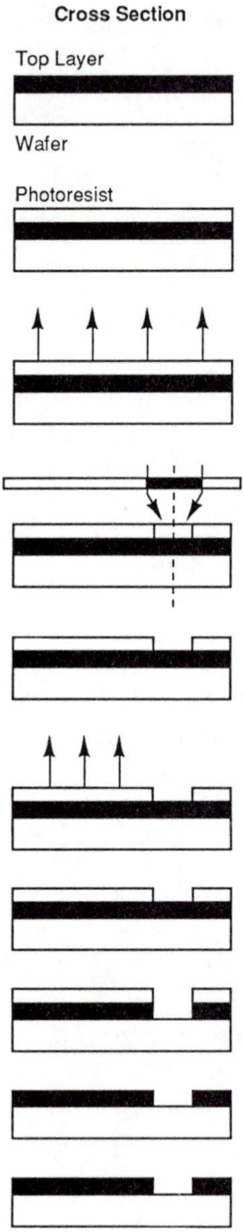

Figure 8.9   Ten-step photomasking process.

and photoresist. Mastering the steps and results using different resist and mask polarities will facilitate understanding the more in-depth explanations in the individual process sections.

Photoresist Chemistry

Photoresists have been used in the printing industry for over a century. In the 1920s photoresists found wide application in the printed circuit board industry. The semiconductor industry adapted this technology to wafer fabrication in the 1950s. Development of photoresists specifically designed for semiconductor use was first supplied by the Eastman Kodak Company. In the late 1950s, they introduced their line of KPR and KMER negative resists. At around the same time, the Shipley Company introduced a line of positive-acting resists. Since then a host of other companies have entered the market with photoresists designed to keep pace with ever-increasing industry demands to print narrower lines with fewer defects and at higher production rates. Today, the photoresist practitioner has available a wide range of products designed to match a variety of needs.

The photoresist is the heart of the masking process. The various steps are fine-tuned to accommodate the particular resist used and the desired results. The selection of a resist and development of a resist process is a detailed and lengthy procedure. Once a resist process is established, it is changed very reluctantly. Rarely will a photoresist engineer risk losing an entire process if only a marginal gain is possible with a new photoresist.

Photoresist composition

Photoresists are manufactured for both general and specific applications. They are tuned to respond to specific wavelengths of light and different exposing sources. They are given specific thermal flow characteristics and formulated to adhere to specific surfaces. These properties come about from the type, quantity, and mixing procedures of the chemical components in the resist. However, all the resists contain four basic ingredients (Fig. 8.10).

Light-sensitive and energy-sensitive polymers. The ingredients that contribute the photosensitive properties to the photoresist are special polymers. Polymers are groups of large, heavy molecules containing carbon, hydrogen, and oxygen that are formed into a repeated pattern. Plastics are a form of polymers.

The techniques of polymer chemistry impart the properties of specific light and radiation sensitivity into the polymer. The most commonly used resists react to some form of light energy: ultraviolet or

Component	Function
Polymer	Changes structure in reaction to energy (polymerization or photosolubilization)
Solvent	Allows spin application of thin layers
Sensitizers	Control of modification chemical reaction when exposed
Additives	Specific Needs

Figure 8.10 Photoresist components.

laser. These are called *optical resists*. Others respond to x rays or electron beams. In each case it is the polymer that determines the property.

Negative resists are based on the polyisopreme polymer. Polyisopreme polymers occur naturally in rubber. In fact, the earlier negative photoresists were based on rubber polymers. (The Hunt Corporation developed the first synthetic polyisopreme polymer structure. See Fig. 8.11.) Before exposure, the negative-resist polymers exist in an unpolymerized condition in which the polymers are not chemically linked to each other. When the resist is exposed to the proper light energy, the polymers become cross-linked or, to use the correct chemical term, become polymerized. This condition can also be achieved when the resist is exposed to heat and/or normal light. To prevent accidental exposure, photomasking areas processing negative resist use yellow filters or yellow lighting. The desirable property of the polymerized resist is its resistance to etchants.

Photoresists respond to many forms of energy. The forms are often referred to by their general category (light, heat radiation, etc.), or by a specific portion of the electromagnetic span (ultraviolet light, deep ultraviolet, I line, etc.). The exposing energies used are detailed in the section on alignment and exposure.

Positive photoresists are based on the phenol-formaldehyde polymer, also called the phenol-formaldehyde novolak resin (Fig. 8.12).

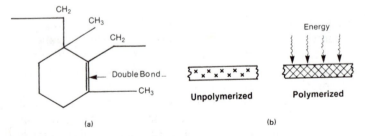

Figure 8.11 Negative resist chemistry.

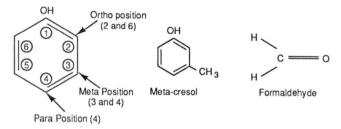

Figure 8.12 Phenol-formaldahyde novolak resin structure. (*After: W. S. DeForest, Photoresist: Materials and Processes, McGraw-Hill, New York, 1975.*)

Within the resist, the polymer is relatively insoluble. After exposure to the proper light energy, the resist converts to a more soluble state. This reaction is called *photosolubilization*.

A discussion of the relative merits and uses of these two types of resists and others are at the end of the process sections. Figure 8.13 contains a list of resist polymers used in semiconductor patterning.

Solvents. The largest ingredient by volume in a photoresist is a solvent. It is the solvent that makes the resist a liquid and allows the resist to be applied to the wafer surface as a thin layer by spinning. Photoresist is analogous to paint, which is composed of the coloring pigment dissolved in an appropriate solvent. It is the solvent that allows the application of the paint onto a surface in a thin layer. For negative photoresist, the solvent is an aromatic type, xylene. In posi-

Resist	Polymer	Polarity	Sensitivity (Coul/cm^2)	Exposure Radiation
Positive	Novolak (M-Cresol-formaldehyde)	+	$3-5 \times 10^{-5}$	UV
Negative	Poly Isoprene	−		UV
PMMA	Poly-(Methyl Methacrylate)	+	5×10^{-5}	E-Beam
PMIPK	Poly-(Methyl Iso-propenyl Ketone)	+	1×10^{-5}	E-Beam/ Deep UV
PBS	Poly-(Butene-1-Sulfone)	+	2×10^{-6}	E-Beam
TFECA	Poly-(Trifluoroethyl Chloroacrylate)	+	8×10^{-7}	E-Beam
COP (PCA)	Copolymer-(α-Cyano Ethyl Acrylate-α-Amido Ethyl Acrylate)	−	5×10^{-7}	E-Beam X-Ray
PMPS	Poly-(2-Methyl Pentene-1-Sulfone)	+	2×10^{-7}	E-Beam

Figure 8.13 Resist comparison table.

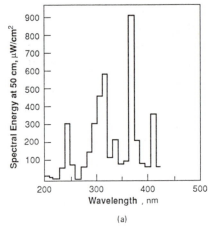

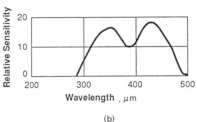

Figure 8.14   Response of a sensitized positive resist. (a) Spectral output of mercury lamp; (b) exposure sensitivity of KTI positive photoresist.

tive resist, the solvent is either ethoxyethyl acetate or 2-methoxyethyl.[2]

Sensitizers. Chemical sensitizers are added to the resists to cause or control certain reactions of the polymer. In negative resists, the untreated polymer responds to a certain range of the ultraviolet spectrum. Sensitizers are added to either broaden the response range or narrow it to a specific wavelength (Fig. 8.14). In negative resists, a compound called bis-aryldiazide is added to the polymer to provide the light sensitivity.[2] In positive resists, the sensitizer is o-naphthaquinonediazide.[3]

Additives. Various additives are mixed with resists to achieve particular results. Some negative resists have dyes intended to absorb and control light rays in the resist film.

Photoresist Performance Factors

The selection of a photoresist starts with the dimensions required on the wafer surface. The resist must first have the capability of producing those dimensions. Beyond that it must also function as an etch barrier during the etching step, a function that requires a certain thickness for mechanical strength. In the role of etch barrier, it must be free of pinholes, which also requires a certain thickness. In addition, it must adhere to the top wafer surface or the etched pattern will be distorted, just as a paint stencil will give a sloppy image if it is not taped tight to the surface. These, along with process latitude and step coverage capabilities, are resist performance factors. In the selection of a resist, the process engineer often must make trade-off decisions between the various performance factors.

Resolution

The smallest opening or space that can be produced in a photoresist layer is generally referred to as its *resolution capability*. The smaller the line produced, the better the resolution capability. Generally, smaller line openings are produced with thinner resist film thicknesses. However, a resist layer must be thick enough to function as an etch barrier and be pinhole-free. The selection of a resist thickness is a trade-off between these two goals.

The capability of a particular resist relative to resolution and thickness is measured by its *aspect ratio* (Fig. 8.15). The aspect ratio is calculated as the ratio of the resist thickness to the image opening. Positive resists have a higher aspect ratio compared to negative resists, which means that for a given image-size opening, the resist layer can be thicker. The ability of positive resist to resolve a smaller opening is due to the smaller size of the polymer. It is a little like the requirement of using a smaller brush to paint a thinner line.

Adhesion capability

In its role as an etch barrier, a photoresist layer must adhere well to the surface layer to faithfully transfer the resist opening into the layer. Lack of adhesion results in distorted images. Resists differ in their ability to adhere to the various surfaces used in chip fabrication.

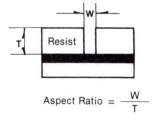

$$\text{Aspect Ratio} = \frac{W}{T}$$

Figure 8.15   Aspect ratio.

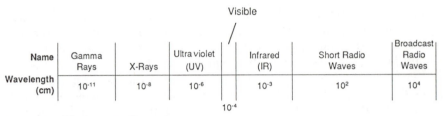

Figure 8.16   Electromagnetic spectrum.

Within the photomasking process, there are a number of steps that are specifically included to promote the natural adhesion of the resist to the wafer surface. Negative resists generally have a higher adhesion capability than positive resists.

Exposure speed and sensitivity

The primary action of a photoresist is the change in structure in response to an exposing light or radiation. An important process factor is the speed at which that reaction takes place. The faster the speed, the faster the wafers can be processed through the masking area. Negative resists typically require 5 to 15 seconds of exposure time while positive resists take three to four times longer.

The sensitivity of a resist relates to the amount of energy required to cause the polymerization or photosolubilization to occur. Further, sensitivity relates to the energy associated with specific wavelength of the exposing source. Understanding this property requires a familiarization with the nature of the electromagnetic spectrum (Fig. 8.16). Within nature we identify a number of different types of energy: light, short and long radio waves, x rays, etc. In reality they are all electromagnetic energy (or radiation) and are differentiated from each other by their wavelengths, with the shorter wavelength radiation having higher energies.

Common positive and negative photoresists respond to energies in the ultraviolet and deep ultraviolet (DUV) portion of the spectrum (Fig. 8.17). Some are designed to respond to particular wavelengths (peaks) within those ranges. When we speak of resist sensitivity, we refer to the specific wavelengths the resist reacts to. This property is also called the *spectral response characteristic* of the resist. Figure 8.18 is the spectral response characteristic of a typical production resist. The peaks in the spectrum are regions (wavelengths) that carry higher amounts of energy. The different light sources used in masking areas are covered in "Alignment and Exposure," p. 194.

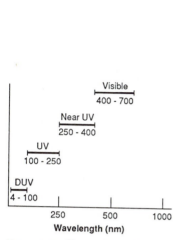

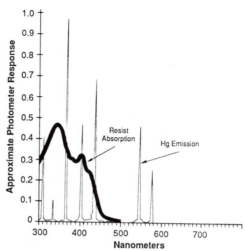

Figure 8.17 Ultraviolet and visible spectrum. (*After Elliott.*[7])

Figure 8.18 Exposure response of Hunt chemical HPR resist. (*Courtesy of Hunt Chemicals.*)

Process latitude

While reading the sections on the individual masking process steps, the reader should keep in mind the fact that the goal of the overall process is a faithful reproduction of the required image size in the wafer layer(s). Every step has an influence on the final image size and each of the steps has inherent process variations. Some resists are more tolerant of these variations, that is, they have a wider process latitude. The wider the process latitude, the higher the probability that the images on the wafer will meet the required dimensional specifications.

Pinhole counts

Pinholes are microscopically small voids in the resist layer. They are detrimental because they allow etchants to seep through the resist layer and etch small holes in the surface layer. Pinholes come from particulate contamination in the environment, the spin process, and from structural voids in the resist layer.

The thinner the resist layer, the more pinholes. Therefore, thicker films have fewer pinholes but they also make the resolution of small openings more difficult. These two factors present one of the classic trade-offs in determining a process resist thickness. One of the principal advantages of positive resists is their higher aspect ratio, which allows a thicker resist film and a lower pinhole count for a given image size.

Step coverage

By the time the wafer is ready for the second masking process, the surface has a number of steps. As the wafer proceeds through the fabrication process, the surface gains more layers. For the resist to perform its etch barrier role, it must maintain an adequate thickness over these earlier layer steps. The ability of a resist to cover surface steps with adequate resist is an important parameter.

Thermal flow

During the masking process there are two heating steps. The second one, hard bake, takes place after the image has been developed in the resist layer. The purpose of the hard bake is to increase the adhesion of the resist to the wafer surface. However, the resist, being a plasticlike material, will soften and flow during the hard bake step. The amount of flow has an important effect on the final image size. The resist has to maintain its shape and structure during the bake.

The goal of the process engineer is to achieve as high a bake temperature as possible to maximize adhesion. This temperature is limited by the flow characteristics of the resist. In general, the more stable the thermal flow of the resist, the better it is in the process.

Comparison of Positive and Negative Resists

Up to the mid-1970s, negative resist was dominant in the masking process. The advent of VLSI circuits and image sizes in the 2- to 5-μm range strained the resolution capability of negative resists. Positive resists had been around for over 20 years, but their poorer adhesion properties were a drawback and their superior resolution capability and pinhole protection were not needed.

By the 1980s, positive resist became the resist of choice. The transition was not easy. To switch a fabrication line from negative to positive resist requires changing the polarity of the masks or reticles from clear-field to dark-field. Unfortunately, it is not a simple matter of reversing the fields in the mask-making process. The dimensions have to be adjusted to accommodate the different characteristics of the positive resist. The determination of the correct mask or reticle dimensions is a lengthy procedure.

Once the mask polarity switch is made, positive-resist processing of holes using dark-field masks offers additional pinhole protection for the wafer (Fig. 8.19). Clear-field masks are prone to small cracks in the glass surface. These cracks, called *glass damage*, block the exposing light, creating in the photoresist layer unwanted holes, which in turn etch into the wafer surface as holes. On dark-field masks, the majority of the surface is covered by chrome, which is hard and less likely to have pinholes. Thus the wafer has fewer unwanted holes.

Another problem with negative resists is oxygenation. This is a reaction of the resist to oxygen in the atmosphere, and it can result in a thinning of the resist film by as much as 20 percent. Positive resists do not have this property. Cost is always an important consideration. Negative resists sell for about one-third the cost of positive resists.

Developing characteristics differ between the two types of resists. Negative resists develop in readily available solvents and possess a wider developer process latitude. Positive resists require carefully prepared developer solutions and require temperature control of the process.

The next-to-last step in the masking process is photoresist removal, which can take place in chemical solutions or in plasma systems. Generally, the removal of positive resists is easier and takes place in chemicals that are more environmentally sound. While positive photoresists are the resists of choice for fabrication areas processing state-of-the-art circuits, there are many lines still producing devices and circuits with image sizes greater than 5 μm. A great many of

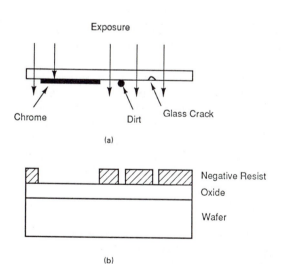

Figure 8.19 (a) Clear-field mask with dirt particle and glass crack, (b) result in negative resist after develop.

Parameter	Negative	Positive
Aspect Ratio (Resolution)		Higher
Adhesion	Better	
Exposure Speed	Faster	
Pinhole Count		Lower
Step Coverage		Better
Cost		Higher
Developers	Organic Solvents	Aqueous
Strippers		
Oxide Steps	Acid	Acid
Metal Steps	Chlorinated Solvent Compounds	Simple Solvents

Figure 8.20   Comparison of negative and positive resists.

these lines use negative resists. Figure 8.20 shows a comparison of properties of the two resists.

Physical Properties of Photoresists

The performance factors just detailed are related to a number of physical and chemical properties of the resist. The properties are discussed in the following text. The influence of a particular property on the process is discussed in the process sections.

Solids content

A photoresist is a liquid that is applied to the wafer by a spinning technique. The thickness of resist left on the wafer is a function of the spin step parameters and several resist properties (two of which are the solids content and viscosity).

Recall that the photoresist is a suspension of polymers, sensitizers, and additives in a solvent. Different resists will contain different amounts of these solids. The amount is referred to as the solids content of the resist and is expressed as the weight percent in the resist. Solids content is measured by evaporating off the solvent and weighing the amount of solids left. Solids contents are in the 20 to 40 percent range.[4]

Viscosity

Viscosity is the quantitative measure of liquid flow. High-viscosity liquids, such as tractor oils, flow in a sluggish manner. Low-viscosity liquids, such as water, flow more readily. In both cases, the mechanism of flow is the same. The molecules in the liquid roll over each other as the liquid is being poured. As the molecules roll about, there exists an attraction between them that acts as an internal friction. Viscosity is the measurement of that friction.

Viscosity is measured by several techniques. One is the falling ball viscosimeter. In this instrument, a ball of measured size and weight is dropped into a tube of the liquid and timed as it passes between two marks. The higher the viscosity of the liquid, the longer it takes the ball to transist the marked distance. Another technique is the Ostwalk-Cannon-Fenske method, which is a timed measurement for a given amount of a liquid to pass through a given orifice.

Most photoresist manufacturers measure viscosity with a rotating vane in the resist. The higher the viscosity, the more force is required to move the vane through the liquid at a constant speed. The rotating-vane apparatus illustrates the force-related character of viscosity.

The unit of viscosity is the centipoise. It is named after the French scientist Poisseulle who investigated the viscous flow of liquids. The poise is equal to one dyne second per centimeter. Photoresist is measured in centipoise. One centipoise is one-hundredth of a poise. The viscosity unit of centipoise is more correctly named the *absolute viscosity*.

Another viscosity unit used by photoresist manufacturers is centistokes. This value is calculated from the absolute viscosity (centipoise) divided by the density of the resist. This value is called the *kinematic viscosity*. Viscosity varies with temperature; therefore, its specified value is stated at a particular temperature, usually 25°C. Viscosity is a major parameter determining the resist thickness during the spin process. Viscosity is closely related to the solids content. The higher the solids content, the higher the viscosity.

Surface tension

The surface tension of a resist also influences the outcome at spin. Surface tension is a measure of the attractive forces in the surface of the liquid (Fig. 8.21). Liquids with high surface tension flow less readily on a flat surface. It is the surface tension that draws a liquid into a spherical shape on a surface or in a tube.

Index of refraction

When the resist-covered wafer surface is being exposed, the transparent resist is part of an optical system. As such, its optical properties must be specified and controlled. One property of transparent films sitting on reflective substrates (resist on a wafer) is that a light ray impinging on the surface at an angle will be bent as it passes through

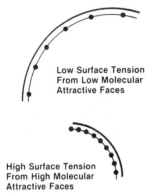

**Low Surface Tension
From Low Molecular
Attractive Faces**

**High Surface Tension
From High Molecular
Attractive Faces**

Figure 8.21 Surface tension.

the film. This comes about as the light ray is slowed up in the material. The index of refraction is a measurement of the speed of light in a material compared to its speed in air, as shown in Fig. 8.22. It is calculated as the ratio of the reflecting angle to the impinging angle. For photoresists, the index of refraction is close to that of glass, approximately 1.45.

Storage and Control of Photoresists

Photoresists are delicate high-technology mixtures. Great care and precision go into their manufacture. Once a photomasking process is developed, its continuing success depends on the day-in, day-out control of the process parameters and a consistent photoresist product.

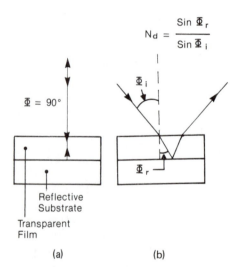

$$N_d = \frac{\sin \Phi_r}{\sin \Phi_i}$$

$\Phi = 90°$

Reflective
Substrate

Transparent
Film

(a)

(b)

Figure 8.22 Index of refraction. (a) 90° incident light; (b) angled light is refracted in the transparent film.

Delivered batch-to-batch consistency is a responsibility of the manufacturer. Maintaining that consistency is the responsibility of the user. Several properties of resists dictate the storage and control conditions required.

Light and heat sensitivity

Both light and heat can activate the sensitive mechanisms in the resist. These mechanisms are intended to be activated in the closely controlled masking process steps. It is imperative that the resist be protected during storage and handling to prevent unwanted reactions that would interfere with the process results. This is the reason that masking areas use yellow light. It is also the reason that resist bottles are brown. The colored glass protects the resist from stray light. Proper transportation and storage of resist require temperature control within limits specified by the manufacturer.

Viscosity sensitivity

Viscosity control is essential for good film thickness control. To maintain the photoresist's viscosity, resist bottles must remain capped prior to use. Opened bottles will allow evaporation of the solvent, which results in a higher solids content, which, in turn, results in a higher viscosity. If photoresist is dispensed from plastic tubing, the material should be tested to ensure that the resist is not leaching plasticizers out of the material. The plasticizers will increase the viscosity of the resist.

Shelf life

A container of photoresist comes with a recommended shelf life. The problem again has to do with the self-polymerization or photosolubilization of the resist. In time, changes to the polymer will take place, altering the resist performance when it reaches the production line.

Cleanliness

Needless to say, any and all equipment used to dispense photoresist must be maintained in the cleanest condition possible. Besides the effects of particular contamination from the system, the resist tubing must be cleaned regularly due to the possible buildup of dried photoresist. Cleaning agents should be checked for their compatibility with the resist. For example, TCE should not be used with negative resist because it can cause bubbles in the resist.

Photomasking Processes

These processes follow the ten photomasking steps just outlined. For each step, the purpose and process considerations are explained, along with the major process options. The choice of a particular process option depends on a number of interrelated factors that include the image size, circuit complexity, transistor type, and the equipment budget.

Surface preparation

Throughout this text, various analogies will be used to aid the reader in understanding the complicated processes. A good analogy to the photomasking process is painting. Even the amateur painter soon learns that to end up with a smooth film of paint that adheres well, the surface must be dry and clean. The same is true in photomasking technology. Ensuring that the resist will stick to the wafer surface requires surface preparation. This step is performed in three stages: particle removal, dehydration, and priming.

Particle removal.   The wafers coming to the photomasking area almost always come in a clean condition from another process such as oxidation, doping, or CVD. However, during storage, loading, and unloading into carriers, they may pick up some particulate contamination that must be removed from the surface. Depending on the level of contamination and/or the process requirements, several particulate removal techniques may be used (Fig. 8.23). In extreme cases, the wafers may be put through a wet chemical cleaning similar to the preoxidation cleaning processes, including an acid cleaning, water rinsing, and drying. The particular acid used must be compatible with the top layer on the wafer surface.

On fabrication lines and in laboratories that employ manual processing, the principal particle removal method is to blow it off using a filtered nitrogen gun, sometimes with an antistatic ionizer. Each wa-

---

- High-pressure nitrogen blowoff
- Wet chemical cleaning
- Rotating brush scrubber
- High-pressure water stream

Figure 8.23 Prespin wafer cleaning methods.

fer is held on a vacuum wand and blown clean. Another method is to blow off all of the wafers in a carrier.

A more effective method is a rotating brush wafer scrubber (Fig. 8.24). These systems are automated, with the wafer being removed from the carrier and placed on a rotating vacuum chuck. The vacuum holds the wafer firmly in place as it rotates. Cleaning takes place as the wafer is flooded with deionized water and/or detergents and a rotating brush is brought into near contact with the wafer. The combined rotations of the wafer and the brush create a high liquid velocity across the wafer surface, effectively removing particles. Scrubber systems are made in standalone units or can be incorporated into other process machines (such as photoresist spinners) to automate the cleaning process.

Another effective cleaning system uses a thin stream of high-pressure water that is swept across the wafer surface. This technique was developed to clean photomasks. Its effectiveness comes from the extremely high pressure, in the range from 2,000 to 4,000 psi. High-pressure water systems are effective in dislodging particles held to the surface by static charges and offer the advantage that only clean deionized water comes in contact with the wafer. This latter aspect minimizes the chance of secondary contamination associated with brush cleaning methods. Often a wetting agent is added to the water stream to minimize the buildup of static on the wafer surface.

Dehydration baking.   It has been mentioned that the wafer surface has to be dry to promote wafer adhesion. As it comes out of most of the processes, the wafer surface is *hydrophobic*. Liquids form into small droplets on a hydrophobic surface, such as water beads on a newly waxed car. A hydrophobic surface is conducive to good photoresist adhesion (Fig. 8.25).

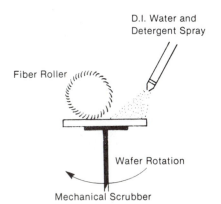

Figure 8.24 Mechanical scrubber.

Unfortunately, when the wafer is exposed to moisture, either from the air or from postcleaning rinses, the surface condition changes to a *hydrophilic* one. This condition is evidenced by a liquid on the surface spreading out in a wide puddle, such as water on a nonwaxed car surface. A hydrophilic surface is also said to be hydrated. Resist does not adhere very well to hydrated surfaces (Fig. 8.25).

Two important ways to maintain a hydrophobic surface are to keep the room humidity below 50 percent and to coat the wafers with photoresist as quickly as possible after being received from a previous process. Often the wafers will be stored in desiccators purged with dry, filtered nitrogen. Additional steps are taken to establish a dehydrated wafer surface. These steps include a dehydration bake and priming with a liquid.

Dehydration bakes use a variety of systems that rely on the three methods of heat transfer of conduction, convection, and radiation. Conduction is the transfer of heat by direct physical contact of the object with a heated surface. Hot plates heat by conduction. In the conduction process the vibrating atoms of the hotter surface cause the object atoms to vibrate also. As they vibrate and collide, the atoms become hotter.

Some dehydration baking is in convection ovens. Systems using convection heating include home forced-air furnaces, hair dryers, air- and nitrogen-fed ovens, and oxidation furnaces. In these systems a unit heats the gas and a blower or pressure pushes the gas to a space where it, in turn, transfers its energy to the object.

The third method is radiation. The term *radiation* describes the travel of electromagnetic energy waves through space. Radiation waves travel in vacuums as well as through gases. The sun transfers heat to the earth by radiation. Heating lamps also transfer heat by radiation. Radiation is the heating method used by RTP systems. When the radiation strikes an object, the energy carried by the wave is transferred directly to the atoms of the object.

Dehydration bakes take place in three temperature ranges to address three different dehydration mechanisms. In the range of 150 to 200°C (low temperature), surface water is evaporated. At 400°C (medium temperature), water molecules loosely attached to the surface

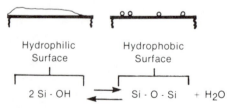

Figure 8.25   Hydrophilic versus hydrophobic surfaces.

will leave. At temperatures above 750°C (high temperature), the surface is chemically restored to a hydrated condition.

In most masking processes, only low-temperature dehydration bake temperatures are used. This is because this temperature range is easily obtainable with hot plates and chest-type convection and vacuum ovens. Another advantage of low-temperature dehydration is that the wafers do not have to wait for a cool-down before the spin process. Systems to perform this step can easily be integrated into a spin-bake system, making them dehydration-spin-bake systems. An explanation on these heating systems is in "Soft Bake," p. 190.

High-temperature dehydration bakes are rare. One reason is that reaching a temperature of 750°C usually requires the use of a tube furnace, and tube furnaces are large and cannot be integrated into the spin-bake processes. A second reason is the temperature level itself. At 750°C, doped junctions in the wafer can move (which is undesirable) and mobile ionic contaminants on the surface can move into the wafer causing reliability and performance problems.

Wafer priming. In addition to dehydration baking, the wafers go through a chemical priming step to ensure good adhesion of the resist. In painting, primers are a subcoat selected for their ability to adhere to the surface and provide a good surface for the finish paint to stick to. In semiconductor photomasking, the role of the primer is to tie up molecular water on the surface, thereby increasing its adhesion property.[5]

A number of chemicals provide priming capabilities, but one, hexamethyldisilazane (HMDS), is in universal use. The use of HMDS is described in U.S. Patent 3,549,368 by R. H. Collins and F. T. Deverse (1970) of IBM. The HMDS is mixed with xylene in solutions of from 10 to 100%. The exact mixture is determined by the process engineers and is based on the particular surfaces and environmental factors in the clean room. Unlike a painting primer, a thickness of only several molecules is sufficient to provide the necessary adhesion promotion.

Immersion priming. The simplest method of applying the primer to the wafer is by dipping the wafers in a beaker of the liquid. This method has the drawbacks of being a manual method and of exposing the wafers to contamination in the HMDS.

Spin priming. In most photomasking areas the primer is applied to the wafer while it is on the resist spinner chuck (Fig. 8.26). The dispensing of the HMDS can be manual from a syringe. Automatic spinners (see next section) have a separate system to dispense HMDS onto the wafer surface just prior to the application of the resist. After the

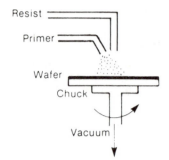

Figure 8.26 Spin dispense of primer.

spinner dispenses the primer onto the rotating wafer, the chuck is ramped to a higher speed to dry the HMDS layer. A major production advantage of spin priming is that it takes place in-line with the spin step.

Vapor priming.   Both immersion and spin priming require the direct contact of the HMDS liquid with the wafer surface. Whenever a liquid is in contact with the wafer, there is the danger of contamination from the liquid. Another consideration is that the HMDS must be dry before the resist is applied. Wet HMDS can dissolve the bottom layer of the resist and interfere with the exposure, development, and etching. Lastly, HMDS is relatively expensive, and in spin priming an excess of HMDS is sprayed on the wafer to ensure adequate coverage. The excess is thrown off the wafer and discarded.

The preceding considerations are overcome by vapor priming techniques. Vapor priming is practiced in three forms. Two are performed at atmospheric pressure and one in a vacuum (Fig. 8.27). One atmospheric system employs a bubbler chamber connected to a desiccator-type chamber. Nitrogen is bubbled through the HMDS and carried into the chamber where it coats the wafers. Another method employs a vapor degreaser with a reserve of liquid HMDS. The HMDS is

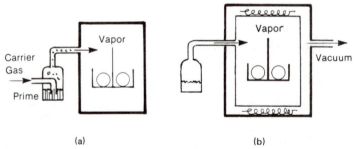

Figure 8.27 Vapor prime methods. (*a*) Atmospheric; (*b*) vacuum bake–vapor prime.

heated to the vapor point and the wafers are suspended in the vapors for coating.

A newer technique is vacuum vapor priming, which uses a sealed flask of HMDS connected to a vacuum oven or single-wafer chamber. The wafers are first heated in the oven in a nitrogen atmosphere. After a temperature of about 150°C is reached, the atmosphere is switched to a vacuum. Once the vacuum level is reached, a valve is opened and HMDS vapors are drawn into the chamber by the low pressure. Within the chamber, the wafers become completely coated as the vapors fill the entire chamber. This method has shown good adhesion longevity even in the presence of high humidity.[6]

Vacuum vapor priming offers the additional advantage of a combined dehydration bake and prime step and a significant reduction in HMDS usage. In a production area that typically uses gallons of HMDS per week, switching to vacuum vapor priming will reduce usage to pints per month. Vacuum vapor priming practiced in a chest-type oven adds an additional step to the process. Many automatic spinner systems offer in-line vapor primers.

Photoresist spinning

The purpose of the photoresist application step is the establishment of a thin, uniform, defect-free film of photoresist on the wafer surface. These qualities are easy to state, but they require sophisticated equipment and stringent controls to achieve. A typical resist layer varies from 0.5 to 1.5 μm in thickness and has to have a uniformity of plus or minus only 0.01 μm (100 Å). This variation is 1 percent of a 1.0-μm thickness.

The usual methods of applying thin layers of liquids to surfaces are brushing, rolling, and dipping. None of these methods is adequate to achieve the quality resist film necessary for photomasking. The method used is spinning, which was briefly described in the section on priming. Spinners are built in manual, semiautomatic, and automatic designs. The systems differ in the degree of automation and are described in the following text. However, the deposit of the film on the wafer is common to each of the systems.

The static spin process.   The wafer that is ready for spinning (after priming) is on the vacuum chuck. Several cubic centimeters (cm^3) of the photoresist is deposited in the center of the wafer (Fig. 8.28) and allowed to spread out into a puddle. The puddle is allowed to spread until it covers the majority of the wafer surface. The size of the puddle is a process parameter that depends on the size of the wafer and the type of resist used. The amount of resist deposited in the puddle is crit-

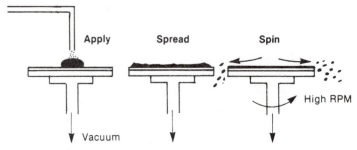

Figure 8.28 Static spin process.

ical only in the extremes. Too small an amount will result in incomplete resist coverage, and too much will cause a buildup of a resist rim or result in resist on the back of the wafer (Fig. 8.29).

When the puddle reaches its specified diameter, the chuck is rapidly accelerated to a predetermined speed. During the acceleration, centrifugal forces spread the resist to the wafer edge and throw off excess resist, leaving a thin uniform layer on the wafer. The high-speed spin continues for some time after the resist is spread to allow drying of the resist.

The final thickness of the film is established as the result of the resist viscosity, the spin speed, the surface tension, and the drying characteristics of the resist. In practice, surface tension and the drying characteristics are properties of the resist, and the viscosity–spin

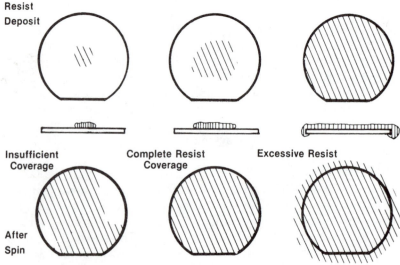

Figure 8.29 Example of resist coverage.

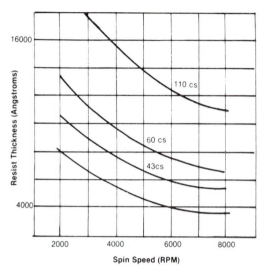

Figure 8.30 Resist thickness versus spin speed. (*Courtesy of KTI Chemicals.*)

speed relationship is determined from curves supplied by the resist manufacturer or established for the particular spin system used (Fig. 8.30).

Although spin speed is specified to control resist thickness, it is actually the acceleration rate that establishes the final resist thickness. The acceleration characteristic of the spinner must be specified, and it is usually maintained as a constant in the spin process.

Dynamic dispense. The need for uniform resist films on larger-diameter wafers led to the development of the dynamic spin dispensing technique in the 1970s (Fig. 8.31). For this technique the wafer is rotated at a low speed of approximately 500 rpm. While the wafer is rotating, the resist is dispensed onto the surface. The action of the rotation assists in the initial spreading of the resist. Less resist is used and a more uniform layer is achieved. After spreading of the resist, the spinner is accelerated to a high speed to complete the spread and thin the resist into a uniform film.

Moving-arm dispensing. An improvement on the dynamic dispense technique is the addition of a moving-arm resist dispenser (Fig. 8.32). The arm moves in a slow motion from the center of the wafer toward its edge. This action creates more uniform initial and final layers. A moving-arm dispenser also saves resist material, especially for larger-diameter wafers.

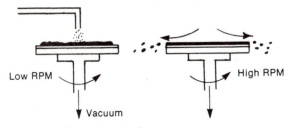

Figure 8.31   Dynamic spin dispense.

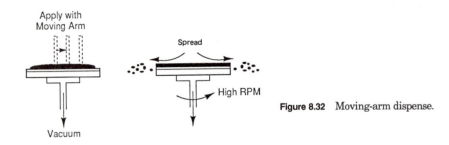

Figure 8.32   Moving-arm dispense.

Manual spinners.   A manual spinner is a simple machine consisting of from one to four vacuum chucks (called *heads*), a motor, a tachometer, and a connection for a vacuum source. Each head is surrounded by a catch cup which serves to collect the excess resist and direct it to a collection vessel. The catch cup also prevents "balls" of resist thrown off the wafers during the acceleration from landing on adjacent wafers. The process starts with the manual removal of particles with a filtered nitrogen gun. The wafer(s) are mounted on the head with tweezers or a vacuum wand and the chuck vacuum is turned on. Next the HMDS is dispensed onto the surface from a syringe or squeeze bottle, and the wafer is spun and dried. Finally, the resist puddle is dispensed from another syringe or squeeze bottle. Most spin processes performed on manual spinners use the static dispense method.

Automatic spinners.   A semiautomatic spinner adds automation to the resist blow-off, resist dispense, and spinning cycles. The nitrogen blow-off is accomplished from a separate tube over the vacuum chuck that is connected to a pressurized nitrogen source (Fig. 8.33). Also in the dispense chamber are a primer resist tube and a resist dispense tube. The resist tube is fed resist from either a nitrogen pressurized vessel or through a diaphragm-type pump. In general the industry has moved away from pressurized resist dispense systems because of problems that arise from the absorption of nitrogen into the resist. The ni-

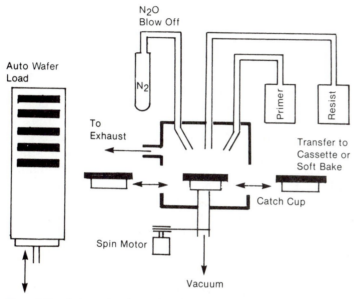

Figure 8.33 Automatic spinner diagram.

trogen comes out of the resist after the dispense cycle, causing voids in the film. Diaphragm pumps eliminate this problem.

Automatic resist dispensers have a negative pressure capability that automatically draws the resist back up into the dispenser tube after each dispensing operation. This drawback (or suck back) minimizes the exposed surface of the resist in the tube from drying into a hard ball that can be deposited on the wafer (Fig. 8.34). In fully automatic systems, all of the events of the spin process are controlled by microprocessors. The systems have mechanisms to extract the wafers from the carriers, place them on the chucks, perform the dispense, per-

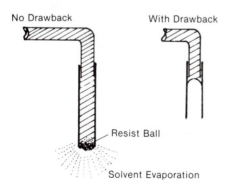

Figure 8.34 Automatic dispense with drawback.

form a soft bake, and place the wafers back in their carriers. The standard system configuration is in a line, referred to as a *track*. Production-level spinners will have from two to four tracks side by side. A complete system may include a wafer cleaning module, dehydration-prime station, spinner unit, and soft bake oven along with wafer load and unload mechanisms.

Backside coating. In some device processes, it is required that the oxide on the back of the wafer remain in place through the masking process. One way this is accomplished is by coating the back of the wafer with photoresist. The requirement for this backside coating is simply a thick-enough layer to survive the etch process. Usually the wafer is backside-coated on a roll coating machine after the front side coat (Fig. 8.35). Another backside etch protection system is provided by adhesive plastic sheets that adhere to the backside of the wafer.

Soft bake

The soft bake step is a heating operation with the purpose of evaporating a portion of the solvents in the photoresist. After the bake, the resist film is still "soft" as opposed to being baked to a varnishlike finish. The solvents are evaporated for two reasons. Remember that the principal role of the solvents is to allow the application of the resist in a thin layer on the wafer. After that role is fulfilled, the presence of the solvent can interfere with the rest of the processing. The first interference occurs during the exposure step. The solvent(s) in the resist can absorb exposing radiation, thus interfering with the proper chemical change in the photosensitive polymers. The second problem is with the resist adhesion. Using the painting analogy, we know that complete drying (evaporation of the solvent) in necessary for good adhesion.

The soft bake parameters of time and temperature are determined by matrix evaluations. A matrix evaluation assesses a number of factors that are varied in a controlled manner to establish the desired results. In photomasking, two major goals are the correct image defini-

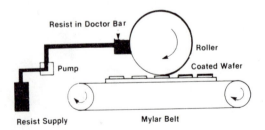

Figure 8.35   Roller coating.

tion and the adhesion of the resist to the wafer during the etch step. Both of these goals are influenced by the degree of the soft bake. In the extreme, an underbaking will result in incomplete image formation at exposure and excessive lifting (poor adhesion) at the etch step. Overbaking will cause the polymers in the resist to polymerize and not react to exposing radiation.

Temperature and time ranges for the soft bake are provided by the resist manufacturer and are fine-tuned by the masking engineer. Negative photoresists have to be baked in a nitrogen atmosphere, whereas positive resists can be baked in air. A number of different methods are used to accomplish the softbake step, and various pieces of equipment incorporating all three heat transfer methods are used.

Convection ovens.   The mainstay baking oven for many semiconductor processes is the convection oven (Fig. 8.36). It is a stainless steel chamber in an insulated enclosure. Either nitrogen or air is supplied by ducts surrounding the chamber and passed through a heater before being directed into the chamber. The inside of the chamber is fitted with racks for the wafer carriers. The carriers stay inside the oven for a predetermined time while the heated gas brings them up to temperature. Convection ovens used for VLSI applications have proportional band controllers and HEPA filters to maintain a clean baking environment.

There are several drawbacks to convection ovens for soft baking. One is batch-to-batch temperature variation, which arises from the amount of time the door is open for loading, the size of the load, and the variable time for all parts of the oven to reach a constant temperature. Within the oven there are different locations that heat up at various rates, depending on the gas flow. A process problem associated with convection heating is the tendency of the top layer of resist to "crust," trapping solvents in the resist (Fig. 8.37).

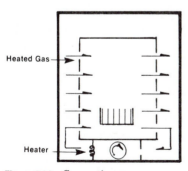

Figure 8.36   Convection oven.

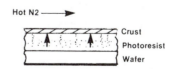

Figure 8.37   Crusting effect of ovens.

Manual hot plates.   In manual and laboratory operations, a simple hot plate is often used for the soft bake. The wafers are placed on an aluminum holder which has a dial thermometer set in a hole drilled into it. The wafers are put on the holder, which is set on the hot plate (Fig. 8.38). The operator monitors the rise in temperature on the thermometer and removes the holder when the proper temperature is reached. With a well-controlled hot plate, an effective soft bake can be achieved. A process advantage to a hot plate is the backside heating. With the bottom of the wafer being heated by conduction, the surface on the resist layer does not crust over, a situation that can trap solvents in the resist. A hot plate process is operator-dependent and has low productivity.

In-line, single-wafer hot plates.   The backside advantage of hot plate heating can be gained in track systems. A single-wafer hot plate is built into the spinner system. Wafers leaving the spinner are positioned on the hot plate and clamped to it with a vacuum. The wafer and resist are heated for a predetermined time, the vacuum released, and the wafer transferred to a carrier. These in-line systems are connected to the facility exhaust system for the removal of the solvent vapors.

Moving-belt hot plates.   A constraint on the productivity of single-wafer hot plates in an integrated system is the total time of the spin step. A typical spin time is 25 to 40 s, which means the soft bake

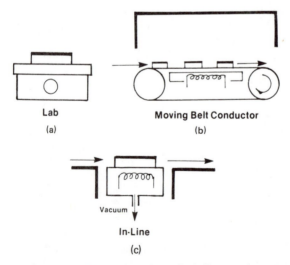

Figure 8.38   (*a*) Manual hot plate; (*b*) in-line continuous hot plate; (*c*) in-line single-wafer hot plate.

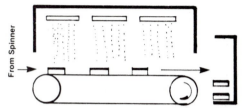

Figure 8.39   Moving-belt infrared (IR) heating.

would have to be completed in that amount of time to keep the wafers moving in a continuous flow. For some resists and for some processes, this is too short a time. A way around the problem is a moving-belt hot plate. The wafers are placed on a heated moving steel belt, and the temperature and belt speed are set to meet the soft bake requirements and process the wafers in a continuous flow.

Moving-belt infrared ovens.   The desire for fast, uniform, and non-crusting soft bake methods led to the use of infrared (IR) radiation sources (Fig. 8.39). Infrared baking is much faster than conduction baking and heats from the "inside out." Inside-out baking is the principle of conduction hot plate baking. The infrared waves pass through the resist layer without heating it, much like sunlight will pass through a window without heating it. The wafer, however, absorbs the energy, gets hot, and in turn heats the resist layer from the bottom.

Microwave baking.   Microwaves as a soft bake heating source have the advantage of infrared heating but at a much faster rate due to the higher energy carried in a microwave. Soft bake temperatures can be well under 1 minute. This brief time lends itself to on-chuck soft baking. Immediately after spinning, a microwave source is directed at the wafer, completing the soft bake (Fig. 8.40).

Vacuum baking.   Vacuum ovens offer several advantages for a number of process steps. A vacuum oven is configured similarly to a convection

Microwave Source

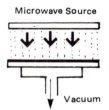

Figure 8.40   Microwave heating.

oven, but is fitted to a vacuum source. Vacuum is particularly efficient for evaporation processes because the reduced pressure aids the evaporation of the solvents, reducing the reliance on the temperature. However, for soft baking the wafers must be heated to a uniform temperature. A problem arises because heating in a vacuum oven is by radiation from the heated chamber walls to the wafers. This heat transfer method is sometimes called line-of-sight, because for uniformity, each wafer must have a clear line of sight to the heat source. In a chest-type vacuum oven packed with carriers of vertically held wafers, this condition cannot be met. The result is poor temperature uniformity in most vacuum ovens.

The benefits of vacuum and uniform hot plate heating can be achieved with in-line, single-wafer systems. Figure 8.41 is a table summarizing the different soft bake methods.

Alignment and exposure. Alignment and exposure (A&E) is, as the name implies, a two-purpose photomasking step. The first part of the A&E step is the positioning or alignment of the required image on the wafer surface. The second part is the encoding of the image in the photoresist layer from an exposing light or other radiation source. It was previously mentioned that the photoresist is the heart of the masking process. More correctly, the photoresist is the "material" heart of the process. A&E is the equipment heart of the process. Correct alignment of the image patterns and establishment of the precise image dimensions in the resist are absolute requirements for the functioning of the devices and circuits. Further, the wafers spend 60 percent of the process time in the lithography area.

Consider the challenge of building a typical VLSI 1-Mbyte RAM circuit. The circuit will contain up to several million individual devices in an area about 0.33 in^2. The individual parts of the devices will have feature sizes in the 1-μm range and have to "fit" or overlay within or next to each other within tolerances of one-third the nominal feature size. This tolerance has to be maintained as each of the level patterns is exposed on the wafer. Advanced processes may require up to 15 separate patterns. The allowable variation in wafer placement must be maintained over all of the pattern alignments.

Aligner system performance capabilities

An aligner system is composed of two major subsystems. One is the subsystem required to correctly position the pattern(s) on the wafer surface. The different aligner types have different alignment subsystems, which are explained in the equipment sections. The second

Method	Bake Time Min.	Temparature Control	Productivity Type	Rate Waf/Hr.	Queing
Hot Plate	5-15	Good	single to small batch	60	Yes
Convection Oven	30	Average - Good	Batch	400	Yes
Vacuum Oven	30	Poor - Average	Batch	200	Yes
I.R. Moving Belt	5-7	Poor - Average	Single	90	No
Conductive Moving Belt	5-7	Average	Single	90	No
Microwave	0.25	Poor Average	Single	60	No

Figure 8.41 Soft bake chart.

- Resolution capability
- Alignment precision
- Contaminaion level
- Productivity

Figure 8.42 Aligner criteria.

part is the exposure subsystem. This subsystem includes an exposure source and a mechanism for directing the radiation rays to the wafer surface.

Aligners are selected and compared by several criteria (Fig. 8.42) that relate to their ability to produce the required images in a consistent and productive manner. Perhaps the most important parameter is the *resolution capability*, or the ability of the machine to produce a particular size image. The higher the resolution capability, the better the machine. In addition to the resolution of the required image size, the aligner must be capable of placing the images in their correct position to each other. This performance parameter is called the *registration capability* of the aligner. These two factors must be performed over the entire wafer, a factor called *dimensional control*. The final performance factor is wafer throughput, which includes the time required to load, align, expose, and unload the wafer and the uptime of the machine. Uptime factors are discussed in Chap. 15 on manufacturing technology.

Alignment and exposure systems

Up to the mid-1970s, the photoresist engineer had the choice of only two A&E systems, contact and proximity aligners. Today the choice has expanded to include both optical and nonoptical aligners (Fig. 8.43). Optical aligners use an ultraviolet light source, while nonoptical systems use exposure sources from other parts of the electromagnetic spectrum. The systems in use today were developed to keep pace with the reduction of feature size, increased circuit density, and required defect reductions of the VLSI era.

Exposure sources

Exposure light sources are chosen to match the spectral response characteristic of the resist and the feature size of the images. Most optical aligner systems use a high-pressure mercury lamp, which produces light as a current is passed through a tube of mercury. A high-pressure atmosphere allows a higher level of stimulation of the mercury without evaporation.

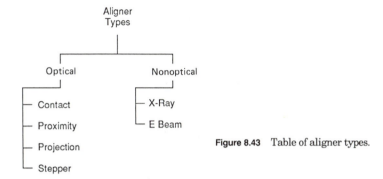

Figure 8.43   Table of aligner types.

The energy from the mercury comes out in bundles of waves grouped in ranges. Some resists are designed to react to the entire range of the wavelengths and some to specific wavelengths. Other resists are designed to respond to the specific high-energy peaks of a mercury lamp. The three peaks are at the 365-, 405-, and 436-nm wavelengths. They are referred to as of the I, H, and G lines, respectively. Steppers used for advanced imaging often have filters to expose the resist to the G line with some available with I-line exposure capabilities.[7]

The advantage of confining the exposure source to a narrow wavelength is to foster complete and faster exposure of the resist. Narrowband exposure limits resolution problems associated with partial exposure of the resist at the clear/opaque edge on the mask. The exposure time is faster for a specific spectral resist since the peaks represent higher-energy portions of the spectrum.

Resolution capabilities of the aligner and the resist are a function of the wavelength of the exposing light. The smaller or narrower the exposing wavelength, the higher the resolution capability. This is due to the bending effects of a wave of energy as it passes the opaque edge of a mask or reticle. As the wave leaves the edge, it becomes bent from a phenomenon known as *diffraction* (Fig. 8.44). The longer the wavelength, the more the diffraction effect, which ultimately acts as a limit on the resolution capability of the system. Shorter wavelengths of light also carry more energy, allowing shorter exposure times and in

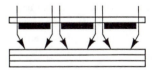

Figure 8.44   Diffracting reduction of image in resist.

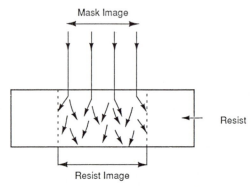

Mask Image

Resist

Resist Image

Figure 8.45   Light scattering in resist film.

turn limiting poor resolution coming from scattering of the light in the resist and from the wafer surface (Fig. 8.45).

Shorter ultraviolet wavelength peaks are in a region of the mercury lamp output spectrum called the mid or deep ultraviolet (DUV). The peaks occur at 313 (mid uv) and at 254 nm (DUV). Deep- and mid-ultraviolet exposing sources are created by the use of filters with a standard ultraviolet source, such as mercury-zenon, zenon, or deuterium lamps.[7]

Alignment criteria

The first mask is aligned by positioning the y axis of the mask at a 90° angle to the major wafer flat on the wafer (Fig. 8.46). Subsequent masks are aligned to a previously patterned mask with the use of alignment marks (also called *targets*). These are special patterns (Fig. 8.47) located in an easily found position on the edge of each chip pattern. Alignment is accomplished by the operator positioning a mark on the mask to a corresponding mark in the wafer pattern. For automatic systems (see "Steppers," p. 202), the alignment marks serve the same purpose. Alignment marks become a permanent part of the chip surface through the etching process. They are then in place for the alignment of the next layer.

Alignment errors, called *misalignment*, fall into several categories (Fig. 8.48). A common one is a simple misplacement in the x-y directions. Another common misalignment is rotational, where one side of the wafer is aligned but the patterns become increasingly misaligned across the wafer. A third rotational misalignment comes about when the die pattern is rotated on the mask or reticle.

Other misalignment problems associated with masks and stepper aligners are run-out and run-in. These problems arise when the chip

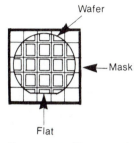

Figure 8.46 First mask alignment.

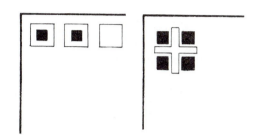

Figure 8.47 Types of alignment marks.

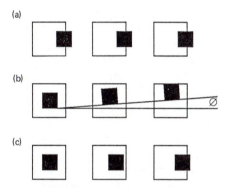

Figure 8.48 Misalignment types. (a) x direction; (b) rotational; (c) run-out.

patterns are not formed on the mask on constant centers or are placed on the chip off center. The result is that only a portion of the mask chip patterns can be properly aligned to the wafer patterns. The pattern becomes progressively misaligned across the wafer.

Contact aligners. Until the mid 1970s, the contact aligner was the workhorse aligner of the semiconductor industry. The alignment part of the system uses a full-wafer-size photomask positioned over a vacuum wafer chuck. The wafer is mounted on the chuck and viewed through a split-field objective microscope (Fig. 8.49). The microscope presents the operator with a simultaneous view of each side of the mask and wafer. The chuck is moved left, right, and/or rotated (x, y, and z movement) by manual controls until the wafer is aligned to the mask pattern.

Once the mask and wafer are aligned properly, the aligner is put into the exposure mode. First the wafer chuck moves up on a piston, pushing the mask into intimate contact with the mask. Once contact is established, a mirror is activated which directs the ultraviolet rays coming from a reflection and lens system through the mask and into

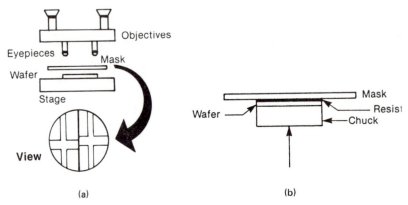

Figure 8.49 Contact aligner system. (*a*) Alignment stage; (*b*) contact stage.

the photoresist. After the proper time, the mirror moves out of the optical path, the chuck retracts, and the vacuum releases the wafer for removal. The lens system spreads the rays coming from the ultraviolet lamp into a uniform front of parallel rays. A properly adjusted exposure ray is referred to as a *collimated* light.

Contact aligners are used for exposing discrete devices and circuits with SSI and MSI densities and feature sizes of approximately 5 μm and above. In this role they rank as an older technology. Actually, a contact aligner is capable of submicron imaging with the proper resist and a well-tuned process. Its replacement by other systems is more related to yield losses associated from the contact of the mask and wafer.

In the contact mode, any particulate contamination will damage the soft resist layer, the mask, or both. Dirt adhering to the clear portions of the mask blocks light during exposure, causing an unwanted defect in negative photoresists or tearing holes in any type of resist. Another defect problem with contact aligners is epitaxial layer spikes on bipolar wafers. The spikes are pointed growths of silicon that damage the masks. With contact aligners, mask damage is so prevalent that masks have to be removed and discarded or cleaned every 15 to 25 exposures, a process that limits the productivity of the method. Dirt between the mask and wafer will cause resolution problems in the immediate area of the piece of dirt, due to the forced local separation of the mask and wafer.

The cost of LSI and higher-density masks and larger mask sizes make an aligner with limited use a very expensive penalty on the overall wafer cost. The high defect levels of contact aligners became unacceptable with the advent of early LSI circuits, leading to development of higher-yield systems.

Proximity aligners. Proximity aligners were a natural evolution of contact aligners. The systems are essentially contact aligners but with mechanisms that either hold the wafer in near or soft contact with the mask. Sometimes proximity aligners are called *soft-contact machines*.

The performance of a proximity aligner is a trade-off between resolution capability and defect density. With the wafer in soft contact with the mask, there is always some scattering of the light, which fuzzes the definition of the image in the resist. On the other hand, the soft contact also greatly reduces the number of defects associated with mask and resist damage. Even with the improved defect density, proximity aligners do not find much use in VLSI photomasking processes.

Scanning projection aligners

The advent of the LSI era, with its smaller feature sizes and higher-density chips, heralded the end of production use of contact and proximity aligners. Actually the end was in sight for years and development work was ongoing in the search for an alternative. The search centered on projection techniques (Fig. 8.50), which are centered on the concept of projection of the mask image onto the wafer surface much as a slide (the mask) is projected onto a screen (the wafer). While simple in concept, the technique requires an excellent optical system to expose the resist in the exact dimensions of the mask. The system became a reality with the introduction in the early 1970s of the Perkin Elmer scanning projection aligner.

Perkin Elmer developed a system under a contract from the U.S. Navy. Their approach to the optical system avoided the distortions introduced with lens systems. They used a mirror system with a slit blocking part of the light coming from the light source. The slit allows a more uniform portion of the light to shine on the mirror system,

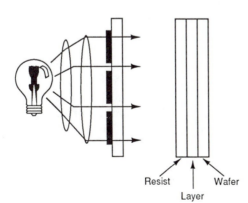

Resist | Wafer

Layer

Figure 8.50 Concept of projection exposure.

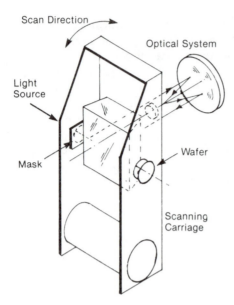

Figure 8.51 Perkin Elmer light projection system. (*Courtesy of Perkin Elmer Corp.*)

which is in turn projected onto the wafer (Fig. 8.51). Since the size of the slit is smaller than the wafer, the light beam is scanned across the wafer. With this system a new parameter, scan speed, became a parameter requiring control.

The early projection systems were sold for up to $100,000 each, at a time when a fully loaded contact aligner could be had for about $20,000. However, the increase in yields gained by the physical separation of the mask and wafer (lower damage-induced defects and reduced mask costs) made payback of the increased cost possible in less than a year. The early scanning projection aligners required manual alignment of the mask and wafer. They are also referred to as 1:1 aligners, since the image dimensions on the mask are the same size as the intended image dimensions on the wafer surface.

Steppers. Scanning projection aligners were a great leap forward over contact aligners for production work, but they still had some limitations, such as alignment and overlay (registration) problems associated with the making of the full-size mask, image distortion, and mask-induced defects from dust and glass damage. These same problems exist to a lesser degree in a step-and-repeat aligner. Recall that a full-size mask is formed on a chrome-covered glass plate from the image projected on it by a chip-size reticle. Each chip is aligned and exposed, and the aligner steps to the next location (Fig. 8.52). The bottom line is that a reticle is of a higher quality than a full-size mask.

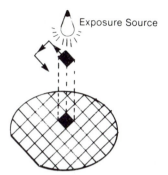

Figure 8.52 Direct step alignment and exposure.

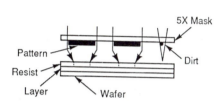

Figure 8.53 5× mask pattern transfer.

By the mid-1970s there were thoughts of stepping images directly from a reticle directly onto the wafer surface. The benefits of such a process are better overlay and alignment because each chip is individually aligned. The procedure of stepping allows precise matching of larger-diameter wafers. Other advantages are resolution improvements because a smaller area is being exposed each time and a lessened vulnerability to dust and dirt. This latter advantage comes from the stepper's reduction option, which uses a reticle that is 5 to 10 times the image size required on the wafer. Making an oversize reticle is easier, and any dirt or small glass distortions are reduced out of existence during the exposure (Fig. 8.53).

The key to production use of steppers required the development of automatic alignment systems. There is no way that an operator could individually align several hundred die on a wafer at a productive rate. Automatic alignment is accomplished by passing low-energy laser beams through alignment marks on the reticle and reflecting them off corresponding alignment marks on the wafer surface. The signal is analyzed and information is fed to the x-y-z wafer chuck controls by a computer, which moves the wafer around until the wafer and reticle are aligned. The images are placed in the photoresist by sequentially exposing each die pattern across and down the wafer (Fig. 8.54).

Most production steppers have G-line exposure sources. I- and H-line exposure systems are available on some systems. There are also development efforts going into excimer laser sources. These sources operate in the deep ultraviolet range and have promise for the resolution of 0.5-μm images.[4] The maintenance of the correct image size during the exposure part of the process requires excellent humidity and temperature control. Most steppers are enclosed in an environmental chamber that controls these important parameters.

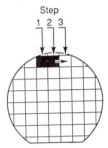

Figure 8.54 Step and repeat die alignment and exposure.

X-ray aligners. The desire for higher-resolution exposure sources inevitably led to the consideration of x-ray sources (Fig. 8.55). X rays are high-energy photons with wavelengths of 4 to 50 Å.[8] This range of wavelengths is capable of image sizes less than 0.5 μm due to the lack of diffraction effects. X-ray aligners are full-mask systems featuring high output through short exposure times. X-ray-exposed wafers show a lower level of defects from dust and organic matter on the mask because the x rays pass through the spots.

A number of difficulties have surfaced with production-level, x-ray aligners. A major problem has been with the development of x-ray blocking masks. Because x rays pass through conventional chrome and glass masks, a process that requires gold as the blocking layer

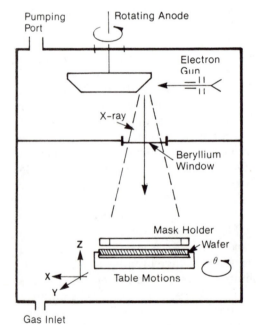

Figure 8.55 X-ray exposure system.

and other materials that stand up to the high energy of the x rays needs to be developed.

While development work goes on in defining x-ray equipment, it also goes on in developing x-ray resists. This work is complicated because there is no standard x-ray source and the resists must show high sensitivity to x rays and also be good etch barriers. These last two factors have proven difficult to balance in resist chemistries.

X-ray sources are still in the development stage. Sources under consideration are standard x-ray tubes, laser-driven sources, and synchrotrons. One proposed solution is to group a number of x-ray aligners around one x-ray source. Whichever source emerges in production systems, it will be expensive.

Electron beam aligners

Electron beam lithography is a mature technology used in the production of high-quality masks and reticles. The system (Fig. 8.56) consists of an electron source that produces a small-diameter spot and a "blanker" capable of turning the beam on and off. The exposure must

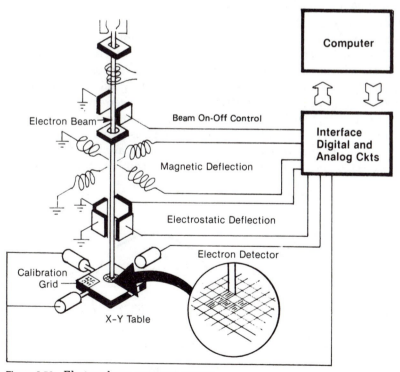

Figure 8.56   Electron beam exposure system.

take place in a vacuum to prevent air molecules from interfering with the electron beam. The beam passes through electrostatic plates capable of directing (or steering) the beam in the x-y direction on the mask, reticle, or wafer. This system is functionally similar to the beam-steering mechanisms of a television set. Precise direction of the beam requires that the beam travel in a vacuum chamber in which there is the electron beam source, support mechanisms, and the substrate being exposed.

There is no mask or reticle used to generate the pattern. The blanking and steering functions are controlled by a computer that has in its memory the wafer pattern directly from the CAD design stage. The beam is directed to specific positions on the surface by the deflection subsystem and the beam turned on where the resist is to be exposed. Larger substrates are mounted on an x-y stage and are moved under the beam to achieve full surface exposure. This alignment and exposure technique is called *direct writing*.

The pattern is exposed in the resist by either raster or vector scanning (Fig. 8.57). Raster scanning is the movement of the electron beam side to side and down the wafer. The computer directs the movement and activates the blanker in the regions where the resist is to be exposed. One drawback to raster scanning is the time required for the beam to scan, since it must travel over the entire surface. In vector scanning, the beam is moved directly to the regions that have to be exposed. At each location, small square or rectangularly shaped areas are exposed, building up the desired shape of the exposed area.

Alignment and overlay parameters are very good with electron beam systems because no distortions are introduced from masks or from optical effects such as diffraction. Resolution is also good, with current machines capable of 0.25-μm feature sizes.[9] Drawbacks to full use of electron beam systems to wafer production are speed and cost. Current electron beam aligners can expose less than ten 6-in-diameter

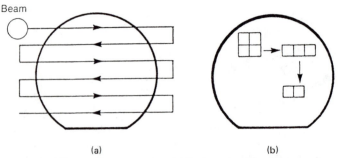

(a)   (b)

Figure 8.57   Electron beam scanning. (*a*) Raster scan; (*b*) vector scanning.

FEATURES	CONTACT	PROJECTION	(DEEP UV)	DIRECT STEP	X-RAY	E-BEAM
Resolution (Microns)	3.0–5.0	1.5–2.0	1.0	1.5–2.0	0.3–0.5	0.75–1.0
Throughput (Wafers/Hr.)	80	50 — 60		30 — 40	20	2
Mask Required	x	x			x	
Reticle Required				x		

Figure 8.58   Alignment and exposure systems comparison.

wafers per shift and cost several million dollars per system. A factor in the slowness of the system is the time required to create the vacuum and release it in the exposing chamber.

Aligner system comparison

The table in Figure 8.58 compares aligner system features and performance parameters.

Key Concepts and Terms

Alignment and exposure
Negative-resist performance
Photoresist components
Photoresist performance factors
Positive-resist performance

Resist spin
Soft bake methods
Surface preparation methods
Ten-step process systems

Review Questions

1. Draw cross-sectional diagrams of the ten-step process with a positive resist and dark-field mask.

2. Make a list of the major components in photoresist and explain the role of each.

3. What type of polymers are in negative resist?

4. What type of polymers are in positive resist?

5. Describe the changes in positive and negative resist during exposure to light.

6. Name the exposure and alignment systems that require a mask or reticle.

7. Which exposure system(s) are used for VLSI processes? Explain why.

8. Name ten factors and their processes that effect control of image size.

9. What is the result of too low a soft bake temperature?

10. What process steps are required to complete the first image transfer?

References

1. S.M. Sze, *VLSI Technology*, McGraw-Hill, New York, 1983, p. 273.
2. Private communication, John Housley, KTI Chemicals Inc.
3. David J. Elliott, *Integrated Circuit Fabrication Technology*, McGraw-Hill, New York, 1976, p. 168
4. David J. Elliott, *Integrated Circuit Fabrication Technology*, McGraw-Hill, New York, 1976, p. 71
5. David J. Elliott, *Integrated Circuit Fabrication Technology*, McGraw-Hill, New York, 1976, p. 116
6. Yield Engineering Systems LP-2 (product bulletin).
7. David J. Elliott, *Integrated Circuit Fabrication Technology*, McGraw-Hill, New York, 1976, p. 370
8. David J. Elliott, *Integrated Circuit Fabrication Technology*, McGraw-Hill, New York, 1976, p. 82
9. David J. Elliott, *Integrated Circuit Fabrication Technology*, McGraw-Hill, New York, 1976, p. 384

Photolithography— Developing to Final Inspection

Overview

In this chapter, the methods used for steps 5 to 10 (developing to final inspection) of the basic process are explained. The end of the chapter examines the processes used for mask making.

Objectives

Upon completion of this chapter, you should be able to:

1. Draw a cross section of a wafer before and after developing.
2. Make a list of the developing methods.
3. Explain the purpose and methods of hard bake.
4. List at least five reasons a wafer can be rejected at develop inspect.
5. Draw a diagram of the develop-inspect-rework loop.
6. Explain the methods and relative merits of wet and dry etch.
7. Make a list of the resist strippers used to strip photoresist from oxide and metal films.
8. Explain the purpose and methods of final inspection.

Development

After the wafer completes the alignment and exposure step, the device or circuit pattern is coded in the photoresist as regions of exposed and

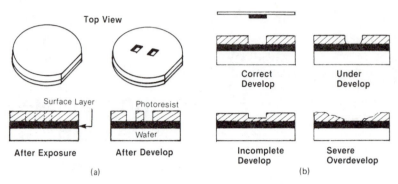

Figure 9.1 Photoresist development. (*a*) Process; (*b*) problems.

unexposed resist (Fig. 9.1). The pattern is "developed" in the resist by the chemical dissolution of the unpolymerized resist regions. Development techniques are designed to leave in the resist layer an exact copy of the pattern that was on the mask or reticle. Problems resulting from a poor developing process (Fig. 9.1) are underdevelopment, which leaves the hole incompletely developed to the correct dimensions, or a coved sidewall. In some cases, the development will not be long enough (incomplete) and will leave a layer of resist in the hole. The third problem is overdevelopment, which removes too much resist from the image edge or top surface.

Negative and positive resists have different developing characteristics and require different chemicals and processes (Fig. 9.2).

Negative resist development

The successful development of the image coded in the resist is dependent on the nature of the resist's exposure mechanisms. Negative resist, upon exposure to light, goes through a process of polymerization which renders the resist resistant to dissolution in the developer chemical. The dissolving rate between the two regions is high enough so that little of the resist layer is lost from the polymerized regions. The chemical preferred for most negative-resist-developing situations

	Positive	Negative
Developer	NaOH Tetramethyl Ammonium Hydroxide	Xylene Stoddard Solvent
Rinse	H_2O	n-Butylacetate

Figure 9.2 Resist developer and rinse chemicals.

is xylene, which is also used as the solvent in negative resists formulas. The development step is done with a chemical developer followed by a rinse. For negative resists, the rinse chemical is usually n-butyl acetate, because it neither swells nor contracts the resist. Either of these actions can change pattern dimensions. Mixtures of alcohol and trichlorethylene (TCE) can function as a negative resist rinse, but they cause some changes in the resist film dimensions. For wafers that have been patterned with a stepper, a milder-acting Stoddart solvent may be used.

The action of the rinse is twofold. First, it rapidly dilutes the developer chemical to stop the developing action. Although the polymerized resist is developer-resistant, there is always a transition region at the exposed edge (Fig. 9.3) that contains partially polymerized molecules. If the developer is allowed to stay in full strength on the wafer, it dissolves into this region and changes the image dimensions. The second action of the rinse is to remove partially polymerized pieces of resist from the open regions in the resist film.

Positive resist development

Positive resists present a different developing condition. The two regions, polymerized and unpolymerized, have a dissolving rate difference of about 1:4. This means that during the developing step some resist is always lost from the polymerized region. Use of developers that are too aggressive or that have overly long developing times may result in an unacceptable thinning of the resist film, which, in turn may cause it to lift or breakdown during the etch step.

Two types of chemical developers are used with positive resist, alkaline-water solutions and nonionic solutions. The alkaline-water solutions can be sodium hydroxide or potassium hydroxide. Since both of these solutions contain mobile ionic contaminants, they are not de-

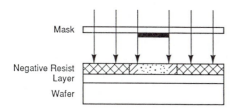

☒ Polymerized Resist

⁄⁄ Partially Polymerized Resist

⚟ Unpolymerized Resist

Figure 9.3 Transition region at resist image edge.

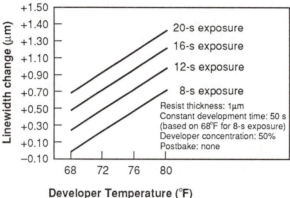

Figure 9.4   Developer temperature and exposure relationship versus line-width change.

sirable in processing sensitive circuits. Most positive resist fabrication lines use a nonionic solution of tetramethylammonium hydroxide (TMAH). The aqueous nature of positive developers makes them more environmentally attractive than the solvent developers required for negative resists.

The positive-resist developing process is more sensitive than negative processes.[1] Factors influencing the outcome are the soft bake time and temperature, degree of exposure, developer concentration, and time, temperature, and method of developing. The development process parameters are determined by matrix testing of all the variables. The effect on line width for a particular process is shown in Fig. 9.4. Tight control of the development and rinse process is critical for dimensional control when using a positive resist. The rinse chemical for positive resist developers is water. It serves the same role as negative resist rinsers but is cheaper, safer to use, and easier to dispose of.

Development methods

There are several methods used to develop resist films (Fig. 9.5). The selection of a method is dependent on the resist polarity, the feature size, defect density considerations, the thickness of the layer to be etched, and productivity.

- Immersion
- Spray
- Puddle

Figure 9.5   Development methods.

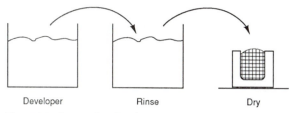

Figure 9.6   Immersion developer steps.

Immersion.   Immersion is the oldest development method. In its simplest form, the wafers, in a chemically resistant carrier, are immersed in a tank of the developer solution for a specific time before being transferred to a second tank of the rinse chemical (Fig. 9.6). The problems associated with such a simple wet procedure are

1. The surface tension of the liquids can prevent the chemicals from penetrating into small openings.
2. Partially dissolved pieces of resist can cling to the wafer surface.
3. The tanks can become contaminated as hundreds of wafers are processed through them.
4. The wafers can become contaminated as they are drawn through the liquid surface.
5. Developer chemicals (especially positive developers) can become diluted through use.
6. Frequent changing of solutions to eliminate problems 1, 2, and 3 raise the cost of the process.
7. Fluctuations in the room temperature can cause changes in the developing rate of the solution.
8. The wafers have to be quickly transferred to a drying process step, which introduces a third step.

Additions are often made to the immersion tanks to improve the development process. Uniformity and penetration of small openings are aided by mechanical agitation of the bath by stirring or rocking mechanisms. A popular stirring system is a Teflon-encapsulated magnet that is coupled to a rotating magnetic field outside the tank.

   Agitation is also achieved by passing ultrasonic waves through the liquid. The waves cause a phenomenon called *cavitation*. The energy in the waves causes the liquid to separate into tiny cavities which immediately collapse. The rapid generation and collapse of the millions of microscopic cavities create a uniformity of development and help the liquid penetrate into small openings. Uniform development rates are also enhanced by the addition of heaters and temperature controllers to the bath.

Spray development.   The preferred method of chemical development is by spray. In fact, spray processing is generally preferred over an immersion system for any wet process (clean, develop, etch) for a number of reasons. For instance, there is a major reduction in chemical use with spray systems. Process improvements include better image definition due to the mechanical action of the spray pressure in defining the resist edge and removing partially polymerized pieces of resist. Spray systems are always cleaner than immersion systems because each wafer is developed (or etched or cleaned) only with fresh chemicals.

Spray processing is done in either single or batch systems. In the single-wafer configuration (Fig. 9.7) the wafer is clamped on a vacuum chuck and rotated while the developer and rinse are sequentially sprayed onto the surface. Drying takes place immediately after the rinse cycle by increasing the rotational speed of the wafer chuck. In appearance and design, a spray developer system for single wafers is the same as a resist spinner but plumbed for different chemicals. Single-wafer spray developers offer the advantage of track automation by integrating the developing and hard bake processes. A major process advantage of these systems is the increased uniformity from the direct impingement of the chemical spray on the wafer.

Spray development had been a standard process for negative resist for years because negative resist is fairly insensitive to the temperature of the developer. For the temperature-sensitive positive resists, spray development is less effective. The problem is the phenomenon of rapid cooling of a fluid dispensed through a orifice under pressure.

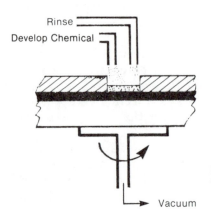

Figure 9.7 Spray development and rinse.

Called *adiabatic cooling*, it is the same phenomenon that causes a can of pressurized household cleaner to cool during dispensing. Spray developers used for positive resist often have a heated wafer chuck or a heated spray nozzle to control the develop temperature. Other problems encountered with the spray development of positive resist are machine deterioration when alkaline developers are used and foaming as the water-based developer comes out of the nozzle under pressure.

Batch developers come in two versions, single-boat and multiple-boat. These machines are the spin-rinse-dryers described in Chap. 7. As developers they require additional plumbing to accommodate the developing chemicals. Batch developing systems are in general less uniform than direct-spray, single-wafer systems because the spray does not impinge directly on each wafer surface and temperature control for positive resist processing is more complicated.

Puddle development. Spray development is very attractive for its uniformity and productivity. A process variation used to gain the advantages of spray for development of positive resist is the *puddle procedure*. This system uses a standard single-wafer spray unit. The difference between regular spray development and the puddle procedure is in the application of the developer chemical to the wafer. The process starts with the deposit of enough developer on the static wafer to cover the surface (Fig. 9.8). Surface tension holds the developer in a puddle on the wafer. The puddle sits there for some required time, usually on a chuck-heated wafer, causing the majority of the development to take place. Puddle development is, in effect, a single-wafer, front-side-only immersion process. After the required puddle time, the wafer surface is sprayed with more developer and rinsed, dried, and passed on to the next step.

Plasma descum. A particularly difficult form of incomplete development is a condition called *scumming*. The scum may be undissolved pieces of resist or dried developer[2] left on the surface. The film is very

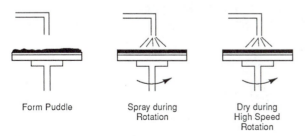

Form Puddle   Spray during   Dry during
                   Rotation   High Speed
                                       Rotation

Figure 9.8 Puddle-spray development.

thin and is hard to detect with visual inspection. In reaction to this problem, advanced VLSI lines with micron-size openings will remove (descum) the film from the wafers in a oxygen-rich plasma chamber after a chemical develop.

Hard Bake

Hard bake is the second heat treatment operation in the masking process. Its purpose is essentially the same as the soft bake step: the evaporation of solvents to harden the resist. For hard bake, however, the goal is exclusively to achieve good adhesion of the resist to the wafer surface. As such, this step is sometimes called preetch bake or prebake.

Hard bake methods

Hard bake is similar to soft bake in the equipment and methods used. Convection ovens, in-line and manual hot plates, infrared tunnel ovens, moving-belt conduction ovens, and vacuum ovens are all used for hard baking. See the "Soft Bake" section in Chap. 8, p. 190.

Hard bake process

The exact time and temperature of the hard bake are determined much the same as in the soft bake process. The starting point is the process recommended by the resist manufacturer. After that, the process is fine-tuned to achieve the adhesion and dimensional control required. Nominal hard bake temperatures are from 130 to 200°C for 30 minutes in a convection oven. Temperatures and times vary in the other methods. The minimum temperature is set to achieve good adhesion of the resist image edge to the surface. The heat-caused adhesion mechanism is dehydration and polymerization. The heat drives water out of the resist, at the same time further polymerizing it, thereby increasing its etch-resistant properties.

The upper temperature limit of the hard bake is set by the flow point of the resist. Resist is a plasticlike material that softens and flows when heated (Fig. 9.9). When the resist flows, the image dimen-

Normal temperature   High temperature

Figure 9.9 Resist flow at high temperature.

sions are changed. Resist flow is first evident as a thickening of the resist edge when viewed with a microscope. Extreme flow exhibits itself as a series of fringe lines around the image. The fringes are an optical effect from the slope left in the resist after the flow.

Hard bake takes place either immediately after the developing step or just before the etching step, as shown in Fig. 9.10. In most production situations, the hard bake is performed in a tunnel oven that is in-line with the developer. When this procedure is used, it is important that the wafers be stored in a nitrogen atmosphere and/or be processed through the develop inspection step as quickly as possible to prevent the reabsorption of water into the resist film.

A goal of process engineering is to have as many common processes as possible. For hard baking that is sometimes difficult due to the different adhesion characteristics of various wafer surfaces. The more difficult surfaces, such as aluminum- and phosphorus-doped oxides, sometimes are given a higher-temperature hard bake or a second hard bake in a convection oven just prior to being etched.

Develop Inspect

The first quality check in the photomaking process is performed after developing and baking. It is appropriately called *develop inspect* or simply DI. The purpose of the inspection is to identify wafers of low-yield wafer sort potential. A well-characterized process will correlate masking defects to wafer sort and final electrical test yields.

This inspection yield, that is, the number of wafers that pass this first quality check, is not factored into the overall yield formula, but it is a much-watched yield for two principal reasons. The critical nature of the photomasking process to the functioning of the circuit has been emphasized throughout this text. It is at develop inspect that the process engineer has the first chance to judge the performance of the process. The second importance of the develop inspect step is related to

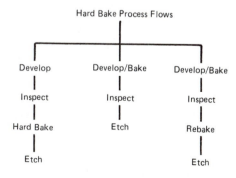

Figure 9.10 Hard bake process flow options.

the two types of rejects made at this inspection. First, some of the wafers will have problems from previous steps that prevent their continuation in the process. These wafers are rejected at develop inspect and discarded. Other wafers have problems associated specifically with the quality of the pattern in the resist film. These wafers can be reworked (Fig. 9.11) by removal of the photoresist and reinsertion into the dehydration process. This is one of the few places in the entire fabrication process where a general rework of mistakes is possible.

Wafers sent back into the masking process are called *reworks* or *redos*. The typical rework rate of a successful fabrication line is around 5 percent. The process engineer has a goal of keeping the rework rate under 10 percent. One reason is the lower sort yield of wafers that have gone through a masking rework. Reworking causes adhesion problems and the additional handling can result in contamination and breakage. If too many wafers are in the rework loop, the overall wafer sort yield will suffer. The second reason for keeping the rework rate under control is related to the additional accounting and identification required to process the rework wafers.

The develop inspect yield and rework rate vary from mask level to mask level. In general, the first levels in the masking sequence have wider feature sizes, flatter surfaces, and lower density, all of which make for a higher yield out of the mask step. By the time the wafers are at the critical contact and metal masks, the rework rate tends to rise.

Develop inspect procedure

The sophistication of techniques used to inspect semiconductor wafers approaches the sophistication of the equipment used to produce the

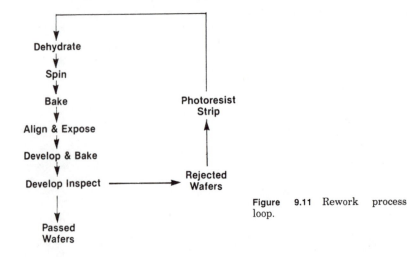

Figure 9.11 Rework process loop.

patterns in the resist. Descriptions of the equipment and methods are in Chap. 14. In this section, the process flow and inspected parameters are explained.

The flow diagram in Fig. 9.12 shows a typical develop inspection sequence. The first step is a naked-eye inspection of the wafer surface. Since no magnification is used, this inspection is sometimes called a 1× (a magnification of 1 is equal to naked-eye viewing) inspection. It may take place in normal room light, but more likely the wafer will be viewed under a collimated white light or a high-intensity ultraviolet light. The wafer is viewed at an angle in the light beam. This method is surprisingly effective in showing film thickness irregularities, gross developing problems, scratches, and contamination, especially stains.

Wafers that pass the 1× visual inspection are placed on a microscope stage for a more detailed inspection. Normal inspection techniques use a system called *random sampling*. Random sampling is based on probability theory that dictates the number of areas on a wafer to be inspected in order to accurately sample the level of defects on the complete wafer. Such an inspection system is based on the operator selecting sites for inspection in a random fashion.

This procedure breaks down because of increasing die sizes and reduced defect densities. As the defect density goes down, as it must for VLSI circuits, the area of inspection must go up if the defects are to be statistically sampled. However, as the die density goes up, the individual parts get smaller, which in turn requires a higher magnification to see them. The increasing magnification narrows the field of view, which in turn increases the time for an operator to inspect a wafer.

The bottom line is that statistically sound inspections are not usually employed for on-line photomasking surface inspections. The usual procedure is to identify a number of specific die for inspection. The chosen die are assumed to geographically represent the wafer surface. This inspection procedure is based on the principle that gross problems with the wafer will show up on the selected die. This method is adequate for gross waferwide problems and to alert the process engineer to problems in the previous masking steps.

Automatic inspection systems using laser and other radiation sources are finding their way onto production lines for both off-line

1. 1× surface inspection
2. 100–400× microscopic inspection
3. Measure critical dimensions

Figure 9.12 Develop inspection steps.

and on-line inspections. One method employs a map of the particular mask level made from the design data. The inspection notes defects on the wafer and compares them to the map. If too many defects have been added to the wafer, it is rejected. Another approach is to make a map of the wafer surface before the particular masking step, and to inspect for newly added defects.

In addition to the issue of the die sampling plan, there is the issue of the wafer sampling plan. The number of wafers inspected varies with the sensitivity and difficulty of the particular mask level. The more sensitive and difficult the layer, the more wafers are likely to be inspected. The number of wafers inspected per batch varies from a few to 100 percent. In addition to process defects, dimensional measurement and control is part of the develop inspection. On each mask level there is a region or set of patterns whose dimensions are critical to the functioning of the whole circuit. A representative pattern is chosen for each level as the *critical dimension* (CD). During the develop inspection the dimensions of these patterns are measured using manual or automatic microscopic techniques.

Causes of develop inspection reject

There are many reasons why a wafer can be rejected or sent for rework at the develop inspection step. Generally the only defects looked for are those added to the wafer during the current photomasking step. Defects from previous masking steps are generally overlooked under the theory that every wafer is passed on with some defects or problems and that the wafer arrived at the current step with acceptable quality. If there is a serious problem with a wafer that somehow escaped previous inspections, it is pulled from the batch.

The operator goes through the inspection on a *first-fail basis*. This means that the operator inspects only until a rejectable level is reached on the wafer. First-fail systems save inspection time but do not give the process engineers a complete evaluation of the wafer image quality. In practice, the operator will have an inspection log that lists the rejectable defects and problems (Fig. 9.13). The information for each wafer is logged on the sheet for data accumulation and analysis by the engineers. Automatic and semiautomatic optical inspection stations have electronic memories for accumulating and correlating this reject data. Some advanced lines have the inspection logs programmed into on-line PCs, which speeds the inspection step and the data analysis.

The reject causes listed have been discussed elsewhere with the exception of bridging. Bridging (Fig. 9.14) is a condition where two patterns are connected by a thin layer of photoresist. The bridge comes

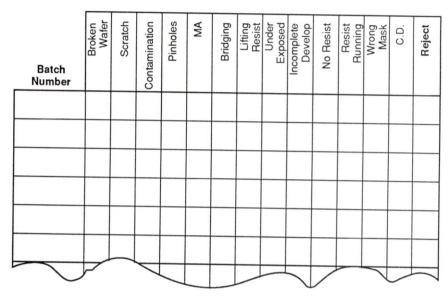

Batch Number	Broken Wafer	Scratch	Contamination	Pinholes	MA	Bridging	Lifting Resist	Under Exposed	Incomplete Develop	No Resist	Resist Running	Wrong Mask	C.D.	Reject

Figure 9.13 Typical develop inspect log.

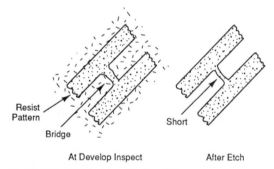

Resist Pattern

Bridge

Short

At Develop Inspect   After Etch

Figure 9.14 Bridged conduction lines.

from an overexposure, poor mask definition, or a resist film that is too thick.

Etch

At the completion of the develop inspect step, the mask (or reticle) pattern is defined in the photoresist layer and is ready for etch. During the etch step the image will be permanently transferred into the surface layer on the wafer. The goal is an exact transfer of the image into the resist layer. The degree of exactness is dependent on several factors which will be explored as a preparation for discussion of the dif-

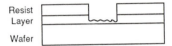

Figure 9.15   Incomplete etch.

ferent etch methods. The factors affecting image transfer are incomplete etch, overetching, undercutting, and selectivity.

Incomplete etch

Incomplete etch is a situation in which a portion of the surface layer still remains in the pattern hole or on the surface (Fig. 9.15). The causes of incomplete etch are too short an etch time, the presence of a surface layer that slows the etching, or an uneven surface layer that results in incomplete etch in the more thickly coated portions of the wafer. If wet-chemical etching is used, a lowered temperature or weak etch solution will cause incomplete etch. If dry plasma etching is used, a wrong gas mixture or an improperly operated system can cause the same effect.

Overetch and undercutting

The opposite condition to incomplete etch is overetch. In any etch process there is always some degree of overetch planned into the process. This is necessary to ensure complete removal of the thickest portions of the the layer and to allow for the etch to break through any slow-etching layers on the top surface.

The ideal etch leaves vertical sidewalls in the layer (Fig. 9.16). Etch techniques that produce this ideal result are said to be *anisotropic*. However, the etching chemical dissolves the top of the sidewall for a longer time than the bottom of the hole. The result is a hole wider at the top than the bottom with a sloped sidewall. Etching techniques that produce this result are called *isotropic*. This action of the etching chemical is called *undercutting* (Fig. 9.17) since the surface layer is undercut below the resist edge. Circuit layout designers take undercutting into account when planning the circuit. Adjacent patterns must be separated a certain distance to prevent shorting. The amount of undercutting must be calculated when the pattern is designed.

An ongoing goal of the etch step is the control of undercutting to an acceptable level. Severe undercutting (or overetch) takes place when the etch time is excessive, the etch temperature is too high, or the etch mixture is too strong. Undercutting is also present when the adhesion bond between the photoresist and the wafer surface is weak. This is a constant worry, and the purpose of the dehydration, prime, soft bake,

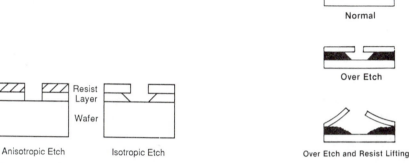

Normal

Over Etch

Over Etch and Resist Lifting

Resist Layer

Wafer

Anisotropic Etch   Isotropic Etch

Figure 9.16 Anisotropic and isotropic etch.

Figure 9.17 Degrees of undercutting.

and hard bake steps is to prevent this type of failure. Failure of the resist bond at the edge of the etch hole can result in severe undercutting. If the bond is very poor, the resist can lift from the wafer surface, causing catastrophic undercutting.

Selectivity

Another goal of the etch step is the preservation of the surface underlying the etched layer. If the underlying surface of the wafer is partially etched away, the physical dimensions and electrical performance of the devices are changed. The property of the etch process that relates to preservation of the surface is *selectivity*. High selectivity implies little or no attack of the underlying surface. In wet etching techniques an etchant acid that will not attack the underlying material is chosen.

Wet Etching

For over 30 years the traditional method of etching has been by immersion techniques using wet etchants. The procedure is similar to the preoxidation clean-rinse-dry process (Chap. 7) and immersion development. The emergence of feature sizes less than 3 μm has seen the shift from wet to dry etching techniques. However, keep in mind that within a circuit whose smallest dimensions are 3 μm or less, there are still mask levels with dimensions well above that level. In many cases dry etching is employed for small dimensions and wet etching for the larger ones.

For wet etching, the wafers are loaded into an etch-resistant boat and immersed in a tank of the etchant. After a predetermined time in the etch tank they are processed through the rinsing and drying steps.

Etching uniformity and process control are enhanced by the addition of heaters and agitation devices, such as stirrers or ultrasonic waves, to the immersion tanks.

Although the basic equipment is simple in concept, a high-production wet "bench" can be very sophisticated,[9] incorporating microprocessor control of the timers and heaters. Many systems have walking beams or robots for the automatic placement of the wafer holders in the etch, rinse, and dry subsystems. The etch tanks of the traditional manual systems are filled by hand, a dangerous and possibly contaminating practice. Newer systems have plumbing to allow the filling of the tanks from reserve tanks by remote control.

The worry about etchant contamination of the wafers is being addressed by *point-of-use filters*. These are special filters fitted to automatic chemical dispensing systems to filter-clean the chemicals just prior to filling the immersion tank. This placement catches particulate contamination from the chemicals, the pumps, and the tubing systems.

Wet etchants are selected for their ability to uniformly remove the top wafer layer without attacking the underlying material (good selectivity).

Etch time variability is introduced by temperature variations as the boat and wafers come to temperature equilibrium in the tank and the continued etching action as the wafers are transferred to a rinse tank. Generally, the process is set at the shortest time compatible with uniform etching and high productivity. The maximum time is limited to the amount of time the resist will continue to adhere to the wafer surface.

Silicon wet etching

Silicon layers are typically etched with a solution of nitric and HF acids mixed in water. The formula becomes an important factor in control of the etch. In some ratios, the etch has an exothermic reaction with the silicon. Exothermic reactions are those that produce heat, which, in turn, speeds up the etch reaction, which, in turn, creates more heat, and so on, resulting in an uncontrollable process. Sometimes acetic acid is mixed in with the other ingredients to control the exothermic reaction.

Some devices require the etching of a trough or trench into the silicon surface. The etch formula is adjusted to make the etch rate dependent on the orientation of the wafer. ⟨111⟩-oriented wafers etch at a 45° angle, while ⟨100⟩-oriented wafers etch with a "flat" bottom.[3]

Other orientations result in different-shaped trenches. Polysilicon films are also etched with the same basic formula.

Silicon dioxide wet etching

The most common etched layer is a thermally grown silicon dioxide. The basic etchant is hydrofluoric acid (HF). HF has the advantage of dissolving silicon dioxide without attacking silicon. However, full-strength HF has an etch rate of about 300 Å/s at room temperature.[4] This rate is too fast for a controllable process (a 3000-Å layer would etch in only 10 s).

In practice, the HF (assay of 49%) is mixed with water or ammonium fluoride and water. The ammonium fluoride (NH_4F) acts as a buffer to the unwanted generation of hydrogen ions which accelerate the etch rate. These solutions are known as *buffered oxide etches* or BOEs. They are mixed in different strengths to create reasonable etch times for the particular oxide thickness (Fig. 9.18). Some BOE formulas include a wetting agent (surfactant such as Triton X-100 or equivalent) to reduce the surface tension of the etch, allowing it to uniformly penetrate into smaller openings.

Aluminum film wet etching

Selective etching solutions for aluminum and aluminum alloy layers are based on phosphoric acid. An unfortunate by-product of the reaction of aluminum and phosphoric acid are tiny bubbles of hydrogen, as

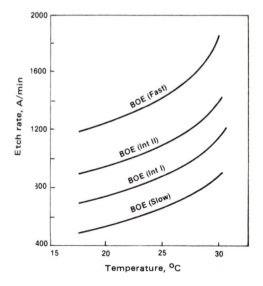

Figure 9.18 Etch rate versus temperature for BOEs.

shown in the reaction in Fig. 9.19. These bubbles cling to the wafer surface and block the etch action. The result is either bridges of aluminum that can cause electrical shorts between adjacent leads or spots of unwanted aluminum, called *snowballs*, left on the surface. Neutralization of this problem is accomplished by use of an aluminum etching solution that contains phosphoric acid, nitric acid, acetic acid, water, and wetting agents. A typical solution of the active ingredients (less wetting agent) is 16:1:1:2.

In addition to the special formulas, a typical aluminum etch process will include wafer agitation by stirring or moving the wafer boat up and down in the solution. Sometimes ultrasonic or megasonic waves are used to collapse and move the bubbles around.

Deposited oxide wet etching

One of the final layers on a wafer is a silicon dioxide passivation film deposited over the aluminum metallization pattern. These films are known as vapox or silox films. While the chemical composition of the films is that of silicon dioxide, the same as thermally grown silicon dioxide, they require a different etch solution. The difference is in the selectivity required of the etchant.

The usual etchant for silicon dioxide is a BOE solution. Unfortunately, the BOE attacks the underlying aluminum pads, causing bonding problems in the packaging process. This condition is called *brown*, or *stained, pads*. The preferred etchant for this layer is a solution of ammonium fluoride and acetic acid mixed in a ratio of 1:2.

Silicon nitride wet etching

Another compound favored for the passivation layer is silicon nitride. It is possible to etch this layer with wet chemical means, but it is not as easy as for the other layers. The chemical used is hot (180°C) phos-

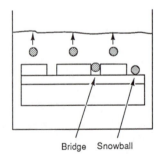

Bridge Snowball

Figure 9.19 Hydrogen bubble blockage of etchant.

⊚ Hydrogen Bubble

	COMMON ETCHANT	ETCH TEMP	RATE Å /MIN	METHOD
SiO₂	HF & NH₄F (1 : 8)	Room	700	Dip & wetting agent predip
SiO₂	HF & NH₄F (1 : 8)	Room	700	Dip & wetting agent predip
SiO₂ (Vapox)	Acetic Acid & NH₄F(2 : 1)	Room	1000	Dip
Aluminum	H₃PO₄ : 16 HNO₃ : 1 Acetic : 1 H₂O : 2 Wetting Agent	40 – 50°C	2000	a) Dip & agitation b) Spray
Si₃N₄	H₃PO₄	150 – 180°C	80	Dip
POLYSi	HNO₃ : 50 H₂O : 20 HF : 3	Room	1000	Dip

Figure 9.20   Summary of wet etching process.

phoric acid. Since the acid evaporates rapidly at this temperature, the etch must be done in a closed reflux container equipped with a cooled lid to condense the vapors. The major problem is that photoresist layers do not stand up to the etchant temperature and aggressive etch rate. Consequently, a layer of silicon dioxide or some other material is required to block the etchant. These two factors have led to the use of dry etching techniques for silicon nitride.

Wet spray etching

Wet spray etching offers several advantages over immersion etching. Primary is the added definition gained from the mechanical pressure of the spray.[5] Spray etching also minimizes contamination from the etchants. From a process control point of view, spray etch is more controllable since the etchant can be instantly removed from the surface by switching the system to a water rinse. Single-wafer spinning-chuck spray systems offer considerable process uniformity advantages.

Disadvantages to spray etching are system cost, safety considerations associated with caustic etchants in a pressurized system, and the requirement of etch-resistant materials to prevent the deterioration of the machine. On the plus side, spray systems are usually enclosed, which adds to worker safety. Figure 9.20 is a table of common semiconductor films and their common etchants.

Dry Etch

The limits of wet etching for VLSI-size patterns has been mentioned in the previous section. For review they are

1. Wet etching is limited to pattern sizes of 3 μm.

2. Wet etching is isotropic, resulting in sloped sidewalls.

3. A wet etch process requires rinse and dry steps.

4. The wet chemicals are hazardous and/or toxic.

5. Wet processes represent a contamination potential.

6. Failure of the resist-wafer bond causes undercutting.

These considerations have led to the use of dry etch processes for the definition of small feature sizes on advanced circuits. Figure 9.21 is an overview of the dry etching techniques used.

Dry etching is a generic term that refers to the etching techniques in which gases are the primary etch medium, and the wafers are etched without wet chemicals or rinsing. The wafers enter and exit the system in a dry state.

Barrel plasma etching

The term dry etching is sometimes used to refer to plasma etching, although there are two other dry etching techniques—ion milling and reactive ion etch. Plasma etching, like wet etching, is a chemical process but uses plasma energy to drive the reaction. Comparison of silicon dioxide etching in the two systems illustrates the differences. In wet etching of silicon dioxide, the fluorine in the BOE etchant is the ingredient that dissolves the silicon dioxide, converting it to water-rinsable components. The energy required to drive the reaction comes from the internal energy in the BOE solution or from an external heater.

A plasma etcher requires the same elements: a chemical etchant and an energy source. Physically, a plasma etcher consists of a chamber, vacuum system, gas supply, and a power supply (Fig. 9.22). The wafers are loaded into the chamber and the pressure inside is reduced

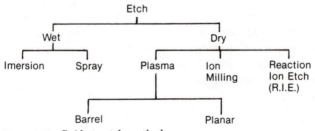

Figure 9.21 Guide to etch methods.

Reactive Gas

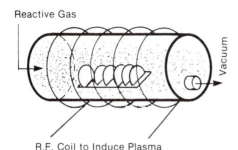

Vacuum

R.F. Coil to Induce Plasma

Figure 9.22   Barrel plasma etch.

by the vacuum system. After the vacuum is established, the chamber is filled with the reactive gas. For the etching of silicon dioxide the gas is usually CF_4 mixed with oxygen. The power supply creates a radiofrequency (RF) field through electrodes in the chamber. The field energizes the gas mixture to a plasma state. In the energized state the fluorine attacks the silicon dioxide, converting it into volatile components that are removed from the system by the vacuum system.

The earlier plasma systems were designed with circular chambers and are called *barrel etchers*. While providing the benefits of dry etching, barrel plasma systems produce isotropic etching. Within the chamber the etching ions are energized by the plasma in a nondirectional manner. The etching ions attack the surface layer from all directions, creating a tapered sidewall. In a barrel system, the wafers are held in a boat and etching relies on the mixing of the etching ions between the wafers. Uniform etching of wafers in a barrel etching system is difficult because it is hard to supply a constant amount of etchant to all the wafers in the system and because of the nondirectionality of the etching ions.

Another consideration of barrel plasma etching is radiation damage resulting from the high-energy plasma field. The high energy causes charges to build up in the wafer surface that compromise the electrical functioning of the circuit. Protection of the wafers from the radiation is provided by perforated metal cylinders that isolate the wafers from the plasma field (Fig. 9.23).

Planar plasma etching

For more precise etching, plasma planar systems are preferred. These systems contain the basic elements of the barrel system, but the wafers are placed on a grounded pallet under the RF electrode (Fig. 9.24). The wafers are actually in the plasma field and the etching ions are more directional than those in a barrel system. The result is a more

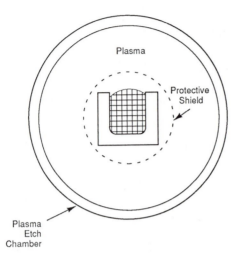

Figure 9.23 Barrel plasma stripper with protective shield.

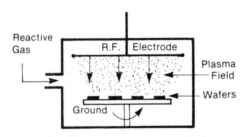

Figure 9.24 Planar plasma etch.

anisotropic etch with almost vertical sidewalls possible. Etching uniformity is increased with the rotation of the wafer pallet in the system.

Etching of silicon dioxide in a planar system takes place in C_4F_8,
which provides good selectivity when etching silicon dioxide over silicon substrates. Selectivity is a major consideration of plasma etching
processes. The three methods used to control selectivity are the selection of the etching gas formula, the dilution of the gas near the end of
the process to slow down the attack of the underlying layer, and
endpoint detectors in the system. An endpoint detector for a silicon dioxide etch over silicon would automatically terminate the etching process as soon as some silicon was detected in the exiting gas stream.

The etch rate of a plasma system is determined by the power supplied to the electrodes, the chemistry of the gas etchants (Fig. 9.25),
and the vacuum (pressure) level in the chamber. Etch rates vary from
600 to 2000 Å/min.[6] Planar plasma etch systems are designed in both
batch and single-wafer chamber configurations. The single-wafer systems are popular for their ability to have the etch parameters tightly

Film	Etchant	Typical Gas Compounds
Al	Chlorine	$BCl_3,CCl_4,Cl_2,SiCl_4$
Mo	Fluorine	CF_4,SF_4,SF_6
Polymers	Oxygen CF_4,SF_4,SF_6	
Si	Chlorine, Fluorine CF_4,SF_4,SF_6	$BCl_3,CCl_4,Cl_2,SiCl_4$
SiO_2	Chlorine, Fluorine	CF_4,CHF_3,C_2F_6,C_3F_8
Ta	Fluorine	''
Ti	Chlorine, Fluorine	''
W	Fluorine	''

Figure 9.25   Plasma etch chemicals.

controlled for uniform etching. Also, with load-lock chambers single-wafer systems can maintain high production rates and are amenable to in-line automation.

Ion beam etching

A second type of dry etch system is the ion beam system (Fig. 9.26). Unlike plasma systems, ion beam etching is a physical process. The

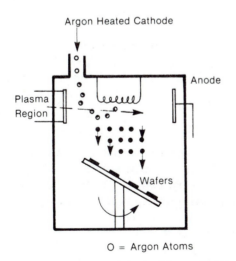

O = Argon Atoms

• = Ionized Argon Atoms

Figure 9.26   Ion beam milling.

wafers are placed on a holder in a vacuum chamber and a stream of argon is introduced into the chamber. Upon entering the chamber, the argon is subjected to a stream of high-energy electrons from a set of cathode-anode electrodes. The electrons ionize the argon atoms to a high-energy state with a positive charge. The wafers are held on a negatively grounded holder which attracts the ionized argon atoms. As the argon atoms travel to the wafer holder they accelerate, picking up energy. At the wafer surface they crash into the exposed wafer layer and literally blast small amounts from the wafer surface. Scientists call this action a *momentum transfer*, a physical process. No chemical reaction takes place between the argon atoms and the wafer material. Ion beam etching is also called *sputter etching* or *ion milling*.

The material removal (etching) is highly directional, resulting in good definition of small openings. Two considerations of ion beam etching is its poor selectivity and radiation damage from the ionization mechanism.

Reactive ion etching

Reactive ion etching (RIE) systems combine plasma etching and ion beam etching principles. The systems are similar in construction to the plasma and the ion beam systems but have a capability of ion milling. The combination brings the benefits of chemical plasma etching along with the benefits of directional ion milling. A major advantage of RIE systems is in the etching of silicon dioxide over silicon layers. The combination etch results in a selectivity ratio of 35:1,[7] whereas ratios of only 10:1 are available with plasma-only etching.

Dry etch etchants

As in wet etching, the selection of a dry etchant is dependent on the layer to be etched, the material under the layer, and the etching method selected. The table in Fig. 9.25 is a list of commonly used etchant gases.

Resist etch barriers in dry etching

For both wet and dry etching processes, a patterned photoresist layer is the preferred etch barrier. In wet etching there is almost no attack of the resist by the etchants. However, in dry etching, residual oxygen in the system attacks the resist layer. An important process parameter is the thickness of a resist destined for a dry etch. The resist must be thick enough to stand up to the etchants without becoming so thin that pinholes are present.

Another resist-related dry etch problem is resist baking. Within the dry etch chamber, the temperature can rise as high as 200°C, a temperature that can bake the resist to a condition that makes it hard to remove from the wafer. Another temperature-related problem is the tendency of resist patterns to flow and distort the images.

Resist Stripping

After etching, the pattern is a permanent part of the top layer of the wafer. The resist layer that has acted as an etch barrier is no longer needed after etching and is removed (or stripped) from the surface. Traditionally, the resist layer has been removed by wet chemical processing. The development of dry etching systems has led to the use of plasma stripping.

Wet chemical stripping of nonmetal surfaces

A number of different chemicals are used for stripping. The choice depends on the wafer surface the resist is being removed from, production considerations, and the polarity of the resist (Fig. 9.27). Generally the strippers are divided into the categories of universal strippers and positive- and negative-only strippers.

Wet stripping is favored for the following reasons:

1. It has a long process history.

2. It is cost-effective.

3. It is effective in the removal of metallic ions.

4. It is a low-temperature process and does not expose the wafers to possibly damaging radiation.

A wet stripping process requires the same steps as a wet etch: strip, water rinse, and dry.

Sulfuric acid and hydrogen peroxide.
Solutions of sulfuric acid and hydrogen peroxide are the most common strippers used for the removal

Stripper	Operating Temperature	Oxide	Surface Metalized	Bathlife	Resist Polarity Negative	Positive	REMOVAL
1) Sulfuric Acid & Oxidant a. exothermic	125°C	X		2-3 Hrs	X	X	Neutralize
b. heated		X		8 Hrs	X	X	Neutralize
2) Organic Acids	90-110°C	X	X	4-8 Hrs	X	X	Remove & bury
3) Chromic/Sulfuric Acid mixture	20°C	X	X	4-8 Hrs	X	X	Remove & bury
4) Solvents	20°C-90°C	X	X	4-8 Hrs	X	X	Remove & recycle

Figure 9.27 Comparison of wet photoresist strippers.

of resist from *nonmetallic* surfaces. A nonmetallic surface is one with either a silicon dioxide, silicon nitride, or polysilicon top layer. This solution is universal in its stripping ability. It is the same chemical process used for pre-tube-cleaning wafers as described in Chap. 7.

Hot sulfuric acid (90 to 150°C) is an effective chemical for removing resist. The problem is that it does not convert the carbon in the resist to carbon dioxide. The carbon can thus build up in the stripper bath and eventually contaminate the wafers. Carbon buildup in the bath is evidenced by the bath turning a black color. The carbon problem is solved with the addition of hydrogen peroxide to the sulfuric acid. The hydrogen peroxide converts the carbon to carbon dioxide, which leaves the bath as a gas. The hydrogen peroxide is mixed into the sulfuric acid by two methods. The traditional method is to mix the room-temperature sulfuric acid and hydrogen peroxide in a ratio of about 2:1. In this ratio the reaction is exothermic, and the solution heats itself rapidly to about 120°C, an effective temperature for resist removal. The drawback to this mixing method is that the temperature of the bath quickly falls below an effective stripping temperature. Typically, a two-bath system lasts only 1 to 2 hours.

An alternative mixing method requires the heating of the sulfuric acid to about 120°C. Small amounts of hydrogen peroxide (50 to 100 cm^3) are then added to the heated sulfuric acid. The advantage of this system is that it provides a long lifetime for the bath. The disadvantage is the requirement of a heating system.

Sulfuric acid and ammonium persulfate. An objection to the use of hydrogen peroxide is its inherent instability. Hydrogen peroxide (H_2O_2) is essentially water with an extra oxygen atom. It is the extra oxygen atom that provides the high reactivity of hydrogen peroxide. The oxygen readily evolves from the solution, a situation requiring vented caps to prevent a dangerous pressure buildup. The cap allows a continual release of oxygen, which over time reduces the reactivity of the solution.

A substitute for the liquid hydrogen peroxide is the crystalline substance ammonium persulfate (AP).[8] AP is added to baths of heated sulfuric acid in amounts of 40 to 80 g. In the sulfuric acid, it oxidizes the carbon from the resist, maintaining a clean bath. AP is less reactive at room temperature than hydrogen peroxide and is therefore safer and can be stored longer. Being less reactive, the AP also ensures a more consistent supply of oxygen to the stripper bath.

Sulfuric acid and nitric acid. Nitric acid is sometimes used as an additive oxidant in a sulfuric acid bath. A mixing ratio of about 10:1 is typical. A drawback to nitric acid is that it turns the bath a light orange color which can mask the buildup of carbon in the bath.

Wet chemical stripping of metallized surfaces

As mentioned in the introduction, stripping from metallized surfaces is a more difficult technology because the metallized layers are subject to attack or oxidation. There are four types of wet chemicals that are used for stripping metallized surfaces. They are

1. Organic strippers
2. Chromic sulfuric acid mixtures
3. Solvent strippers

Phenolic organic strippers. Organic strippers contain a combination of sulfonic acid (an organic acid) and chlorinated hydrocarbon solvents such as duodexabenzene. The formula requires phenol to create a rinsable solution. This class of stripper was introduced in the early 1960s with the invention of the stripper J-100 by Industri Chem. It was the first stripper to provide universal stripper capabilities and nonattack of metal films. Stripping the photoresist required heating the solution to the 90 to 120°C range and a series of solvent rinses, followed by a water rinse and drying step.

Since the introduction of J-100, a number of suppliers have developed similar formulas, some designed for direct water rinsing. These were the workhorse strippers for over two decades. However, in the 1970s concern over the toxic ingredients in these formulas led to the development of sulfonic acid, nonphenolic, nonchlorinated[9] resist strippers.

Chromic and sulfuric acid mixtures. These strippers are a mixture of chromium trioxide in sulfuric acid. The strippers are used at room temperature, which simplifies the process. In fact, these strippers cannot be heated or an attack of metal films will take place. The mixtures must also be kept water-free to prevent metal attack. Like the phenolic organic strippers, the chromic and sulfuric acid strippers represent an environmental problem due to the chromium trioxide, which is unacceptable in the food chain. The environmental considerations require that this type of stripper be removed from the site rather than rinsed and dumped into the city drain.

Solvent positive-only strippers. One of the advantages of positive resists is their ease of removal from the wafer surface. A positive resist layer that has not been hard-baked is easily removed from the wafer with a simple acetone soak. In fact, acetone has been the traditional positive resist stripper. Unfortunately, acetone represents a fire hazard and its use is discouraged.

Several chemical manufacturers supply positive-only strippers based on the solvent N-methyl pyrrolidine (NMP).[10] These strippers are effective, water-rinsable, and drain-dumpable. For processes that include a positive resist hard bake, some of the solvent strippers can be heated to increase the removal rate. Figure 9.27 is a table of the most common wet resist strippers and their uses.

Dry stripping

Like etching, the dry plasma process can also be applied to resist stripping. The wafers are placed in a barrel chamber and oxygen is introduced (Fig. 9.28). The plasma field energizes the oxygen to a high-energy state, which, in turn, oxidizes the resist components to gases that are removed from the chamber by the vacuum pump.

$$C_xH_y \text{ (resist)} + O_2 \text{ (plasma energized)} \rightarrow CO \text{ (gas)} + CO_2 \text{ (gas)} + H_2O$$

Figure 9.28   Resist removal by plasma oxygen.

The major advantage of plasma resist stripping is the elimination of wet hoods and the handling of chemicals. The principal disadvantage is its ineffectiveness in the removal of metallic ions. There is not enough energy in the plasma field to volatilize the metallic ions. Another consideration of plasma stripping is radiation damage to the circuits from the high-energy plasma field. This problem is reduced with system designs that have the plasma chamber removed from the stripper chamber. They are called *downstream strippers* because the plasma is created downstream from the wafers. MOS wafers are more sensitive to radiation effects during stripping. Some companies use a dry plasma strip followed by a wet strip to remove the metallic ions.

Final Inspection

The final step in the basic photomasking process is a visual inspection. It is essentially the same procedure as develop inspect, with the exception that the majority of the rejects are fatal. The one exception is contaminated wafers that may be recleaned and reinspected. Final inspection certifies the quality of the outgoing wafers and serves as a check on the effectiveness of the develop inspection. Wafers that

Possible Process Cause	Contamination	Misalign	Undercut	Incomplete Etch	Wrong Mask	Pin Holes	C.D.'s	Visual Reject
Contaminated Etch	X		X	X				
Contaminated Stripper	X							
Contaminated H$_2$O	X							
Insufficient Rinse	X		X					
No Wet Agent				X				
Under Etch				X			X	
Over Etch			X				X	
Wrong Etch			X	X			X	
Hard Bake Too High			X	X			X	
Poor Develop				X			X	
P$_2$O$_5$ & SiO$_2$			X				X	
B$_2$O$_3$ & SiO$_2$				X			X	
Low Hard Bake			X					
Develop Inspect Escapes		X	X	X	X	X		

Figure 9.29 Final inspect rejects and process causes.

should have been identified and pulled from the batch at develop inspect are called *develop inspect escapes*.

The wafers receive a first surface inspection in incident white or ultraviolet light for stains and large particulate contamination. This inspection is followed by a microscopic inspection and a measurement of the critical dimensions for the particular mask level. Of primary interest is the quality of the etched pattern with underetching and undercutting being two parameters of concern. The table in Fig. 9.29 is a list of typical causes of wafer rejection found in the final inspection.

Mask Making

In Chap. 5, the steps of circuit design were detailed. In this section, the process used to construct a photomask or reticle is examined. The mask-making process is almost identical to the basic wafer-patterning operation (Fig. 9.30). In fact the goal is the same, the transfer of a pattern into the thin chrome layer on the glass mask surface. The two processes differ primarily in the exposure step. For mask making the image must be very precise or any image distortion will be translated to the wafer surface. For circuits with very demanding geometries, an

- Cut glass (or quartz) blank
- Clean glass
- Deposit chrome
- Inspect
- Clean
- Deposit resist
- Expose
- Etch
- Resist strip

Figure 9.30   Chrome mask process steps.

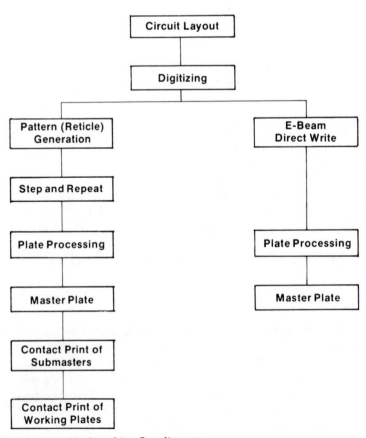

Figure 9.31   Mask-making flow diagrams.

electron beam exposure system is preferred, especially for the master plates (Fig. 9.31).

For most production plates and reticles, a pattern generator is used to establish the image in the photoresist. A pattern generator consists of a light source and a series of motor-driven shutters. The chrome-covered mask is moved under the light source as the shutters are moved and opened to allow precisely shaped patterns of light to shine onto the resist, exposing it.

After imaging by either electron beam or a pattern generator, the reticles or masks go through development, inspection, etch, strip, and inspection steps that transfer the pattern permanently into the chrome layer. Inspections are very critical since any undetected mistake or defect has the potential of creating thousands of scrap wafers. If the process calls for a lot of identical plates, called *working plates*, a special mask contact printer may be used to transfer the master plate pattern to a number of working plates.

With the advent of VLSI-level circuits a considerable amount of development had gone into the mask-making process. The masks and reticles must be virtually defect-free. There are now procedures to eliminate unwanted chrome spots and pattern pertusions with laser "zapping" techniques. There is also a technology to fill in missing pattern parts. Advanced circuit masks are made on low-expansion glass and quartz to minimize pattern distortions from temperature variations in the photolithography exposure and alignment steps.

Key Concepts and Terms

Develop inspect and rework	Negative resist developers
Dry etch methods	Plasma descum
Dry stripping	Positive resist developers
Etch process	Puddle develop
Final inspect	Resist development
Hard bake methods	Resist stripping
Hard bake process	Spray develop
Immersion develop	Wet etch methods
Mask making	Wet strip chemicals

Review Questions

1. Name the major methods of resist development.

2. What are the chemicals used to develop negative and positive resist?

3. What is the purpose of the hard bake step?

4. Name three methods used for hard bake.

5. What problems arise if the hard bake temperature is too low? Too high?

6. Name the preferred wet etchants for etching: silicon dioxide layers, silicon nitride layers, and aluminum layers.

7. Name two advantages of dry etching over wet etching.

8. List the three principal dry etching techniques.

9. Write a wet stripping flow diagram.

10. Name the purpose and methods of final inspect.

References

1. D. Elliott, *Integrated Circuit Fabrication Technology*, McGraw-Hill, New York, 1976, p. 216.
2. S. Wolf, R. Tauber, *Silicon Processing for the VLSI Era*, Lattice Press, Newport Beach, Calif., 1986, p. 530.
3. S. Wolf, R. Tauber, *Silicon Processing for the VLSI Era*, Lattice Press, Newport Beach, Calif., 1986, p. 532.
4. S. Wolf, R. Tauber, *Silicon Processing for the VLSI Era*, Lattice Press, Newport Beach, Calif., 1986, p. 532.
5. C. Murray, "Wet etching update," *Semiconductor International*, May 1986, p. 82.
6. D. Elliott, *Integrated Circuit Fabrication Technology*, McGraw-Hill, New York, 1976, p. 275.
7. D. Elliott, *Integrated Circuit Fabrication Technology*, McGraw-Hill, New York, 1976, p. 282.
8. EKC Technology Inc., Technical Bulletin SA-80.
9. EKC Technology Inc., Technical Bulletin—Nophenol 922.
10. EKC Technology Inc., Technical Bulletin—Posistrip Series.

10

Advanced Photolithographic Processes

Overview

The demands of patterning feature sizes less than 5 microns and maintaining low defect densities have led to the development of patterning processes that are variations on the basic ten-step process. These processes are explained in their application and execution.

Objectives

Upon completion of this chapter, you should be able to:

1. Describe four exposure-related effects that cause image distortion.
2. Draw a cross-sectional flow diagram of a dual-layer resist process.
3. Describe the uses and process of a lift-off procedure.
4. List two planarization techniques.
5. State an advantage of an image-reversal process.
6. Describe how antireflective layers, contrast enhancement layers, and resist dye additives improve resolution.
7. Identify the parts of a pellicle and the advantages it offers to a resist process.

Introduction

The ten-step patterning process detailed in Chaps. 8 and 9 is a core process. As presented, it would be sufficient for the production of MSI circuits. It could also be a basic process for some LSI circuits. However, the basic process undergoes a number of variations for the production of VLSI-level circuits. The reasons for the variations are due to physical and chemical limits of the materials and equipment available. In the mid-1970s it was widely accepted that optical photoresist processes had a lower resolution limit of about 1.5 μm. This wisdom gave rise to the interest in x-ray and E-beam exposure systems. However, manipulation and improvements on the basic processes have resulted in optical resist processes capable of 0.5-μm images.

In this chapter some of the limits of optical imaging are discussed and the major advanced process variations are detailed.

Contrast effects

The goal of the patterning process is the faithful reproduction of the pattern dimensions on the mask in the resist layer and ultimately in the top layer of the wafer. This job becomes difficult in regions of the mask where an opaque line is surrounded by a large clear area. The large amount of radiation coming around the opaque line tends to shrink the dimension of the line in the resist layer (Fig. 10.1). This problem is called a *proximity effect*.

Another contrast effect is called *subject contrast* (Fig. 10.2). This situation comes about when some exposure radiation penetrates the opaque region of a mask or reflects off the wafer surface into the resist. The result is a partially exposed region that leaves a distorted image after the development step. This is more a problem with negative resist than with positive resist.

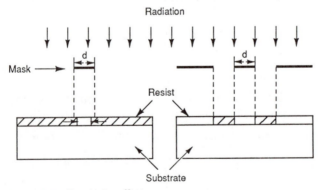

Figure 10.1 Proximity effects.

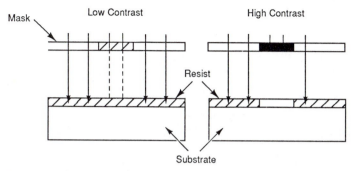

Figure 10.2 Subject contrast.

Resist light scattering

In addition to light radiation reflecting off the wafer surface, the radiation tends to diffuse into the resist causing poor image definition. The amount of diffusion is in proportion to the resist thickness. Some additives put in the photoresist to increase radiation absorption also increase the amount of radiation diffusion, thus reducing image resolution.

Subsurface reflectivity

The high-intensity exposing radiation ideally is directed at a 90° angle to the wafer surface. When this ideal situation exists, exposing waves reflect directly up and down in the resist, leaving a well-defined exposed image (Fig. 10.3). In reality some of the exposing waves are traveling at angles other than 90° and expose unwanted portions of the resist.

This subsurface reflectivity varies with the surface layer material and the surface smoothness. Metal layers, especially aluminum and aluminum alloys, have higher reflectivity properties. A goal of the deposition processes is a consistent and smooth surface to control this form of reflection.

Reflection problems are intensified on wafers with many steps, also called a varied topography. The sidewalls of the steps reflect radiation

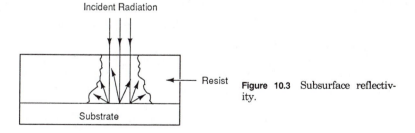

Figure 10.3 Subsurface reflectivity.

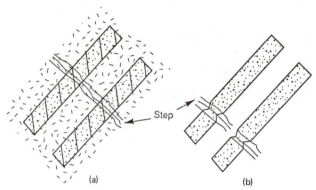

Figure 10.4 Metal line "notching" over step. (*a*) Before etch; (*b*) after etch.

at angles into the resist, causing poor image resolution. A particular problem is light interference at the step that causes a "notching" of the pattern as it crosses the step (Fig. 10.4).

Standing waves

In "Subsurface Reflectivity" it was mentioned that the ideal exposure situation is when the radiation waves are directed to the wafer surface at 90°. This is true when reflection problems are under consideration. However, 90° reflection causes another problem in positive photoresists, the creation of standing waves. As the radiation wave reflects off the surface and travels back up through the resist, it interferes with the incoming wave. The combination of the two waves creates regions of energy that vary from low to high (Fig. 10.5). In the high-energy regions the resist receives more radiation, which causes a rippled sidewall and a loss of resolution. A number of solutions are used to moderate standing waves, including dyes in the resist, separate antireflective coatings directly on the wafer surface, and a bake step between exposure and development of the resist layer.

Double Masking

For the patterning of smaller geometries the resist engineer prefers thinner resist layers and shorter exposure times. Unfortunately, these solutions give rise to higher densities of pinholes in the resist. A process variation that resolves this problem is double masking.

Double masking starts with a thin layer of resist that is processed through the development step (Fig. 10.6). Within the thin resist, small image sizes can be resolved to specification. After the first resist layer,

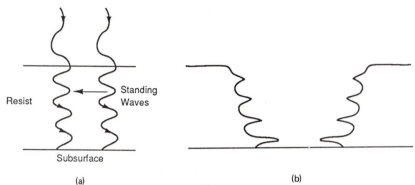

Figure 10.5 Standing-wave effect. (*a*) During exposure; (*b*) after develop.

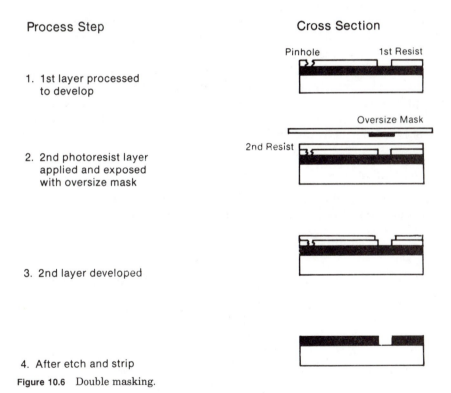

Figure 10.6 Double masking.

a second layer of resist is spun onto the wafer. The wafer goes through another alignment and exposure, but with an oversize mask. The oversize mask makes for an easy alignment and protects the previously patterned image from being decreased by the light. The second resist layer fills in any pinholes in the first layer, providing protection during the etch step. The drawback to this process is the requirement of two masks and the longer time to process the wafer.

Lift-off Process

The final dimensions of the images in the surface layer are the result of variations in both the exposure step and the etch step. In processes where etch undercutting (resist adhesion) is a problem, such as aluminum etching, the etch component of the dimensional variation can be the dominant one.

A patterning variation that eliminates the etch variation component is lift-off (Fig. 10.7). In this variation the wafer is processed through the development step, leaving a hole in the resist layer where a deposited layer is to be located. The exposure and development steps are adjusted to create a negative slope in the sidewall of the hole.

Next, the wafer receives the deposited layer, which covers the entire wafer and fills in the hole. Definition of the pattern comes when the wafer is processed through a photoresist removal step that lifts off the resist and unwanted metal layer. Usually the removal step is assisted by ultrasonic agitation. This helps form a clean break of the deposited film at the resist edge. After resist and film removal, the desired pattern is left on the wafer surface.

Planarization

The advancement of circuits to the VLSI levels has required the addition of more layers to the wafer surface. These layers have in turn cre-

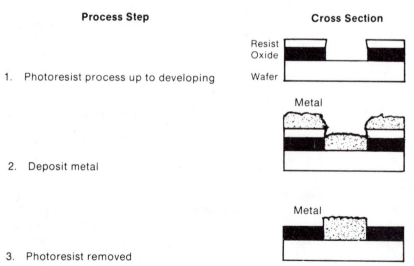

Process Step	Cross Section
1. Photoresist process up to developing	Resist / Oxide / Wafer
2. Deposit metal	Metal
3. Photoresist removed	Metal

Figure 10.7 Lift-off.

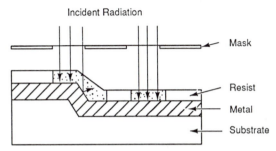

Figure 10.8 Light reflection at steps.

ated more steps on the wafer surface, which in turn make the resolution of small image sizes more difficult due to light reflection and the thinning of resist layers over the steps (Fig. 10.8). A number of techniques are used to offset the effects of a varied wafer topography. The techniques of multilayer resist processing, planarization layers, and reflow are collectively known as *planarization techniques*.

Multilayer resists

There are several multilayer resist processes. In production situations both dual layer and trilayer processes are used. The choice of a process depends on the size of the resist opening and the severity of the surface topography. While multilayer resist processes are longer and require more yield-limiting steps, in some situations they are the only reliable means of creating the desired images.

A dual multilayer resist process uses two layers of photoresist, each with a different polarity. The process is suited to resolve small geometries on wafers with a varied topography. First a relatively thick layer of resist is applied and baked to the thermal flow point (Fig. 10.9). A typical thickness is three to four times the highest step height on the wafer. The goal is a planar top resist surface. A typical multilayer process will use a positive-acting polymethylmeth acrylate (PMMA) resist sensitive to deep ultraviolet radiation.

Next, a thin layer of positive resist sensitive to just ultraviolet radiation is spun on top of the first layer and processed through the development step. The thin top layer allows the resolution of the pattern without the adverse effects encountered with thick resist layers or reflections from steps in the surface. Since the top layer conforms to the shape of the bottom layer, it is referred to as a *conformal layer* or *portable conformal layer*. This top layer of resist acts as a radiation block, leaving the bottom layer unpatterned. Next, the wafer is given a blanket or flood (no mask) deep ultraviolet exposure, which exposes the underlying positive resist through the holes in the top layer, thus ex-

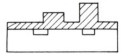

Start

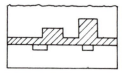

1. Apply 1st layer.

2. Bake 1st layer to cause slight flow.

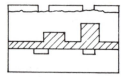

3. Apply 2nd layer and process
   to develop.

4. Flood expose 1st layer.

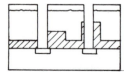

5. Develop 1st layer.

6. Etch and strip.

Figure 10.9   Dual-layer photoresist processing.

tending the pattern down to the wafer surface. A development step completes the hole resolution and the wafer is ready for etch.

Considerations in the choice of photoresists are compatibility of the two resists through the process, reflection problems from the subsurface, standing waves, and sensitivity problems with PMMA resists.[1] In addition, the two resists must have compatible bake processes and independent developing chemistries.

Variations on the basic dual-level resist process include dyes in the PMMA and the use of antireflection layers under the first resist layer. Many variations on the basic dual-level process are practiced. One use of a dual-layer resist process is as a lift-off technique. By adjusting the development of the bottom layer, an overhang can be created that assists in the clean definition of the metal line on the surface (Fig. 10.10).

A trilevel resist process (Fig. 10.11) incorporates a "hard" layer between the two resist layers. The hard layer may be a deposited layer of silicon dioxide or other developer-resistant material. As in the dual-layer process, the image is formed in the top photoresist layer. Then

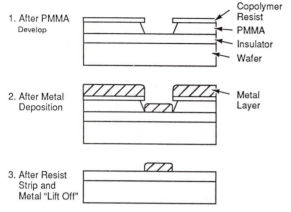

Figure 10.10   Dual-layer resist lift-off process.

the image is transferred into the hard layer by a conventional etching
step. The finishing step is the formation of the pattern in the bottom
layer, using the hard layer as an etch mask. The use of a hard intermediate layer makes possible the use of nonphotoresist bottom layers,
such as a polyimide layer.

Polyimide planarization layers

Along with the introduction of VLSI-level processing has come a host
of new materials and processes. One popular material is polyimide solutions. Polyimides have been used for years in printed circuit board
manufacturing. For semiconductor use, the polyimides offer the dielectric strength of deposited silicon dioxide films and the process advantage of application to the wafer with the same spinning equipment
used for photoresist.[2]

Once applied to the wafer, the polyimides flow over the surface,
making it more planar. After application and flow, the polyimide can
be covered with a hard layer and patterned with chemicals much like
a photoresist. A popular use of polyimide layers is as an interdielectric
layer between two layers of conducting metal. The planarizing effect
of the polyimide makes the definition of the second metal layer easier.

Reflow

Some device schemes use a hard planarizing layer or layers. A popular
layer is a deposited silicon dioxide doped with about 4 to 5% boron,
called boron silicate glass (BSG). The presense of the boron causes the

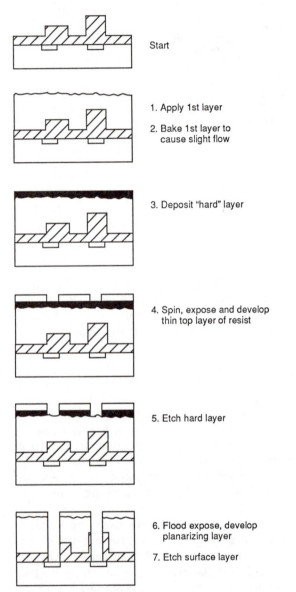

Start

1. Apply 1st layer

2. Bake 1st layer to cause slight flow

3. Deposit "hard" layer

4. Spin, expose and develop thin top layer of resist

5. Etch hard layer

6. Flood expose, develop planarizing layer

7. Etch surface layer

Figure 10.11   Trilevel resist process.

glass to flow at a relatively low temperature (less than 500°C), creating a planarized surface.

Another hard planarizing layer used is a spin-on-glass (SOG) layer. The glass is a mixture of silicon dioxide in a solvent that evaporates quickly. After spin application the glass film is baked, leaving a planarized silicon dioxide film. The glass as spun is brittle[3] and some formulas contain between 1 and 10% carbon to increase its resistance to cracking.

Image Reversal

The preference for positive resists over negative resists for small-geometry patterning has been discussed. One of the advantages with positive resists is the use of dark field masks for the imaging of holes. Dark-field masks offer a lower-defect process because the majority of the surface is covered by hard chrome which does not damage like glass. However, some mask levels require the printing of islands rather than holes. Metal mask levels are island patterns. Unfortunately, the printing of an island with a positive resist requires the use of a clear field mask with its glass damage potential.

A process that allows the printing of islands with positive resists and dark-field masks is image reversal, which involves the formation of the image in the resist with a dark-field mask by conventional masking steps (Fig. 10.12). At the conclusion of the exposure step there is an image in the resist that is reversed from the desired image. That is, if the resist was developed, a hole rather than an island would be formed.

The image reversal process involves exposing the resist-covered wafer to amine vapors in a vacuum oven. The vapors penetrate the resist, reversing its polarity. Upon removal of the wafers from the oven, they are given a flood exposure which completes the reversal process. The effect of the amine bake and flood exposure is to change the relative

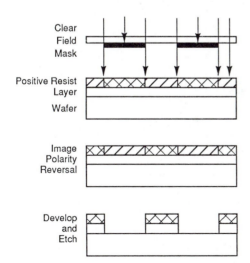

Figure 10.12   Image reversal.

dissolution rates of the exposed and unexposed regions, thus reversing the original image when the resist is finally developed. This process is capable of the same resolution capabilities as a nonreversed positive process.[4]

Contrast Enhancement Layers

Optical projection system resolution is approaching the limits imposed by the constraints of the lens and the wavelength of the exposing radiation. The two set up a condition where the resist *contrast threshold* becomes the limiting factor. This is because at short exposure times and ultraviolet and deep ultraviolet energies, the energy of the exposing wave varies in intensity. Thus the image formed in the resist is fuzzy.

A method used to decrease this threshold is a contrast enhancement layer (CEL), which is a layer spun on top of the resist that is initially opaque to the exposing radiation (Fig. 10.13). During the exposure cycle, the CEL becomes bleached (transparent) and allows the radiation to pass into the underlying photoresist. The CEL responds first to the higher intensities before turning transparent, in effect storing the

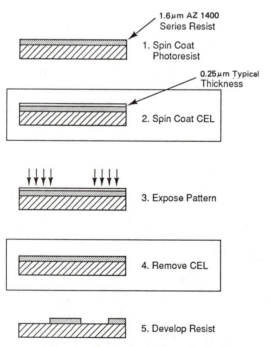

Figure 10.13 Contrast enhancement layer process flow.

lower intensities before turning transparent. The result is that the resist receives a uniform exposure of high-intensity radiation which improves its resolution capability. Another way to imagine the role of the CEL is as the top layer of a dual-layer resist system with the image being formed in the thin top layer.

Before the resist is developed, the CEL is removed by a chemical spray and development of the resist proceeds by normal processing. Positive resist processes normally capable of 1.0-μm resolution can achieve a 0.5-μm image with a CEL.

Antireflective Coatings

Antireflective coatings (ARCs) spun onto the wafer surface before the resist (Fig. 10.14) can aid the patterning of small images. The ARC layer brings several advantages to the masking process. First is a planarizing of the surface, which makes for a more planarized resist layer. Second an ARC cuts down on light scattering from the surface into the resist, which helps in the definition of small images. An ARC can also minimize standing wave effects and improve the image contrast. The latter benefit comes from increased exposure latitude with a proper ARC.

An ARC is spun onto the wafer and baked. After the resist is spun on top of the ARC, the wafer is aligned and exposed. The pattern is developed in both the resist and the ARC. During the etch, the ARC acts as an etch barrier. To be effective, an ARC material must transmit light in the same range as the resist. It must also have good adhesion properties with the wafer surface and the resist. Two other requirements are that the ARC must have a refractive index that matches the resist, and that the ARC must develop and be stripped in the same chemicals as the resist.

There are several penalties associated with the use of an ARC. One is an additional layer requiring a separate spin and bake. The resolution gains offered by an ARC can be offset with poor thickness control and/or with an ill-controlled developing step. The time of exposure can increase 30 to 50 percent, increasing the wafer throughput time. ARC layers may also be used as the intermediate layer in a trilayer resist process.

Dyed Resists

Various dyes may be added into the resist during manufacture. A dye may have one or several effects during the exposure step. One possible effect is the absorption of radiation, thereby attenuating the reflected radiation and minimizing standing wave effects. Another is a change

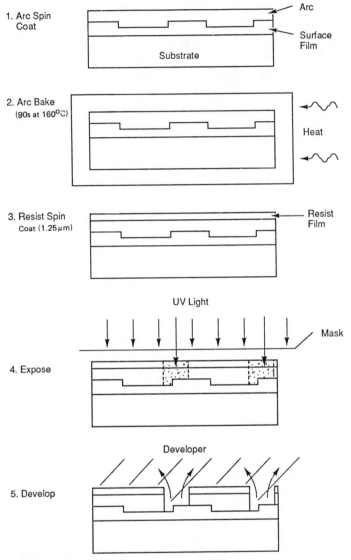

Figure 10.14   Antireflective process sequence.

of the dissolution rate of the resist polymer during development. This effect creates a cleaner developed line (increased contrast).[5] An important use of dyes is the elimination of the notching that occurs in thin lines of deposited material crossing over surface steps. Addition of a dye to a resist can cause an increase in exposure time of 5 to 50 percent.[6]

Pellicles

The development of projection exposure systems (projection aligners and steppers) brought with them an increased mask and reticle lifetime. With the increased lifetime came the incentive to make higher-quality masks and reticles. In a production line where masks are used for a long time, wafer-sort yield loss comes from dirt and scratches picked up during handling and use. One source of damage comes from mask and reticle cleaning steps. This situation is a "Catch-22." The masks and reticles become dirty during the process and require cleaning. The cleaning procedure then itself becomes a source of contamination, scratches, and breakage.

A solution to these problems is a pellicle (Fig. 10.15). A *pellicle* is a thin layer of an optically neutral polymer stretched onto a frame. The frame is designed to fit onto the mask or reticle. The pellicle is fitted to the mask or reticle after the mask is made and cleaned. Once in place, the pellicle membrane is the surface collecting any dirt or dust in the environment. The height of the membrane above the mask surface holds the dirt particles out of the focal plane of the mask. In effect, the particles are transparent to the exposing rays.

Another benefit of a pellicle is the elimination of scratches from the mask surface since the surface is covered. A third benefit is that once pelliclized, a mask or reticle does not need in-process cleaning. For some applications, the pellicle membrane is given an antireflective coating which assists in the imaging of small geometries, especially on reflective wafer surfaces. These benefits translate into wafer-sort yield increases of 5 to 30 percent.[7]

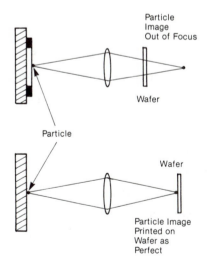

Figure 10.15   Pellicle.

Pellicle membranes are made from either nitrocellulose (NC) or cellulose acetate (AC). NC films are used in exposure systems with broadband exposure sources (340 to 460 nm)[8] while AC films are used in mid-ultraviolet applications. The membranes are thin (0.80 to 2.5 μm) and must show a high transmission rate for the rate exposure wavelengths. A typical pellicle will exhibit over a 99 percent transmission rate for the peaks of the exposing wavelengths. Pellicle effectiveness requires stringent thickness control, on the order of ± 800 Å, and control of particles to less than 25 μm in diameter.

A pellicle membrane is made by a spin casting technique. The pellicle material is dissolved in a solvent and spun onto a rigid substrate, such as a glass plate. This is the same technique used to spin photresist onto wafers. Thickness of the membrane is controlled by the viscosity of the solution and the spin speed of the spin coater. The membrane is removed from the substrate and fixed onto the frame. Frame shapes are determined by the size and shape of the mask or reticle. Cleanliness control requires class 10 or better clean rooms and antistatic packaging.

Key Concepts and Terms

Antireflective layers	Light scattering
Contrast enhancement	Pellicle
Image contrast	Planarization
Image reversal	Standing waves
Lift-off	Subsurface reflectivity

Review Questions

1. What is a contrast effect and how does it affect an image?
2. Name two problems resulting from subsurface reflection.
3. What is the purpose of the top layer of a multilayer resist process?
4. What is the purpose of the bottom layer of a multilayer resist process?
5. What is flood exposure and at what step is it used in a multiresist process?
6. What basic resist process step is eliminated in a lift-off process?
7. What advantages come from the use of a planarization layer?
8. How does a contrast enhancement layer improve resolution?
9. Draw a cross section of a reticle with a pellicle attached. Identify all parts.
10. How does a pellicle improve the wafer sort yield?

References

1. C. H. Ling and K. L. Liauw, "Improved DUV multilayer resist process," *Semiconductor International*, Nov. 1984, p. 102.
2. K. Skidmore, "Techniques for planarizing device topography," *Semiconductor International*, Apr. 1988, p. 116.
3. K. Skidmore, "Techniques for planarizing device topography," *Semiconductor International*, Apr. 1988, p. 117.
4. E. Aling and G. Stauffer, "Image reversal of positive photoresist," *SPIE Journal*, vol. 539, Mar. 11–12, 1985, p. 194.
5. J. F. Boland, H. F. Stanford, and S. A. Fine, "Dye effects on exposure and development of positive resists," *Micro Manufacturing and Test*, Aug. 1985, p. 18.
6. J. Housley, R. Williams, and I. Horiuchi, "Dyes in photoresists: today's view," *Semiconductor International*, Apr. 1988, p. 142.
7. R. Ixcoff, "Pellicles 1985: an update," *Semiconductor International*, Apr. 1985, p. 111.
8. Micropel product data sheet, Micropel Division, EKC Technology, Hayward, Calif., 1988.

11

Doping

Overview

Conductive layers and N-P junctions in the wafer surface are essential parts of the semiconductor devices formed in and on the wafer surface. In this chapter the definition of a junction and the two techniques used to create them, diffusion and ion implantation, are described. The principles of the two techniques and the actual processes are detailed.

Objectives

Upon the completion of this chapter, you should be able to:

1. Define an N-P junction.
2. Draw a flow diagram of a complete diffusion process.
3. Describe the differences between the deposition and drive-in steps.
4. List the three types of deposition sources.
5. Draw a typical concentration vs. distance curve for a deposition and drive-in.
6. List the major parts of an ion implanter.
7. Describe the principle of an ion implanter.
8. Compare the advantages and disadvantages of diffusion and ion implant processes.

Formation of a Junction

In Chap. 5, the steps for the formation of an MOS metal gate device were presented. Step 3 showed the formation of an N-P junction in the

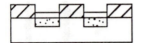

Figure 11.1   Doping and layering. N-type doping and reoxidation of source drain.

wafer surface (Fig. 11.1) by doping. The cross section shows only that a localized region of the wafer surface was changed from P type to N type. In this section the conversion mechanism and the nature of the N-P interface is detailed.

Formation of a Doped Region

The formation of a doped region is illustrated by examining the doping of a wafer in a diffusion process. The starting condition is depicted in Fig. 11.2. The wafer illustrated is from a P-type crystal. The +'s in the diagram represent the P-type dopants that were incorporated into the crystal during the crystal-growing process. They are uniformly distributed throughout the wafer.

In many texts and articles dopant atoms are referred to as *impurities*. This term comes from the field of chemistry which contributed the mathematics of diffusion. Unfortunately, the term impurity is confusing since the industry spends so much time and energy eliminating contaminating impurities from the process. In this text, the terms *dopant atoms* or *dopants* are used in reference to the creation of the needed N- and P-type regions and *impurities* is reserved for unwanted dopants and contaminants.

The wafer receives a thermal oxidation and a patterning process that leaves a hole in the oxide layer. In a diffusion tube, the wafer is exposed to a concentration of N-type dopants at a high temperature (the −'s in Fig. 11.3). The N-type dopants diffuse through the hole in the oxide layer.

The effect in the wafer is illustrated by examining what happens at different levels in the wafer. The conditions in the diffusion tube are set such that the number of N-type dopant atoms that diffuse into the wafer surface are greater than the number of P-type atoms in layer 1. In the illustration there are seven more N-type atoms than P-type atoms and that level is electrically an N-type layer. This process has converted the top layer from P type to N type.

The diffusion process proceeds with N-type atoms diffusing from the first level down to the second level (Fig. 11.4). At the second level there are again more N-type dopants than P-type dopants, converting level 2 to N type. In the table (Fig. 11.5) is an accounting of the num-

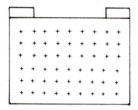

Figure 11.2   P-type wafer ready for diffusion.

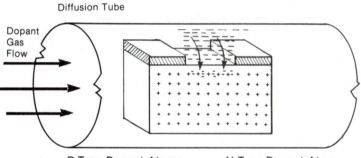

\+ = P Type Dopant Atoms   − = N Type Dopant Atoms

Figure 11.3   Start of diffusion process.

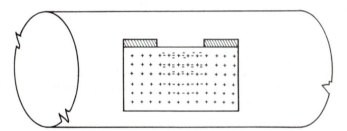

Figure 11.4   Cross section of wafer at conclusion of diffusion.

Layer	# N's (−)	# P's (+)	Net (N − P)	Layer
1	12	5	7	N
2	10	5	5	N
3	8	5	3	N
4	5	5	0	Jct
5	3	5	−1	P
6	0	5	−5	P

Figure 11.5 Dopant amounts and level conductivity type.

ber of N-type and P-type atoms at each level. This process goes deeper into the wafer.

The N-P junction

At level 4 there are exactly the same number of N-type and P-type atoms. This level is the location of an N-P junction. A junction is the interface between N- and P-type regions in a semiconductor wafer. The definition of an N-P junction is the location in the wafer where the number of N-type and P-type dopant atoms are equal. Note that below the junction at level 5 there are three N-type atoms, which are not enough to convert that layer to N type.

The term N-P junction indicates that there is a higher concentration of N-type dopants in the doped region. A P-N junction would indicate a higher P-type concentration in the doped region. The behavior of electrical currents across semiconductor junctions gives rise to the particular performance of individual semiconductor devices and is the subject of Chap. 14. In this chapter the emphasis is on the formation and character of doped regions in a wafer.

Lateral diffusion

The diffusion doping process shown in Fig. 11.4 shows that the incoming dopant atoms travel straight down into the wafer. This is an ideal situation. In reality the dopant atoms move in all directions. An accurate cross section (Fig. 11.6) would show that some of the atoms have moved in a lateral direction, forming a junction under the oxide barrier. This movement is also called *lateral* or *side diffusion*. The amount of the lateral or side diffusion is approximately 85 percent of the vertical junction depth. The effects of side diffusion on circuit density are discussed in the introduction to ion implantation.

Same-type doping

Some devices call for a doping with a dopant type the same as the host region. In other words, an N-type dopant will be put into an N-type

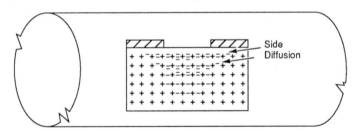

Figure 11.6 Side-diffused N-type dopants.

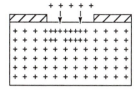

Figure 11.7   P-type doping of P-type wafer.

wafer (Fig. 11.7) or a P-type dopant will be put into a P-type wafer. When this situation happens, the added dopant atoms simply increase the concentration of the dopant atoms in the localized region. No junction is formed.

Graphical Representation of Junctions

In a cross section (Fig. 11.8) of a semiconductor device, the N-P junctions are indicated simply as regions in the device. There is no graphical convention to represent N- or P-type areas. The drawings just show the relative location of the doped region and the junction. This type of graphical representation gives little information about the concentration of the dopant atoms and only approximates the actual dimensions of the regions. A drawing of a 2-μm-deep junction in a 20-mm-thick wafer, scaled to an 8-ft wafer, would have the junction depth only be 0.4-in thick!

Concentration versus depth graphs

Another two-dimensional graphical representation of a doped region is the concentration versus depth graph. This type of graph has the concentration on the vertical axis and the depth into the wafer on the horizontal axis. An example of such a graph is illustrated in Fig. 11.9. The illustration uses the data from the doping example shown in Fig. 11.6. First the P-type dopant concentration is plotted. In the example there are exactly five P-type dopant atoms at every level resulting in a straight horizontal line on the graph (Fig. 11.9b). Next the number of N-type dopant atoms is plotted. Since the number of atoms decreases deeper into the wafer, the plotted line slopes down and to the right. At level 4, the number of N- and P-type dopants is equal and the

Ideal   Lateral Diffusion

Figure 11.8   Cross-sectional representation of diffused junctions.

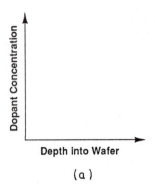

(a)

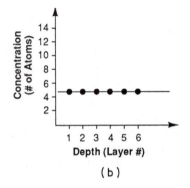

(b)

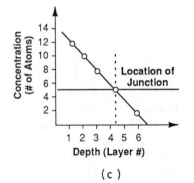

(c)

Figure 11.9 Construction of concentration versus depth curve. (a) Axes; (b) P-type dopant; (c) N-type and P-type dopant.

lines cross. This is a graphical representation of the location of the junction.

The graph of an incoming dopant concentration versus depth profile for an actual process is not a straight line. They are curved lines. The shape of the curve is determined by the physics of the dopant tech-

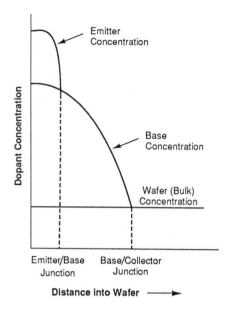

Emitter
Concentration

Base
Concentration

Wafer (Bulk)
Concentration

Dopant Concentration

Emitter/Base
Junction

Base/Collector
Junction

Distance into Wafer

Figure 11.10 Concentration versus depth graph example.

nique. The actual shapes are discussed in the appropriate doping section.

In some devices, a third doping will take place into the same region as the first one shown in Fig. 11.10. A cross-sectional and concentration versus depth representation is shown. When the dopant types are alternated, two different junctions are created.

Concept of Diffusion

In the early days of semiconductor processing, junctions were formed in the wafer surfaces by alloying dots (Fig. 11.11) of an opposite-conductivity type of metal into the wafer surface. This method produced junctions that varied in concentration, depth, and area. The procedure was definitely not a planar process.

A major advance in junction formation was the development of diffusion doping techniques. Diffusion, the movement of one material through another, is a natural chemical process with many examples in everyday life. Two conditions are necessary for a diffusion to take place. First, one of the materials must be at a higher concentration than the other. Second, there must be sufficient energy in the system for the higher-concentration material to move into the other.

An example of a gas-state diffusion is the action of a common pressurized spray can (Fig. 11.12). Inside the can is some material, a room

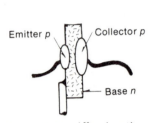

Figure 11.11 Alloy junction transistor.

Figure 11.12 Examples of diffusion.

deodorant or such, under pressure. The pressure allows the storage of the material at a higher concentration than it exists in the air outside the can. When the nozzle is depressed, the material leaves the can and moves into the surrounding air. Initially the movement out of the can is driven by the pressure. Thereafter, movement of the gas into the room proceeds by the process of diffusion. The movement takes place while the nozzle is depressed and continues after it is closed. The diffusion will continue as long as the advancing spray is at a concentration higher than that in the air. As the material moves away from the can the concentration of the material is progressively less. This is a characteristic of a diffusion process. Diffusion will continue until the concentration is even throughout the room.

An example of a liquid-state diffusion is represented when a drop of ink is dropped into a glass of water. The ink is more concentrated than the surrounding water and immediately starts diffusing into the glass of water. The diffusion will continue until the whole glass of water is the same color. This example can be used to illustrate the influence of energy on the diffusion process. If the water in the glass is heated (giving the water more energy) the ink will spread into the water faster.

Solid-State Semiconductor Diffusion

The industry has mastered the technology to diffuse dopant atoms into solid semiconductor wafers. The technique is called *solid-state diffusion*. A complete diffusion process requires two steps, called *deposition* and *drive-in–oxidation*. Both steps take place in a tube furnace. The reader should read or review the equipment section in Chap. 7 on oxidation tube furnaces.

The goals of a diffusion process are threefold:

1. The creation of a specific number (concentration) of dopant atoms in the wafer surface

2. To create an N-P (or P-N) junction at a specific distance below the wafer surface

3. To create a specific distribution (concentration) of dopant atoms in the wafer surface

Open-Tube Deposition

The first step of a diffusion process is called *deposition*; it is also called *predeposition*, *dep*, or *predep*. It takes place in a tube furnace, with the wafers placed in the flat zone of the tube. A source of dopant atoms is located in the source cabinet and their vapors are transferred into the tube at a required concentration (Fig. 11.13).

In the tube, the dopant atoms proceed to diffuse into the exposed wafer surface as explained in the "Formation of a Doped Region," p. 260. Within the wafer, the dopant atoms move by two different mechanisms. In the vacancy model (Fig. 11.14a) the dopant atoms move by filling empty crystal positions, the vacancies. The second (Fig. 11.14b) model relies on interstitial movement of the dopant.[1] In this model the dopant atom moves through the spaces between the crystal sites, that is, intersite.

A deposition process is controlled or limited by several factors. One is the *diffusivity* of the particular dopant. Diffusivity is a measure of the rate of movement of the dopant through the particular wafer material. The higher the diffusivity, the faster the dopant moves through the wafer. Diffusivity increases with temperature.

Another factor is the *maximum solid solubility* of the dopant in the wafer material. Maximum solid solubility is the maximum concentration of a particular dopant that can be put into the wafer. A familiar analogy is the maximum liquid solubility of sugar in coffee. The coffee

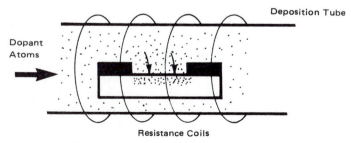

Figure 11.13   Deposition.

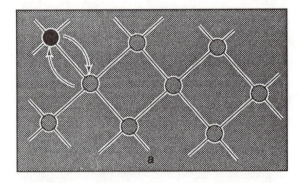

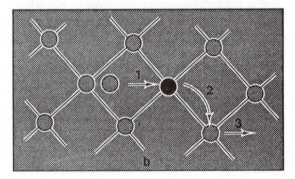

Figure 11.14 Diffusion models. (*a*) Vacancy model; (*b*) interstitial model.

can only dissolve a certain amount of sugar before collecting in the bottom of the cup as a solid. The maximum solid solubility limit increases with increasing temperature.

A semiconductor deposition is performed with a concentration of dopant in the tube that exceeds the maximum solid solubility of the dopant in the wafer material. This situation adds to the control of the deposition process by eliminating the dopant concentration as a variable. The amount of dopant entering the wafer surface is a function of the temperature only, and the deposition is said to take place at solid solubility. Maximum solubility levels for various dopants in silicon are shown in Fig. 11.15.

The concentration of dopant atoms at each level in the wafer is an important factor in the performance of junction diodes and transistors. A dopant concentration versus depth curve for a deposition is shown in Fig. 11.16. The shape of the curve is specific; it is known in mathematics as an *error function*. An important factor in device performance is the concentration of the dopant at the wafer surface. This is called the *surface concentration* and is the quantity indicated where the error

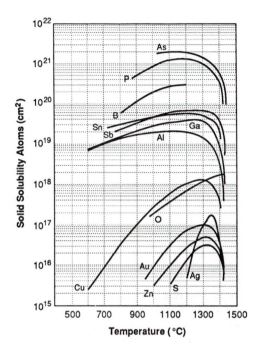

Figure 11.15 The solid solubility of impurities in silicon.

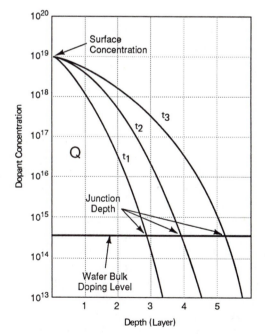

Figure 11.16 Typical deposition (error function) dopant profile for three different deposition times.

function curve intersects the vertical axis. Another deposition parameter is the total quantity of atoms diffused into the wafer. This amount increases with the time of the deposition. Mathematically, the quantity of atoms (Q) is represented by the area under the error function curve.

Deposition steps

A deposition process requires four steps, they are

1. Preclean and etch
2. Tube deposition
3. Deglaze
4. Evaluation

Preclean and etch. Wafers coming to the deposition station first receive a preclean to remove particulates and stains. The chemicals and processes are the same as in the preoxidation process cleaning. After the preclean, the wafers will be chemically etched in an HF or HF and water solution to remove any oxide that may have grown on the exposed silicon surface. An oxide can form on the silicon from exposure to the air and from the chemical preclean process. Removal of the oxide is essential to allow the dopants to diffuse unimpeded into the wafer surface. The etch time and concentration must be balanced to prevent the surface-blocking oxide from being removed or thinned too much.

Deposition. The deposition process, like oxidation, requires a minimum of three cycles. The first cycle is the loading cycle, which takes place in a nitrogen atmosphere. The second cycle is the actual doping cycle. The third cycle is the exit cycle, which also takes place in a nitrogen atmosphere.

Wafers are positioned on the boats at right angles (Fig. 11.17) or parallel to the tube axis. Right-angle placement is the highest packing density, but it can cause uniformity problems since the wafers act as baffles to the gas flow. For uniform doping the gas must mix uniformly between all of the wafers. Parallel placement offers the advantage of more uniform doping since the doping gas proceeds unimpeded through the wafer boat. The disadvantage is a lower loading density. In both placement systems dummy wafers are placed on the outsides and/or the front and back positions of the boat to create uniform doping of the interior device wafers.

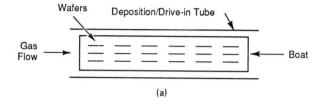

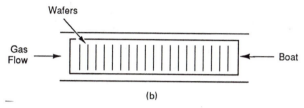

Figure 11.17 Boat loading patterns. (*a*) Parallel; (*b*) perpendicular.

Deglaze . During the deposition cycle, a thin oxide can form on the exposed wafer surface. This oxide is doped and can act as an unwanted source of dopant during the drive-in–oxidation step. Also, a deposition-created oxide can be unetchable, causing incomplete etch in a subsequent masking process. The oxide is removed from the surface by immersion in a diluted HF solution, followed by a water rinse and a drying step.

Evaluation. Test wafers are added to the boat of device wafers that go into the deposition tube. A deposition test wafer has no patterns and has a conductivity type opposite that of the dopant. They are placed on the deposition boat in locations to sample the deposition distribution throughout the wafer batch. After deglazing the test wafers are evaluated. The primary in-line test is of the sheet resistance with a four-point probe or contactless apparatus. The concept of sheet resistance and the measurement techniques are explained in Chap. 14. The junction depth after deposition is very thin and is generally not measured at this point.

Certification of process and tube cleanliness is determined by oxidizing test wafers and performing a capacitance-voltage measurement for mobile ionic contaminants (Chap. 14). Device wafers are 100 percent or sample-inspected for surface contamination and/or stains with high-intensity ultraviolet lamps or microscopes.

Dopant sources

A deposition depends on the presence of a concentration of dopant atom vapors in the tube. The vapors are created from a dopant source

Type	Element	Compound Name	Formula	State	Diffusion Reactions*
N	Antimony	Antimony Trioxide	Sb_2O_3	Solid	
	Arsenic	Arsenic Trioxide	As_2O_3	Solid	$2AsH_3 + 3O_2 \longrightarrow As_2O_3 + 3H_2O$
		Arsine	AsH_3	Gas	
	Phosphorus	Phosphorus Oxychloride	$POCl_3$	Liquid	$4POCl_3 + 3O_2 \longrightarrow 2P_2O_5 + 6Cl_2$
		Phosphorus Pentoxide	P_2O_5	Solid	$2PH_3 + 4O_2 \longrightarrow P_2O_5 + 3H_2O$
		Phosphine	PH_3	Gas	
P	Boron	Boron Tribromide	BBr_3	Liquid	$4BBr_3 + 3O_2 \longrightarrow 2B_2O_3 + 6Br_2$
		Boron Trioxide	B_2O_3	Solid	$B_2H_6 + 3O_2 \longrightarrow B_2O_3 + 3H_2O$
		Diborane	B_2H_6	Gas	
		Boron Trichloride	BCl_3	Gas	$BCl_3 + 3H_2 \longrightarrow 2B + 6HCl$
		Boron Nitride	BN	Solid	
	Gold	Gold	Au	Solid (Evap.)	
	Iron		Fe		
	Copper		Cu		
	Lithium		Li		Undesirable impurities
	Zinc		Zn		from contamination
	Manganese		Mn		
	Nickel		Ni		
	Sodium		Na		

*Note: Only selected diffusion reactions are listed

Figure 11.18 Deposition source table.

located in the source cabinet of the tube furnace and passed into the tube with a carrier gas. Dopant sources are either in liquid, gaseous, or solid states. Several dopant elements are available in more than one state (Fig. 11.18).[2]

Liquid sources. Liquid sources of dopants are chlorinated or brominated compounds of the desired element. Thus, a boron liquid source is boron tribromide (BBr_3) and a phosphorus liquid source is phosphorus oxychloride ($POCl_3$). The liquid sources are held in temperature-controlled quartz flasks (Fig. 11.19). An inert gas, such as nitrogen, is bubbled through the heated liquid, and the gas becomes saturated with the dopant vapors. The flask is connected to a gas manifold whose valves are controlled by a microprocessor. The nitrogen carries the dopant vapors into the tube and maintains a gas volume sufficient to create a laminar flow condition in the tube. Laminar flow is required to prevent the gas from spiraling in the tube, creating nonuniform doping. Another device used to create laminar

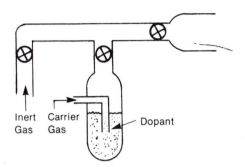

Figure 11.19 Liquid dopant deposition source.

flow is a baffle (Fig. 11.20) at the tube inlet. Baffles break up the incoming gas stream into a laminar form.

Also connected to the manifold is a reaction gas that is required to create the elemental dopant form in the tube. For BBr_3, the reaction gas is oxygen, which creates the boron oxide (B_2O_3), as shown in Fig. 11.21. At the wafer surface, a boron trioxide layer deposits on the silicon, and the boron diffuses from the oxide into the surface. Liquid sources offer the advantages of low to moderate cost and consistent doping. The disadvantages are uniformity problems (especially for larger-diameter wafers), safety considerations, and the potential of contamination associated with the opening of the flasks for recharging. Several vendors supply liquid dopants in pluggable sealed ampules that minimize contamination and safety problems.

Gas sources. Many wafer manufacturers prefer gas dopant sources. These are hydrated forms of the dopant atom such as arsine (AsH_3)

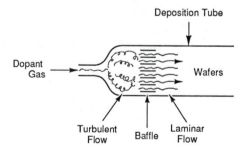

Figure 11.20 Use of baffle to create laminar flow.

$$4BBr_3 + 3O_2 \rightarrow 2B_2O_3 + 6Br_3$$

Figure 11.21 Reaction of liquid source BBr_3 and oxygen in a deposition tube.

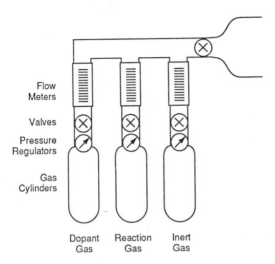

Figure 11.22   Gas source manifold.

and diborane (B_2H_6). The gases are mixed in different dilutions in pressurized containers and connected directly to the gas manifold (Fig. 11.22). Gas sources offer the advantage of precise control through pressure regulators and are favored for deposition on larger-diameter wafers. The processes are in general cleaner, since the pressurized sources last longer than liquid sources. On the downside, unwanted chemical reactions in the manifold can create silica dust that can contaminate the tube and wafers.[2]

Solid sources.   The original deposition sources were solid. An oxide powder of the desired dopant was placed on a quartz holder, called a *spoon*, and placed in a source tube furnace attached to the main deposition tube (Fig. 11.23a). In the source furnace, the oxide gives off source vapors that are carried into the deposition tube where the diffusion takes place. This setup is called a *remote solid source*. Remote solid sources are economical but nonuniform. They are used primarily in the doping of discrete devices where high precision is not required.

A more popular solid source is the planar source wafer (Fig. 11.23b). These are wafer-size "slugs" that contain the desired dopant. Boron slugs are a compound of boron and a nitride (BN). Slugs are also available for arsenic and phosphorus diffusions.

The slugs are stacked on the deposition boat, with one slug between every two device wafers. This arrangement is called a *solid neighbor source*. In the tube the dopant diffuses out of the slug, crosses the short distance to the wafer, and diffuses into the surface. This system provides good uniformity for larger-diameter wafers since the slug is the

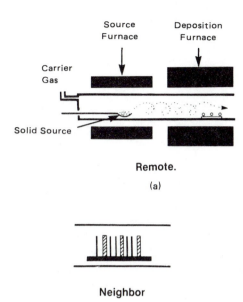

Remote.

(a)

Neighbor

(b)

Figure 11.23 Solid source. (a) Remote furnace; (b) neighbor source.

same size as the wafer. Slugs are safer to use, having no toxic vapors at room temperature. Drawbacks to the slug process include breakage of the slugs, lower productivity (the slugs take up space in the tube), and the necessity of cleaning the slugs. Some slug compounds require bake steps to maintain the dopant activity.

The third solid dopant source is a conformal layer spun directly on the wafer surface. The sources are powdered oxides (same as remote sources) mixed in solvents. They are spun onto the wafer surface using photoresist-type spinners and baked to evaporate the solvents. Left on the surface is a layer of doped oxide which conforms to the wafer surface. The wafers are placed on a boat and positioned in the deposition tube where the heat drives the dopant out of the oxide and into the wafer.

Spin-on dopants have the potential of high uniformity and offer high productivity, and like the slugs are safe to handle. Problems with the system are the distribution of the dopant in the oxide layer; thickness variations, especially over surface steps; and the costs of the additional spin and bake steps.

Closed-Tube Deposition

Some manufacturers use the closed-tube system, which was developed by IBM (Fig. 11.24). The wafers and a powdered dopant source are placed in a quartz capsule (or ampule) that is evacuated of air to a

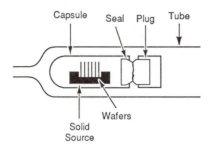

Figure 11.24   Capsule diffusion.

very low pressure and sealed. The capsule is placed in a tube furnace where the heat causes the source powder to give off dopant vapors that diffuse into the wafers. The diffusion is uniform due to the absence of an atmosphere in the capsule. This system has not been widely accepted due to the high cost of the capsule (which is broken open and discarded after each use), and the expensive vacuum pumps required.

Drive-in–Oxidation

The second major part of the diffusion process is the drive-in–oxidation step. It is also variously known as *drive-in, diffusion, reoxidation,* or *reox*. The purpose of this step is twofold:

1. The redistribution of the dopant deeper into the wafer. During the deposition a high-concentration layer and shallow layer of dopant are diffused into the surface. In the drive-in there is no dopant source. The heat alone drives the dopant atoms deeper and wider into the wafer just as material from a spray can will continue to spread into the room after the nozzle is released. During this step the total amount of atoms (Q) from the deposition step remains constant. The surface concentration is reduced and the distribution of atoms takes a new shape. The distribution after the drive-in is described by mathematicians as a gaussian distribution (Fig. 11.25). The junction depth increases. Generally the drive-in–oxidation step takes place at a higher temperature than the deposition step.

2. The second purpose of drive-in–oxidation is the oxidation of the exposed silicon surfaces. The atmosphere in the tube is oxygen or water vapor, which performs the oxidation simultaneously as the drive-in of the dopants is taking place.

The setup, process steps, and equipment for the drive-in–oxidation step are the same as an oxidation process. After the completion of the drive-in, the wafers are again evaluated. The test wafers (from the deposition step) are again measured for surface concentration with a

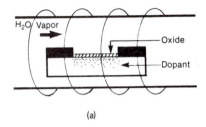

(a)

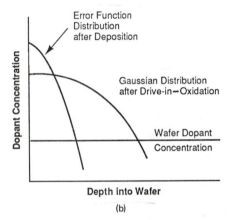

Depth into Wafer

(b)

Figure 11.25 Drive-in–oxidation. (*a*) Cross section of wafer; (*b*) dopant concentration in wafer.

four-point probe and the device wafers checked for cleanliness. The test wafers are measured for junction depth and perhaps for mobile ionic contamination. After some diffusion steps, a test structure in the engineering die may be electrically probed for junction parameters.

Oxidation Effects

The oxidation of the silicon surface affects the final distribution of the dopants.[3] The effects are related to the relocation of the top-level dopants after the oxidation. Recall that the silicon in the silicon dioxide film is consumed from the wafer surface. The question to ask is, What happened to the dopants that were in the top level? The answer to that question depends on the conductivity type of the dopant.

If the dopant is an N-type, an effect called *pile-up* (Fig. 11.26*a*) occurs. As the oxide-silicon interface advances into the surface, the N-type dopant atoms segregate into the silicon rather than the oxide. The effect is to increase the number of these dopants in the new top

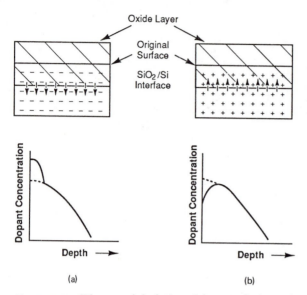

Figure 11.26 Pile-up and depletion of dopants during oxidation. (*a*) Pile-up of N-type dopants; (*b*) depletion of P-type dopants.

layer of the silicon. In other words, the N-type dopants pile up in the wafer surface and the surface concentration of the dopant is increased. Pile-up changes the electrical performance of the devices.

If the dopant is the P-type boron, an opposite effect occurs. The boron atoms are more soluble in the oxide and are drawn up into it (Fig. 11.26*b*). The effect on the wafer surface is a lowering of the concentration of boron atoms, which lowers the surface concentration and also affects the electrical performance of the devices. A summary of the deposition and drive-in–oxidation steps are listed in Fig. 11.27.

Introduction to Ion Implantation

Semiconductor technology has advanced when device requirements have strained the capabilities of existing processes. With the advent of VLSI-based and MOS-based circuits and their smaller feature sizes, diffusion as a doping technique became a processing limitation. Side diffusion placed a limit on the density of circuits. Four problems are side diffusion, low doping control, surface doping, and dislocation generation. Not only do the dopant layers diffuse in a lateral direction during the diffusion step, but this lateral diffusion continues whenever the wafer is processed at high temperatures in other diffusion steps. The situation is illustrated in Fig. 11.28. The circuit designer must leave enough room between adjacent regions to prevent the lat-

	Deposition	**Drive-In**
Goals	Introduction of Dopant	1. Redistribution of Dopant
		2. Reoxidation
Variables		1. Surface Composition
		2. Junction Depth
		3. Time
		4. Diffusivity
		5. Temperature
		6. Quantity of Atoms
Source Conditions	Continuous Source	No Source
Temperature Range	900 – 1100°C	1050 – 1200°C
Oxidation	No	Yes

Figure 11.27   Summary of deposition and drive-in steps.

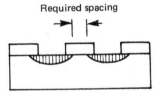

Required spacing

Figure 11.28   Side diffusion.

erally diffused regions from touching and shorting. The accumulative effect for a dense circuit can be a largely increased die area.

The diffusion-related limit encountered in MOS transistors was related to the difficulty of creating doped regions of low concentration (below 10^{15} atoms/cm^2). Gate regions of this low concentration are required to make efficient transistors. The third problem encountered is imposed by the physics of a diffused region. As illustrated in Figs. 11.16 and 11.25, the majority of the dopant atoms are located near the wafer surface. When the devices are operating, this means that the majority of the current is traveling near the surface, exactly where the presence of contamination can interfere with the current flow.

A fourth problem comes from the high temperatures required of the diffusion process. Every time a wafer is heated and cooled, crystal damage from dislocations occurs. A high concentration of these dislocations can cause device failure from leakage currents.

Ion implantation overcomes these limits of diffusion and also adds additional benefits. During the ion implant process there is no side diffusion, the process takes place at close to room temperature, the dopant atoms are placed below the wafer surface, and a wide range of doping concentrations are possible. With ion implantation there is greater control of the location and number of dopants put in the wafer. Also, photoresist and thin metal layers can be used as doping barriers along with the usual silicon dioxide layers. Given the benefits, it is not surprising that the majority of doping steps for advanced circuits are done by ion implantation. Diffusion still finds use in the doping of less critical layers and will continue to be used for lower-integration-level circuits.

Concept of Ion Implantation

Diffusion is a chemical process. Ion implantation is a physical process, that is, the act of implanting does not rely on chemical interaction between the dopant and the wafer material. An analogy that demonstrates the concept of ion implantation is a cannon firing balls into a wall (Fig. 11.29). Given enough momentum from the powder in the cannon, the balls will penetrate the wall and come to rest below the surface of the wall. The same events take place in an ion implantation machine. Dopant atom sources are ionized, selected, accelerated (gain momentum), formed into a beam, and swept across into the wafer. The dopant atoms enter the wafer and come to rest below the surface (Fig. 11.30).

Ion Implantation Equipment

An ion implanter is a collection of very sophisticated subsystems. Ion implanters come in a variety of designs used for advanced research and high-volume production. An in-depth understanding of ion implantation is enhanced by examining each of the subsystems. Figure 11.30a is a detailed diagram of a typical implanter.

Implant sources

The same dopant elements diffused into a silicon wafer are required in an implant process. In diffusion processes the dopants originate in ei-

Figure 11.29   Ion implantation analogy.

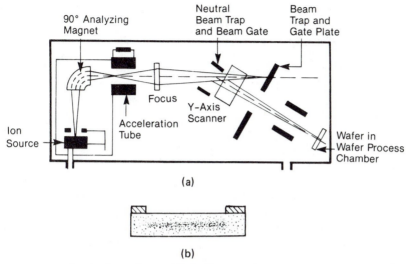

Figure 11.30 Ion implantation. (*a*) Block diagram of ion implanter; (*b*) distribution of implanted atoms in wafers.

ther liquid, gas, or solid sources. For ion implantation, only gas and solid sources are used.

Gases are favored for ion implantation because of their ease of use and higher control. The gases are mostly fluorine based, such as PF_5, AsF_5, $B^{11}F_3$, SbF_3, and PF_3. The gas cylinders are connected to the ion source subsystem through mass flow meters, which offer more control of the gas flow than normal flow meters.

Certain solid sources are used for special applications. One is phosphorus pentoxide (P_2O_5). In this system the powder is heated to create vapors that are carried into the ion source chamber.

Ionization chamber

The name "ion implant" implies that ions are a part of the process. Recall that ions are atoms or molecules with a negative or positive charge. The ions implanted are ionized atoms of the dopants. The ionization occurs in a chamber that is fed by the source vapors. The chamber is maintained at a low pressure (vacuum) of about 10^{-3} torr. Inside the chamber is a filament that is heated to the point where electrons are created from the filament surface. The negatively charged electrons are attracted to an oppositely charged anode in the chamber. During the travel from the filament to the anode, the electrons collide with the dopant source molecules and create a host of positively charged ions from the elements in the molecule. The results of the ionization of the source BF_3 are shown in Fig. 11.31.

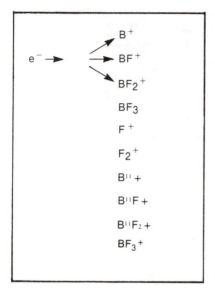

Figure 11.31 Ionization species of BF_3.

Another ionization method uses a cold-cathode technique to generate the electrons. A high-voltage electric field is created between a cathode and anode, which creates the electrons in a self-sustaining process.

Ion selection

At the top of the list in Fig. 11.31 is a lone boron ion. This is the atom desired in the wafer surface. The other species resulting from the ionization of the boron trifluoride cannot be implanted in the wafer. The boron ion must be *selected* from the group of positive ions. This process is called *analyzing, selection*, or *separation*.

Selection is accomplished in a mass analyzer. This subsystem was first developed during the Manhattan Project for the atomic bomb. The analyzer (Fig. 11.32) creates a magnetic field. The species leave the ionization subsystem with voltages of 15 to 40 keV (thousand electron-volts). In other words, they are traveling at a relatively high speed. In the field, each of the positively charged species is bent into an arc with a specific radius. The radius of the arc is dictated by the mass of the individual species, its speed, and the strength of the magnetic field. At the exit end of the analyzer is a slit that will allow only one species to exit. The magnetic field is adjusted to match the path of the boron ion to the exit slit position. Thus only the boron ion leaves the analyzing subsystem.

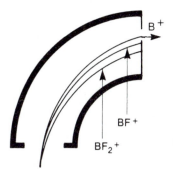

Figure 11.32 Analyzing magnet.

Acceleration tube

Upon leaving the analyzing section, the boron ion moves into an acceleration tube. The purpose is to accelerate the ion to a high enough velocity such that it will have the momentum to penetrate the wafer surface. Momentum is the product of the mass of the atom multiplied by its velocity.

The required velocity is achieved by taking advantage of the fact that negative and positive charges attract each other. The tube is a linear design with annular anodes along its axis. Each of the anodes has a negative charge. The charge amount increases down the tube. As the positively charged ion enters the tube, it immediately starts to accelerate down the tube. The voltage value is selected based on the mass of the ion and the momentum required at the wafer end of the implanter. The higher the voltage, the higher the momentum, and the faster and deeper the dopant ion can be implanted. Voltages range from 5 to 10 keV for low-energy implanters to 0.2 to 2.5 MeV (million electron-volts)[4] for high-energy implanters.

The stream of positive ions exiting the tube is actually an electric current. Ion implanters are classified into medium- and high-current machines. The current level translates into the number of ions implanted per minute. The higher the current, the more atoms implanted. The amount of atoms implanted is called the *dose*. Medium-current machines produce currents in the 0.5 to 1.7 mA (milliampere) range at energies from 30 to 200 keV. High-current machines generate beam currents of about 10 mA at energies up to 160 keV.[4] The relationships of current and energy for production-level implanters is shown in Fig. 11.33.

A successful ion implant relies on the implantation of only the de-

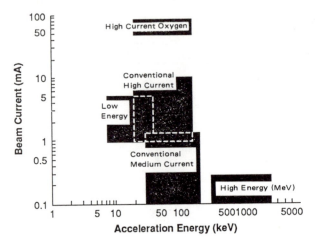

Figure 11.33 Ion implanters. The conventional descriptions of ion implanters were based on applications. However, some of today's more advanced implanters cannot be classified easily; several systems are capable of a broader process window than the conventional descriptions indicate. (*From Ref. 4.*)

sired dopant atom. Single-dopant implantation requires that the system be maintained at a low pressure, greater than 10^{-6} torr. The danger is that any residue molecules in the system (such as air) can become accelerated and end up in the wafer surface. Either oil-diffusion or cryogenic high-vacuum pumps are employed to reduce the pressure. The operation of these systems is described in Chap. 12.

Beam focus

Upon exiting the acceleration tube, the beam separates due to repulsion of like charges. The separation (or defocus) causes uneven ion density and nonuniform layers in the wafer. For successful implantation the beam must be focused. Electrostatic or magnetic lenses are used to focus the ions into a 1-cm-diameter beam.

Neutral beam trap

Despite the vacuum removal of the majority of the air in the system, there are still some residual gas molecules in the vicinity of the ion beam. Collisions between the ions and the residual gas atoms result in a neutralization of the dopant ion.

$$P^+ + N_2 \rightarrow P^0 \text{ (neutral)} + N_2^+$$

In the wafer, the "neutrals" cause nonuniform doping and, since they

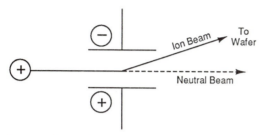

Figure 11.34   Deflection of ion beam from neutral beam.

cannot be "counted" by evaluation equipment, they result in incorrect counting of the amount of dopants in the wafer. Suppression of the neutral beam is accomplished by bending the ion beam (Fig. 11.34) with electrostatic plates, leaving the neutral beam to travel straight ahead away from the wafers.

Beam scanning

Given a 1-cm-diameter beam and a 4- to 8-in-diameter wafer, a procedure must be provided to uniformly dope the entire wafer. Three methods are used: beam scanning, mechanical scanning, shuttering, alone or in combination.

A beam-scanning system (Fig. 11.35) has the beam pass between a number of electrostatic plates. Negative and positive charges can be controllably changed on the plates to attract and repel the ionized beam. By manipulating the charges in two dimensions, the beam can be swept across the entire wafer surface in a raster scan pattern.

Beam sweeping is used primarily in medium-current machines for single-wafer implanting. The procedure is fast and uniform. A drawback is the requirement that the beam be moved completely off the wafer to make the turns. For a large-diameter wafer, this procedure can lengthen the implant time by 30 percent or more. Another problem occurs with high-current machines where the high density of ions causes discharges (called *space charge forces*) that destroy the electrostatic plates.

Mechanical scanning approaches the scanning problem by holding the beam in one position and moving the wafer(s) in front of it. Mechanical scanning is used primarily on high-current machines. One advantage is that there is no wasted time to turn the beam and the beam speed is constant. A number of different designs are used (Fig. 11.36). The procedure is essentially a batch process which is efficient but there are higher maintenance and alignment chores. If the wafer is at an angle to the beam, nonuniform implant depths can result. Beam

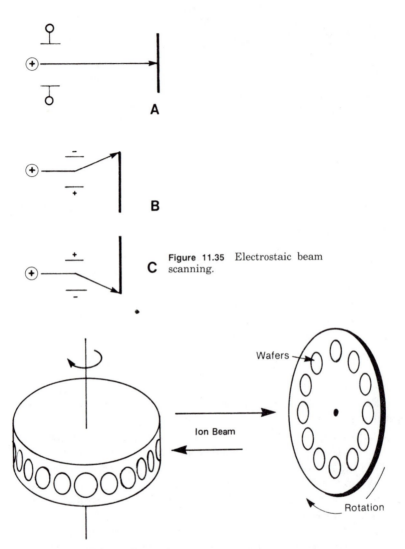

Figure 11.35 Electrostaic beam scanning.

Figure 11.36 Mechanical scanning.

shuttering employs either an electronic field or a mechanical shutter to turn the beam on when it is on the wafer and off when it is not.

Target chamber

The actual implantation takes place in the target chamber, sometimes called an end chamber. It includes the scanning system and the entrance and exit mechanisms. There are several tough requirements

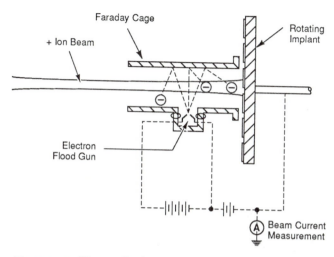

Figure 11.37 Electron flood gun.

for end stations. The wafers must be entered, loaded on the holder, implanted, removed from the holder, and removed from the station with a very low breakage rate. These mechanical motions take longer than the implant itself and represent the focus for improving wafer throughput time.

Wafer breakage can cause contamination from wafer chips and dust, which in turn requires time-consuming cleaning. Contamination on the wafers causes shadowing that blocks the ion beam. Production speed must be maintained with a system that quickly evacuates the chamber for implanting and returns it to room pressure for exit and reloading. The target chamber may house a detector (called a *Faraday cup*) to "count" the number of ions impacting the surface. These detectors can automate the process by allowing beam contact with the wafer until the correct dose is achieved.

High-current implantation can cause the wafer to heat up, and these machines often have cooling mechanisms on the wafer holders. These machines may also have an *electron flood gun* (Fig. 11.37) designed to minimize a buildup of charge on the wafer surface that can electrostatically attract contamination.

Ion Implant Masks

A major advantage of ion implantation is the variety of masks that are effective blocks to the ion beam. In diffusion processing the only effective mask is silicon dioxide. Most films employed in the semiconductor process can be used to block the beam, including photoresist, silicon dioxide, silicon nitride, aluminum, and other thin metal films.

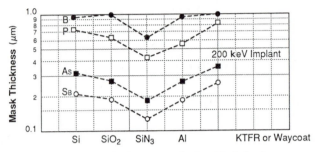

Figure 11.38 Barrier thickness required to block ion beam.

Figure 11.38 compares the thicknesses required to block a 200-keV implant for various dopants.

The use of resist films as a beam block rather than an etched opening in an oxide layer offers the same dimension control advantage as the lift-off process; the etch step and its variability are eliminated. Use of resist layers also is more productive. Options to the use of silicon dioxide increase overall yield by minimizing the number of heating steps the wafers undergo.

Atomic Distribution in Implanted Regions

The distribution of the ions in the wafer surface is different than the distribution after a diffusion process. The number and location of the dopant atoms in a diffusion process is determined by the diffusion laws and time and temperature. In an ion implant process, the number of atoms (dose) implanted is determined by the beam current density (ions per square centimeter) and the implant time.

The location of the ions in the wafer is a function of the incoming energy of the ions, the orientation of the wafer, and the stopping mechanism of the ion. The first two factors are physical ones. The heavier the incoming ion and/or the higher its energy, the deeper into the wafer it will move. The wafer orientation influences the stopping position because different crystal planes have different atom densities and the incoming ions are stopped by the wafer atoms.

Within the wafer the ions are slowed and stopped by two mechanisms. The positive ions are slowed by electronic interactions with the negatively charged electrons in the crystal. The other interaction is the physical collision with the nucleus of the wafer atoms. All of the stopping factors are variable; the ions have a distribution of energies, the crystal is not perfect, and the electronic interactions and collisions vary. The net result is that the ions come to rest over an area in the wafer (Fig. 11.39). They are centered about a depth called the *pro-*

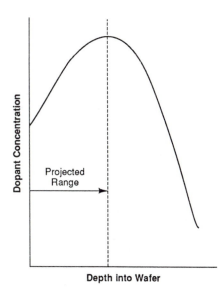

Figure 11.39 Dopant concentration profile after ion implant.

Depth into Wafer

jected range and fall off in density on each side of it. Projected ranges for different dopants are shown in Fig. 11.40. The mathematical shape of the ion distribution is a gaussian curve. A junction between the implanted ions and the bulk doping in the wafer takes place where the ion concentration equals the bulk doping concentration.

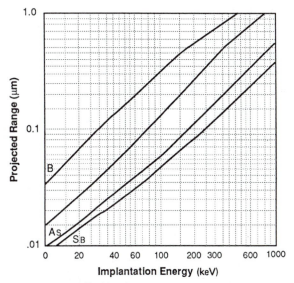

Figure 11.40 Projected range of various dopants in silicon. (*After Blanchard, Trapp, and Shepard.*)

Crystal Damage

During the process of implantation the wafer crystal structure is damaged by the colliding ions. There are three types of damage: lattice damage, damage cluster, and vacancy-interstitial.[5]

Lattice damage occurs when the ions collide with host atoms and displace them from their lattice site. A damage cluster occurs when displaced atoms in turn displace other substrate atoms, creating a cluster of displaced atoms. The most common implant-produced defect is a vacancy-insterstitial. This defect comes about when an incoming ion knocks a substrate atom from a lattice site and the displaced atom comes to rest in a nonlattice position (Fig. 11.41).

Light atoms, such as boron, produce a small number of displaced atoms. The heavier atoms, phosphorus and arsenic, generate a large number of displaced atoms. With prolonged bombardment, the regions of dense disorder may change to an amorphous (noncrystal) structure. In addition to the structural damage to the wafer from ion implantation, there is an electrical effect. The damaged regions do not have the required electrical characteristics because the implanted atoms do not occupy lattice sites.

Annealing

Restoration of the crystal damage and electrical activation of the dopants can be achieved by a thermal heating step. The temperature of the anneal is below the diffusion temperature of the dopant to prevent lateral diffusion. A typical anneal in a tube furnace will take place between 600 and 1000°C in a hydrogen atmosphere.

RTP techniques are also used for postimplant annealing. RTP offers fast surface heating that restores the damage without the substrate temperature rising to the diffusion level. Additionally, the anneal can take place in seconds while a tube process takes 15 to 30 minutes. If the wafer has a heavy subsurface amorphous layer, the damage may be restored with a second implant of a light atom, such as oxygen or neon.

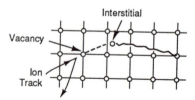

Figure 11.41 Vacancy-interstitial damage mechanism.

Channeling

The crystalline structure of the wafer presents a problem during the ion implantation process. The problem comes about when the major axis of the crystal wafer is presented to the ion beam. Ions can travel down the channels, reaching a depth as much as 10 times the calculated depth. An ion concentration profile of a channeled cross section (Fig. 11.42) shows an significant amount of additional dopants. Channeling is minimized by several techniques: a blocking amorphous surface layer, misorientation of the wafer, and creating a damage layer in the wafer surface.

The usual blocking amorphous layer is simply a thin layer of grown silicon dioxide (Fig. 11.43). The layer randomizes the direction of the ion beam so that the ions enter the wafer at different angles and not directly down the crystal channels. Misorientation of the wafer 3 to 7° off the major plane also has the effect of preventing the ions from entering the channels (Fig. 11.44). Predamaging the wafer surface with

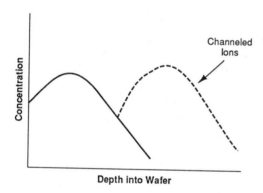

Figure 11.42   Effect of channeled ions on total dose.

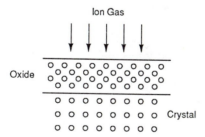

Figure 11.43   Implant through an amorphous oxide layer.

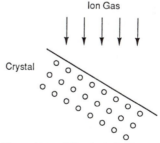

Figure 11.44   Misorient the beam direction to all crystal axes.

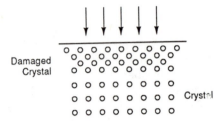

Figure 11.45 Predamage on the crystal surface.

a heavy silicon or germanium implant creates a randomizing layer in the wafer surface (Fig. 11.45). The method increases the use of the expensive ion-implant machine. Channeling is more of a problem with low-energy implants and heavy ions.[6]

Evaluation of Implanted Layers

The evaluation of implanted wafers is essentially the same as for diffused layers. Four-point probes are used to determine the sheet resistance of the layer. Spreading resistance techniques and capacitance-voltage techniques determine the concentration profile, dose, and junction depth. Junction depths can also be determined by bevel and decoration methods. These procedures are explained in Chap. 14.

For implanted layers, a special structure, called the *Van Der Pauw structure*, is sometimes used in place of a four-point probe (Fig. 11.46). The structure allows a determination of sheet resistance without the contact resistance problems of a four-point probe. Variation in the implanted wafers can come from many sources: the beam uniformity, variations in voltage, scanning variations, and problems in the mechanical systems. These potential problems have the possibility of causing a wider sheet resistance surface variation than a diffusion process. To detect and control the sheet resistance across the wafer surface, mapping techniques are popular and, for critical implants, required. Wafer surface maps (Fig. 11.47) are drawn from four-point probe measurements that are computer-corrected for proximity and edge effects.

A measurement technique unique to ion implantation is optical dosimetry. The technique requires spinning a glass disk with a layer

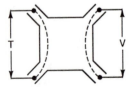

Figure 11.46 Van Der Pauw test pattern.

Run 2   Run 4

Figure 11.47 Four-point probe surface measurement pattern. (*Courtesy of Prometrics.*)

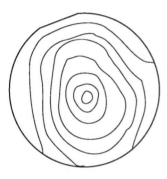

Figure 11.48 Contour map of ion dose.

of photoresist. Before being placed in the ion implanter, the resist film is scanned with a dosimeter that measures the absorption of the film. The information is stored in a computer. The wafer and film receive the same implant as the device wafers. The resist absorbs the ion dose and darkens. After the implant, the film is again scanned. The computer subtracts the before-implant value for each location and prints out a contour map of the surface. The spacing of the contour lines is an indication of the uniformity of the dopants in the surface (Fig. 11.48).

Uses of Ion Implantation

An ion implantation can be substituted for any deposition process. The greater control and lack of side diffusion make it the preferred doping technique for dense and small-feature-size circuits. A predeposition application in CMOS devices is the creation of the deep P-type wells (see Chap. 16).

Perhaps the most important use of ion implantation is for MOS transistor threshold adjustment (Fig. 11.49). An MOS transistor consists of three parts: a source, a drain, and a gate. During operation, a voltage is applied between the source and drain regions. However, no current can flow between the regions until the gate becomes conduc-

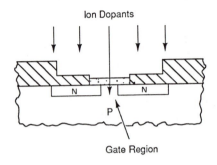

Ion Dopants

N   N

P

Gate Region

Figure 11.49   Ion doping of MOS gate region.

tive. The gate becomes conductive when a voltage applied to it causes a conductive channel to form in the surface and connects the source and drain. The amount of voltage required to first form the connecting channel is called the *threshold voltage* of the device. The threshold voltage is very sensitive to the dopant concentration in the wafer surface under the gate. Ion implantation is used to create the required dopant concentration in the gate region. Also in MOS technology, ion implantation is used to alter the field dopant concentration. However, in this use the purpose is to set a concentration level that prevents current flow between adjacent devices. In this application the implanted layer is part of a device *isolation* scheme.

In bipolar technology, ion implantation is used to create all of the various transistor parts. The "custom" dopant profiles available from implantation can improve device performance. One particular application is arsenic buried layers. When the buried layer is diffused, the high concentration of arsenic atoms affects the quality of a subsequent epitaxial layer deposited on the surface. By using an arsenic ion implant, a high concentration of arsenic is possible and an anneal process restores the damage, allowing a higher-quality epitaxial layer to be deposited.

Resistors for both MOS and bipolar circuits are good candidates for ion implantation. Diffused resistors vary in uniformity from 5 to 10 percent, whereas ion-implanted resistors have variations of less than 1 percent or better.

Summary of Ion Implantation

Since the mid-1970s the uses and applications of ion implantation have grown steadily. For advanced circuit fabrication, it has become the preferred method of doping for the following reasons:

- Precise dose control from 10^{10} to 10^{16} atoms/cm^2
- Uniform topping of large areas
- Dopant profile through energy selection
- Relative ease of implanting all dopant elements
- Minimal side diffusion
- Implanting of nondopant atoms
- Can dope through surface layers
- Choice of dopant barriers for selective doping

Ion implantation also has its drawbacks. The equipment is expensive and complex. Training and maintenance take longer than for comparable diffusion processes. The machines present new dangers in the form of high voltages and more toxic gas use. From the process perspective, the biggest worry is over the ability of the annealing processes to completely eliminate the implant-induced damage. However, despite the drawbacks, ion implantation is the preferred doping process for advanced circuits. In addition, many new structures are only possible with the unique advantages offered by ion implantation.

Key Concepts and Terms

Channeling	Drive-in–oxidation
Closed-tube deposition	Implant analogy
Crystal damage	Implant mask types
Deposition	Implant sources
Diffusion	Implanted dopant profiles
Diffusion dopant profiles	Implanted layer uses
Diffusion process steps	Implanter subsystems
Dopant source states	Ion implantation
Dopant source types	Open-tube deposition

Review Questions

1. Explain the difference between a deposition and a drive-in–oxidation.
2. Name the three types of sources used in a diffusion process.
3. List two evaluation measurements made after the drive-in–oxidation.

4. Is the junction depth deeper after deposition or the drive-in–oxidation?

5. What type of source is BCl_3?

6. Name the subsystems in an ion implantation.

7. In what subsystem is the dopant ion selected?

8. What is the difference between a diffusion and an ion-implanted dopant profile?

9. Name two materials used as an ion implant doping barrier.

10. Why is an anneal required after ion implantation?

References

1. P. B. Griffin and J. D. Plummer, "Advanced diffusion models for VLSI," *Solid State Technology*, May 1988, p. 171.
2. K. T. Robinson, "A guide to impurity doping," *Micromanufacturing and Test*, Apr. 1986, p. 52.
3. P. Guise and R. Blanchard, *Modern Semiconductor Fabrication Technology*, Reston Books, Reston, Va., 1986, p. 46.
4. P. Burggraaf, "Ion implanters: major trends," *Semiconductor International*, Apr. 1986, p. 78.
5. J. Hayes, P. Van Zant, *Doping Today Seminar Manual*, Semiconductor Services, San Jose, Calif., 1985.
6. D. Zrudsky, "Channeling control in ion implantation," *Solid State Technology*, July 1988, p. 73.

12

Deposition

Overview

One of the important layering techniques is deposition. Many different types of layers are added to the surface by deposition and a variety of deposition techniques are available. This chapter will describe the deposition techniques, layer materials and their use in basic device structures.

Objectives

Upon completion of this chapter, you should be able to:

1. Name the parts of a CVD reactor.
2. Describe the principle of chemical vapor deposition.
3. List the conductor, semiconductor, and insulator materials deposited by CVD techniques.
4. Know the difference between atmospheric CVD, LPCVD, hot wall, and cold wall systems.
5. Explain the difference between epitaxial and polysilicon layers.

Introduction

Much attention is given the advances in photomasking technology that have allowed the fabrication of VLSI circuits. But in the 1950s it was the development of chemically deposited silicon epitaxial layers that allowed the formation of high-performance bipolar transistors and a practical scheme for isolating circuit components from each other (Fig. 12.1). Coincident with the growth in circuit density has been a growth in the number of layers on a chip. Most of the new lay-

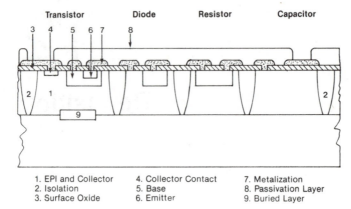

Figure 12.1   Cross section of bipolar circuit showing epitaxial layer and isolation.

ers are deposited layers that function in the devices as either semiconductors, insulators, dielectrics, or conductors. The growth is illustrated by a comparison of an MOS transistor circa 1972 to its 1980s version (Fig. 12.2).

Not surprisingly, the growth in the number and kind of deposited films has resulted in a number of new deposition techniques. Where the process engineer of the 1960s had a choice of only atmospheric chemical vapor deposition (CVD), today's engineer has many more options (Fig. 12.3). (Chapter 12 will detail the basics of film deposition, as well as the techniques used and the films deposited by these important processes. The uses of the particular films, while indicated in this chapter, are detailed in Chaps. 16 and 17.)

Film Parameters

The particular films incorporated into device structures are examined in "Low-Pressure Chemical Vapor Deposition," p. 308. While each film material has a set of specific parameters, there are several general criteria that all films must meet for semiconductor use. Most of the criteria come about from the choice of source chemistries used and the system design and operating parameters.

Thickness control is of paramount importance, especially as the films are getting thinner. Epitaxial films have shrunk from 5-μm levels to submicron thicknesses. Surface flatness is as important as the thickness. In Chap. 10 the effect of steps and surface roughness on image formation was detailed. Deposited films must be flat and as smooth as the material will allow to minimize steps, cracking, and subsurface reflections.

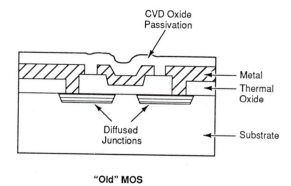

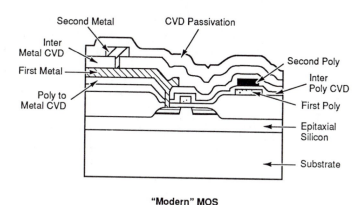

Figure 12.2 Evolution of MOS layers.

Atmospheric Pressure	Low Pressure
Cold wall	Hot wall
• Horizontal	Plasma enhanced
• Vertical	Vertical isothermal
• Pancake	
Hot wall	
Photochemical	
VPE	
MOCVD	

Figure 12.3 Overview of CVD systems.

The films must be of the desired uniform composition. Some of the reactions are complex and it is possible that films will be deposited with different than the intended composition. The process engineer will make stoichiometric calculations to determine the proper amount of reactants required in the chamber to create the film composition.

Stoichiometry is the methodology by which the quantities of reactants and products in chemical reactions are determined. Stress-free films are another requirement. A stressed film can relieve its stress while on the wafer surface and crack or cause cracking of a second film deposited on top of it. Cracked films cause surface roughness and can allow contamination to pass through to the wafer.

Purity, that is, no unwanted chemical elements or molecules in the film, is required for the film to carry out its intended function. For example, oxygen contamination of an epitaxial film will change its electrical properties. Purity also includes the exclusion of mobile ionic contaminants and particulates. On surfaces with existing steps, it is imperative that the deposited film lay over the step without thinning or cracking. Failure of deposited films over steps is responsible for poor current flow in conductors and poor dielectric properties. Last, but not least, deposited films must be continuous and free of pinholes to prevent the passage of contamination and to prevent shorting of sandwiched layers.

Chemical Vapor Deposition

The deposition of the epitaxial layers on the earlier devices was by a process known as chemical vapor deposition (CVD). At the time CVD shared deposition duties with the evaporation of aluminum films. CVD technology has expanded into a number of system designs, generally pushed by the needs to achieve the film qualities and the shrunken dimensions needed for VLSI circuit fabrication.

CVD basics

Thus far the terms *deposition* and *CVD* have been used without explanation. In semiconductor processing, deposition refers to any process in which a material is deposited intact on the wafer surface, as opposed to the formation of films by surface chemical reaction such as silicon dioxide. The majority of films are deposited by CVD techniques. In concept the process is simple (Fig. 12.4). Chemicals containing the atoms required in the final film are mixed and reacted in a deposition chamber. The reaction forms the proper film elements or molecules in a vapor state. The elements or molecules deposit on the wafer surface and build up to form a film. Figure 12.4 illustrates the reaction of silicon tetrachloride ($SiCl_4$) with hydrogen to form a deposited layer of silicon on the wafer. Generally, CVD reactions require the addition of energy to the system, such as heating the chamber or the wafer.

The chemical reactions that take place fall into the four categories of

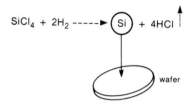

$$SiCl_4 + 2H_2 \longrightarrow \boxed{Si} + 4HCl \uparrow$$

wafer

Figure 12.4 Chemical vapor deposition of silicon from silicon tetrachloride.

Pyrolysis	$SiH_4 = Si + 2H_2$
Reduction	$SiCl_4 + 2H_2 = SI + 4HCL$
Oxidation	$SiH_4 + O_2 = SiO_2 + 2H_2$
Nitridation	$3SiH_2Cl_2 + 4NH_3 = Si_3N_4 + pH + 6H_2$

Figure 12.5 Examples of CVD reactions.

pyrolysis, reduction, oxidation, and nitridation (Fig. 12.5). Pyrolysis is the process of chemical reaction driven by heat alone. Reduction causes a chemical reaction by reacting a molecule with hydrogen. Oxidation is the chemical reaction of an atom or molecule with oxygen. Nitridation is the chemical process of forming silicon nitride.

Deposited film growth proceeds in several distinct stages. The first stage, *nucleation*, is very important and critically dependent on substrate quality. Nucleation occurs as the first few atoms or molecules deposit on the surface. These first atoms or molecules form islands that grow into larger islands. In the third stage the islands spread, finally coalescing into a continuous film. This is the transition stage of the film growth with a typical thickness of several hundred angstroms (Fig. 12.6). The transition region film has chemical and physical properties much different than the final, thicker "bulk" film.[1]

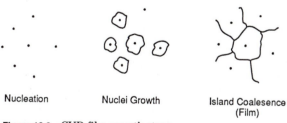

Nucleation   Nuclei Growth   Island Coalesence
(Film)

Figure 12.6 CVD film growth steps.

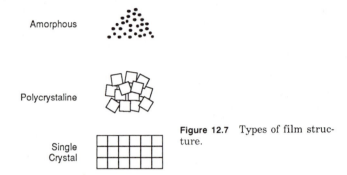

Amorphous

Polycrystaline

Single
Crystal

Figure 12.7 Types of film structure.

After the transition film is formed, the growth of the bulk begins. Bulk films can take three forms: amorphous, polycrystalline, and single crystal (Fig. 12.7). These terms have been defined before. In general a process is set up to produce a particular film structure. A poorly defined or controlled process can result in a film ending up with the wrong structure. For example, attempting to grow a single-crystal epitaxial film on a wafer with islands of unremoved oxide will result in regions of polysilicon in the bulk film.

Basic CVD system design

CVD systems come in a wide variety of designs and options. Understanding the many variations is helped by an examination of the basic subsystems required in all CVD systems (Fig. 12.8). In most respects, a CVD system has the same basic parts as a tube furnace. In some cases the entire CVD system is a tube furnace identical to those used for oxidation and diffusion. The source chemicals are housed in a source section. Vapors are generated from pressurized gas cylinders or liquid source bubblers. Gas flow control is maintained by pressure regulators and mass flow meters.[2]

A *mass flow meter* is preferred in place of a standard flow meter for its inherent superior control. Stoichiometric considerations require that the same amount of material, as measured by its mass, be delivered into the reaction chamber. Standard flow meters measure the volume of material, and equal volumes of the same material may contain different amounts of material due to pressure and temperature differences. Semiconductor processes use a thermal-type of mass flow meter. The system consists of a heated gas passage tube with two temperature sensors. When no gas is flowing, the temperature sensors are at the same temperature. With the introduction of a gas flow the downstream sensor reads higher. The difference between the two sen-

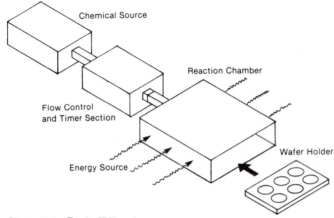

Figure 12.8 Basic CVD subsystems.

sors is related to the amount of heat mass (not volume) that has moved downstream. The meter has a feedback mechanism to control the gas flow such that a steady amount of material flows through the meter. The source section also contains the microprocessor-controlled valves that meter the gas into the reaction chamber in the right sequence for the right amount of time.

The actual deposition takes place in a reaction chamber which serves the dual role of isolating the reaction from the atmosphere and holding the wafers. In most systems the reaction chamber or the wafers inside are supplied energy as required by the reaction. Energy sources are either heat (conduction-convection), induction RF, radiant, plasma, or ultraviolet. The particular source is chosen for particular reasons as explained in the sections on particular systems. Temperatures range from room temperature to 1250°C depending on the reaction, film thickness required, and the growth parameters.

The fourth basic part of the system is the wafer holder. Different chamber configurations and heat sources dictate the style and material of the holders. Most production-level systems for VLSI work are automated, with the "one-button" hands-off design preferred.

CVD process

A CVD process follows the same steps as an oxidation or diffusion process. For review, the steps are preclean (and etch, if required), deposition, and evaluation. Cleaning processes are those already described to remove particulates and mobile ionic contaminants. Evaluation of the films are for thickness, purity, cleanliness, and composition. Evaluation techniques are explained in Chap. 14.

Systems Overview

CVD systems (Fig. 12.3) are divided into two primary types: atmospheric pressure (AP) and low pressure (LP). Within each type a system may be either cold wall or hot wall. Cold-wall systems use induction or radiant heating (see "Continuous-Conduction-Heated APCVD," p. 307). to directly heat the wafer holder or wafers through a quartz reaction chamber. Hot-wall types supply the reaction energy by a method that heats the wafers, the wafer holder, and the chamber walls. The advantage of cold-wall CVD is that the reaction occurs only at the heated wafer holder. In a hot-wall system the reaction occurs throughout the chamber, leaving reaction products on the inside chamber walls. The reaction products build up, necessitating rigorous and frequent cleaning to prevent contaminating the wafers. In some systems, a plasma field is created in the chamber to enhance the film deposition. The plasma adds additional energy to the reaction, resulting in more uniform depositions and/or lower temperatures. These are called *plasma-enhanced* CVD systems, or PECVD.

Atmospheric Pressure Systems

As the name implies, atmospheric CVD system reactions and deposition take place at atmospheric pressure. There are a number of system designs that fall under this category.

Horizontal tube–induction-heated APCVD

The first use of CVD was for the deposition of silicon epitaxial films for bipolar devices. The same basic system is still used today for epitaxial deposition (Fig. 12.9). It is essentially a horizontal tube furnace, but with some significant differences. First the tube has a

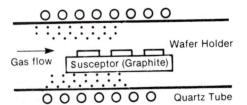

Figure 12.9 Cold-wall induction APCVD with horizontal susceptor.

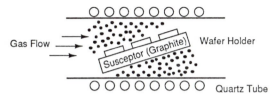

Figure 12.10   Cold-wall induction APCVD with tilted susceptor.

square cross section. The major difference, however, is in the heating method and wafer holder.

Wafers are arranged on a flat graphite slab and positioned in the tube. Surrounding the tube are copper coils that are connected to an RF generator. The RF waves traveling in the coils pass through the quartz tube and the flowing gas in the tube without heating them. This is the cold-wall aspect of the system. When the radiant waves reach the graphite wafer holder, they "couple" with the molecules of the holder causing the graphite to heat up. This heating method is called *induction*.

The heat of the holder is passed to the wafers by conduction. The film deposition takes place at the wafer surface (and at the holder surface). One problem with this type of system is downstream depletion of the reactants in the gas flow. The problem arises from the laminar nature of the flow in the tube. Laminar flow is required to minimize turbulence. But as the reaction takes place in the plane directly above the wafers, that plane is depleted of reactants. This in turn results in successively thinner films along the wafer holder. The standard fix for the problem is to tilt the holder in the tube so that each line of wafers is exposed to a different layer of the gas stream (Fig. 12.10).

Barrel radiant–induction-heated APCVD

Large-diameter wafers have imposed uniformity and productivity limits on horizontal induction-heated tube systems. Larger-diameter wafers laid horizontally on a holder have a low packing density. And loading more wafers on a larger holder strains the uniformity capabilities of the system.

The solution to these problems came with the development of the barrel radiant-heated system (Fig. 12.11). The reaction chamber of the system is a cylindrical stainless steel barrel with high-intensity quartz heaters placed about the inside surface. The wafers are placed on a graphite holder that rotates in the center of the barrel. The rotation of the wafers in the reaction gases in the barrel produces a more uniform film thickness compared to the horizontal systems.

Gas Flow

Graphite Wafer Holder

Radiant Heaters

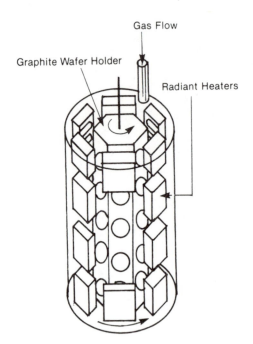

Figure 12.11 Cylindrical or barrel system.

Radiant heat from the lamps heats the wafer surface, causing the deposition reaction. While some heating of the chamber walls occurs, the system is close to a cold-wall deposition. The nature of the radiant heat produces a very controlled and even film growth. In an induction-heated system, the wafers are heated from the bottom, and as the film grows there is some small but measurable drop in temperature at the film surface. In the barrel system the wafer surface is always facing the lamps and receives a more uniform film growth rate.

In 1987, Applied Material introduced a jumbo barrel system for large-diameter wafers featuring an induction heating system.[3] A principal advantage of the barrel reactor is an increased productivity based on the increased number of wafers per cycle. This design is the current production standard for deposition of epitaxial silicon in the 900 to 1250°C range.

Pancake induction-heated APCVD

The pancake or vertical-flow APCVD system has been a favorite for small fabrication lines and R&D labs (Fig. 12.12). The wafers are held on a rotating holder of graphite and heated by induction-conduction from an RF coil below the holder. The reaction gases are fed through a tube exiting above the wafers. Vertical gas flow offers the advantage of a continuous supply of fresh reactants to the wafers, thus minimiz-

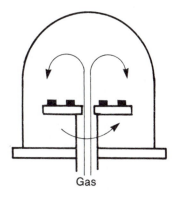

Figure 12.12 Rotating pancake APCVD.

ing downstream depletion. The combination of the rotation and vertical flow of the gases produces good film uniformity. Productivity in smaller systems is restricted, as in the horizontal tube system, by the number of wafers that the pancake can accommodate.

A production-level variation of the pancake design is produced by Gemini Research. The reactor features radiant-resistance heating and a large-capacity holder with robot autoloading.[4]

Horizontal conduction-heated APCVD

The horizontal conduction-heated APCVD system (Fig. 12.13) is also one of the original CVD designs. These systems were used to deposit the first silicon dioxide passivation layers from silane. The reaction chamber is a stainless steel box with inlets for the gases. The wafers are mounted on a movable hot plate and placed in the chamber. The hot plate heats the wafers and chamber walls, which classifies the technique as a hot-wall system.

Continuous-conduction-heated APCVD

A production-level variation on the horizontal conduction-heated APCVD system features a heated hot plate wafer holder that moves

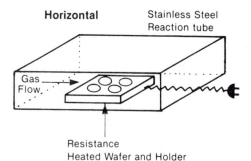

Figure 12.13 Hot-plate APCVD.

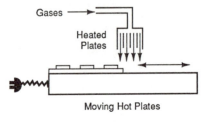

Figure 12.14 Moving hot-plate APCVD with shuttle.

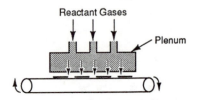

Figure 12.15 Continuous hot-plate APCVD.

back and forth (Fig. 12.14) under a series of nozzles that dispense a vapor of the desired material. The reaction takes place in a mixing chamber above the nozzles. Another version of the system (Fig. 12.15) has a continuous belt that is heated from below by a hot plate.

Low-Pressure Chemical Vapor Deposition (LPCVD)

Uniformity and process control within atmospheric pressure CVD systems rely on temperature control and the flow dynamics in the system. A factor influencing film uniformity and step coverage is the mean free path of the molecules in the reaction chamber. The mean free path is the average distance a molecule will travel before colliding with an object in the chamber be it another molecule, the wafers, or wafer holder. Collisions change the direction of the particles. The longer the mean free path, the higher the uniformity of the film deposition. A major determiner of the length of the mean free path is the pressure in the system. Lowering the pressure in the chamber increases the mean free path and the film uniformity. Decreasing the pressure also allows a lowering of the deposition temperature.

These benefits became available to the industry in 1974 when Unicorp, under license to Motorola Inc., introduced an LPCVD system that operated at a few hundred millitorr.[5] The complete list of LPCVD system advantages includes:

Lower chemical reaction temperature

Good step coverage and uniformity

Vertical loading of wafers for increased productivity and lower exposure to particles

Less dependence on gas flow dynamics

Less time for gas phase reaction particles to form

Can be performed in standard tube furnaces

A vacuum pump must be added to the system to reduce the pressure in the chamber. A discussion of the types of pumps used with LPCVD systems is in Chap. 13.

Horizontal conduction–convection-heated LPCVD

A standard production-level LPCVD system is an adaption of the horizontal tube furnace (Fig. 12.16). The system is a standard horizontal tube system with three major exceptions. The tube is connected to a vacuum pump that pulls the system down to a pressure range of 0.25 to 2.0 torr.[6] A second change is a ramping of the temperature in the center zone to offset reaction depletion down the tube. The third change may be special injectors at the gas inlet end to improve gas mixing and deposition uniformity. In some systems, the injectors positioned in the deposition zone directly over the wafers. Disadvantages of this system design are particles formed on the inside wall surface (hot-wall reactions), uniformity along the tube axis, the use of cages around the wafers to minimize particle contamination, and the higher downtime required for frequent cleaning.

These systems are most often used for polysilicon, silicon dioxide, and silicon nitride films with typical thickness uniformities of ±5 percent. The primary deposition variables are temperature, pressure, gas flow, gas partial pressure, and wafer spacing. These variables are carefully balanced for each deposition process. The deposition rates are somewhat lower (100 to 500 Å/min) than AP systems, but productivity is enhanced by the vertical wafer-loading densities that can approach 200 wafers per deposition.

Vertical isothermal LPCVD

A novel LPCVD system was introduced by Anicon in the mid-1980s. Called a *vertical isothermal reactor* (Fig. 12.17), the reaction takes

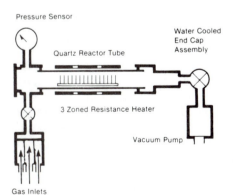

Figure 12.16 Horizontal hot-wall LPCVD system.

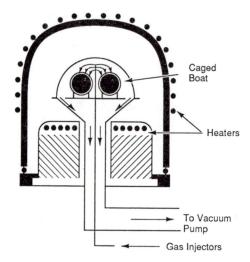

Figure 12.17 Vertical flow isothermal LPCVD. *(Courtesy of Anicon/SVG.)*

Labels in figure: Caged Boat, Heaters, To Vacuum Pump, Gas Injectors

place in a reduced-pressure bell-jar-type reaction chamber. The wafers, which are held vertically in a quartz boat, are heated from below, creating an isothermal (uniform) temperature at each wafer. The system features vertical gas flow.

Gases are metered separately into a tube that allows the mixing just above the wafers. The deposition takes place in the vertical plane. Film uniformity is improved since there is no downstream depletion of the gas and there is a continuous supply of equal amounts of "fresh" reactants. Reported uniformities are typically ±2 percent. Wafer cleanliness is improved since the gas supply is fresh and the cage protects the wafers from reaction particles. Cleanliness is also improved because the wafers are loaded from the bottom and not through the deposition zone.

Plasma-enhanced CVD (PECVD)

The desire to replace silicon dioxide metal protection layers with silicon nitride led to the development of PECVD techniques. The problem came from the approximate 660°C temperature required for a thermal CVD silicon nitride process. This temperature causes enough problems (see Chap. 13) with underlying aluminum layers to be unusable. The answer to this problem was a plasma enhancement of the deposition energy to allow a reduction temperature under the 450°C level for deposition over aluminum layers. The plasma in such a system is established by a radiofrequency-induced glow discharge. PECVD depositions are performed at low pressure and lower temperature with the combination of the two features providing good film uniformity and throughput.

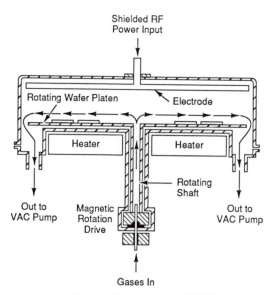

Figure 12.18   Vertical-flow pancake PECVD.

PECVD reactors have the capability of also using the plasma for etching and cleaning the wafer prior to the deposition step. This step is the same as the dry etch processes described in the patterning chapter. This in situ cleaning prepares the deposition surface, eliminating the problem of added contamination picked up during the loading step.

Horizontal vertical flow PECVD. This system follows the design of a bottom-heated pancake vertical-flow CVD. The plasma region is created in the top of the chamber by a radiofrequency (rf) feed to an electrode (Fig. 12.18). Wafer heating comes from radiant heaters mounted below the wafer holder, creating a cold-wall deposition system. With PECVD systems, there are several more critical parameters to control. In addition to those in a standard LPCVD reactor, RF power density, the RF frequency, and the duty cycle are added. Film deposition speed is generally increased, but it must be controlled to prevent film stress and/or cracking.

The need for close parameter control is addressed in the single-wafer chamber PECVD system (Fig. 12.19) where the chamber is small and each successive wafer is exposed to identical conditions. Single-wafer systems are generally slower than batch processes. The productivity trade-off with the larger-chamber batch machines lies in fast mechanisms to feed wafers in and out of the chamber and how quickly the vacuum can be established and released. Load-lock systems enhance the productivity by moving the wafers into an ante-

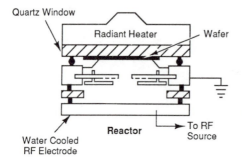

Quartz Window
Radiant Heater   Wafer

Reactor   To RF Source

Water Cooled RF Electrode

Figure 12.19 Single-chamber planar PECVD.

chamber, pumping it down to the required pressure, and moving the wafers into the deposition chamber as it becomes available.

Barrel radiant-heated PECVD.   This system is a standard barrel radiant-heated system with low pressure and plasma capabilities. It is favored for the deposition of tungsten silicide.

Horizontal-tube PECVD.   This system is based on a standard tube furnace design. The plasma region is created in the specially designed wafer boat. The boat consists of slabs of graphite, alternately connected to an RF power supply (Fig. 12.20). Each pair of slabs is in effect a mini-parallel-plate plasma generator. Wafers are placed between the plates and the deposition gases passed over and through the holder where the localized plasma enhances the deposition.

Primary advantages of this approach to PECVD are the economic savings from adapting a tube furnace and high productivity. Disad-

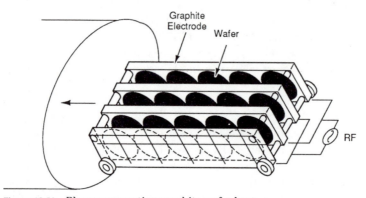

Graphite Electrode   Wafer

RF

Figure 12.20 Plasma generating graphite wafer boat.

vantages are in the more difficult control of the film stoichiometry and the large, heavy wafer carrier.

Photochemical (Ultraviolet) LPCVD

Photochemical systems utilize an ultraviolet light source to dissociate the reactant gases into the desired film material. The major advantage of the photochemical LPCVD approach is a lower reaction temperature. These systems can deposit both silicon dioxide and silicon nitride at temperatures between 50 to 200°C. The lower temperature can result in a deposited film with fewer stress cracks and very good step coverage and with less thermal damage to the wafer surface. Heating of the wafers is provided by a hot plate and operates the system at the lower end of the LPCVD pressure range, 0.3 to 1.0 torr. While offering advantages to the CVD arsenal, these systems are still under development and are not in use as a mainstream LPCVD technique. Figure 12.21 is a summary table of CVD systems by temperature range.

Deposited Films

The types of films deposited by CVD techniques are divided into their electrical classifications of semiconductors, dielectrics, and conductors. In the following sections the deposition of the various films is examined. For each film, its principal use(s) in semiconductor devices is presented along with the particular film properties for the described use. The uses are treated in general. (For a more detailed explanation of their roles in particular devices, see Chap. 16.)

Deposited semiconductors

So far in this text we have discussed the nature of semiconductor materials, their important doping characteristics, and the formation of wafers as the base of semiconductor devices and circuits. There are

Level	Temperature Range	Methods
High Temp.:	600–1250°C	R.F. Induction (Cold Wall) Radiant Heat (Cold Wall) Resistance Coils (Hot Wall)
Mid Temp.:	200–600°C	Hot Plates Plasma Enhanced LPCVD
Low Range:	22–200°C	Hot Plates P.E. CVD Photochemical

Figure 12.21 Summary table of CVD methods.

several drawbacks to crystal quality, doping ranges, and doping control achievable in wafer form. Early on these factors placed a limit on the fabrication of high-performance bipolar transistors. The development of a deposited silicon layer, called an *epitaxial layer*, is one of the major advances in the industry. Epitaxial layers were a part of the technology as early as 1950.[7] Since then deposited silicon layers have found additional uses in advanced bipolar device design and as a high-quality base for CMOS circuits. The technology has expanded to include the deposition of gallium arsenide and other 111-1V films.

Epitaxial silicon

The term *epitaxial* comes from the Greek word meaning "arranged upon." In silicon layer deposition it refers to the single crystalline structure of the film. The structure comes about when silicon atoms are deposited on a bare silicon wafer in a CVD reactor (Fig. 12.22). When the chemical reactants are controlled and the system parameters set correctly, the depositing atoms arrive at the wafer surface with sufficient energy to move around on the surface and orient themselves to the crystal arrangement of the wafer atoms. Thus an epitaxial film deposited on a $\langle 111 \rangle$-oriented wafer will take on a $\langle 111 \rangle$ orientation.

If, on the other hand, the wafer has a silicon dioxide layer, an amorphous surface layer, or contamination and is put in the same reactor under the same conditions, the depositing atoms have no structure to align to. The resulting film structure is polysilicon. This condition is useful for some applications, such as MOS gates, and is unwanted if the goal is to grow a single-crystal film structure.

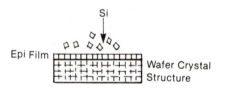

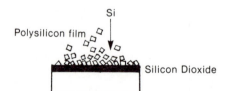

Figure 12.22 Epitaxial and polysilicon film growth.

Silicon tetrachloride	$SiCl_4 + 2H_2 \leftrightarrow Si + 4HCl$
Silane	$SiH_4 + heat \rightarrow Si + 2H_2$
Dichlorosilane	$SiH_2Cl_2 \leftrightarrow Si + 2HCl$

Figure 12.23   Epitaxial silicon chemical sources.

Silicon tetrachloride source chemistry.   A number of different sources are used for the deposition of epitaxial silicon (Fig. 12.23). Deposition temperature, film quality, growth rate, and compatibility with a particular system are factors in choosing a silicon source. An important process parameter is the deposition temperature. The higher the temperature, the faster the growth rate. Faster growth rates create more crystal defects and film cracking and stress. Higher temperatures also cause higher levels of autodoping and out-diffusion. (These effects are described in the following text.)

The favored source of silicon for deposition is silicon tetrachloride ($SiCl_4$). The reasons are the high formation temperature (growth rate) and the reversible nature of the reaction. Note that in the first reaction in Fig. 12.23 there is a double-headed arrow, which indicates that the reaction creates silicon atoms in one direction and removes (etches) silicon in the other direction. Within the reactor these two reactions compete with each other.

Initially, the silicon surface is etched preparing it for the deposition reaction. In the second stage, the deposition of silicon is faster than the etch, with the net result of a deposited film.

The graph in Fig. 12.24 shows the effect of the two reactions. With an increasing percentage of $SiCl_4$ molecules in the gas stream, the deposition rate first increases. At the 0.1 ratio, the etching reaction starts to dominate and slows down the growth rate. This latter reaction is actually one of the first events in the reactor. Hydrogen chloride (HCl) gas is metered into the chamber where it etches away a thin layer of the silicon surface, preparing it for the silicon deposition.

Silane source chemistry.   The second-most-used silicon source chemistry is silane (SiH_4). Silane offers the advantage of not requiring a second reaction gas. It forms silicon atoms by decomposing when heated. The reaction takes place several hundred degrees lower than a silicon tetrachloride deposition, which is attractive from an autodoping and wafer warping perspective. Also silane does not produce pattern shift (see "Epitaxial Film Quality," p. 317). Unfortunately the reaction occurs at all locations in the system, creating a powdery film inside the reactor which, in turn, contaminates the wafers. Silane finds more use as a source for polysilicon and silicon dioxide depositions.

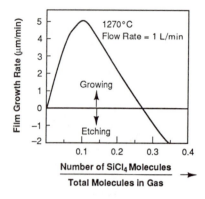

Figure 12.24 Growth-etch characteristic of SiCl$_4$ epitaxial deposition.

Dichlorosilane source chemistry. Dichlorosilane (SiH$_2$Cl$_2$) is also a lower-temperature silicon source that is used for thin epitaxial films. The lower temperature reduces autodoping and solid-state diffusion from previously diffused buried layers and provides a more uniform crystal structure.

Epitaxial film doping. One of the advantages of an epitaxial film is the precise doping and doping range available by the process. Silicon wafers are manufactured in a concentration range of approximately 10^{13} to 10^{19} atoms/cm^3. Epitaxial films can be grown from 10^{12} to 10^{20} atoms/cm^3. The upper limit is close to the solid solubility of phosphorus in silicon.

Doping in the film is achieved by the addition of a dopant gas stream to the deposition reactants. The sources of the dopant gases are exactly the same chemistries and delivery systems used in deposition doping furnaces. In effect the CVD deposition chamber is turned into a doping system. In the chamber the dopants become incorporated into the growing film where they establish the required resistivity. Both N- and P-type films can be grown on either N- or P-type wafers. The classic epitaxial film in bipolar technology is an N-type epitaxial film on a P-type wafer.

Epitaxial film quality. Epitaxial film quality is a prime concern of the process. In addition to the usual considerations over contamination, there are a number of faults specifically associated with epitaxial growth. Contaminated systems can cause a problem called *haze*.[8] Haze is a surface problem that varies from a microscopic disruption to severe cases that are observable as a dull matte finish. Haze comes about from residual oxygen in the reactant gases or from leaks in the system.

Contaminants on the surface at the start of the deposition result in

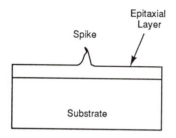

Figure 12.25 Epitaxial growth spike.

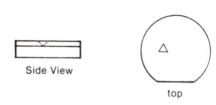

Figure 12.26 Stacking fault on ⟨111⟩ Si.

an accelerated growth known as *spikes* (Fig. 12.25). The spikes can be as high as the film thickness. They cause holes and disruption in photoresist layers and other deposited films.

During the growth a number of crystal problems can occur. One is *stacking faults*. A stacking fault is due to the inclusion of an extra atomic plane with a corresponding "dislocation" of the atoms around the plane. A stacking fault begins at the surface and "grows" to the surface of the film. The shape of the stacking fault depends on the orientation of the film and wafer. Faults in ⟨111⟩-oriented films have a pyramidal shape (Fig. 12.26), while ⟨100⟩-oriented wafers form rectangular-shaped stacking faults. The faults are detected by either x-ray or etching techniques.

A growth problem associated with ⟨111⟩ wafers is *pattern shift*. This problem occurs when the deposition rate is too high and the film planes grow at an angle to the surface. Pattern shift is a problem when alignment to a subsurface pattern relies on locating it from a film surface step (Fig. 12.27). Another major growth problem is *slip*. This condition comes about from poor control of the deposition parameters and results in a "slippage" of the crystal along plane interfaces (Fig. 12.28).

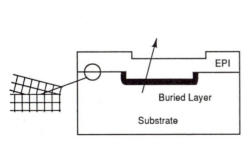

Figure 12.27 Epitaxial pattern shift.

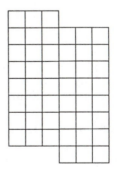

Figure 12.28 Crystal slip.

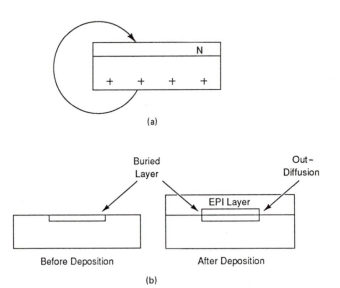

(a)

(b)

Figure 12.29 (*a*) Epitaxial autodoping and (*b*) out-diffusion.

There are two issues associated with the temperature of the deposition: autodoping and out-diffusion. Autodoping of the growing film occurs when dopant atoms from the back of the wafer diffuse out from the wafer (Fig. 12.29), mix in the gas stream, and become incorporated into the growing film. In the film they change the resistivity and the conductivity level. Autodoping in a P-type film, grown over an N-type wafer, will be less P-type than intended and have a lower P-type concentration as the autodoped atoms neutralize a number of the P-type atoms in the film.

Out-diffusion causes the same effect but at the epitaxial layer–wafer interface. The source of the out-diffused atoms is doped regions diffused into the wafer before the epitaxial deposition. In bipolar devices the regions are called *buried layers* or *subcollectors*. In the usual format the buried layer is an N-type region in a P-type wafer, over which is grown an N-type epitaxial layer. During the deposition the N-type atoms diffuse out and become incorporated into the bottom of the epitaxial film, changing the concentration. In the extreme, the buried layer can out-diffuse up into the bipolar device structure causing electrical malfunctions.

CMOS epitaxy. Until the late 1970s, the dominant use of epitaxial films was as the collector region of bipolar transistors. The technique provided a quality substrate for device operation and a clever means of isolating adjacent devices (see Chap. 16). A newer and perhaps

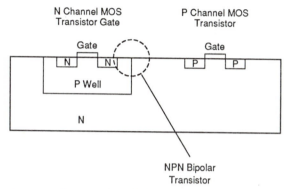

Figure 12.30   Lateral NPN bipolar transistor in CMOS circuit.

more dominant use of silicon epitaxial films is for CMOS circuit wafers. The need for an epitaxial layer was driven by a CMOS circuit problem called *latch-up*. Figure 12.30 shows a cross section of a portion of a CMOS chip. The side-by-side MOS transistors form a lateral bipolar transistor. During circuit operation the lateral bipolar transistor can function as an unwanted amplifier, increasing its output to a point where it causes the memory cell to go into a state where it cannot be switched. This condition is known as *latch up*. In this state the cell cannot be addressed for its information. The solution to preventing latch-up is a low-resistivity epitaxial layer that shunts (shorts) the emitter of the bipolar transistor so that it does not "turn on." This technique is now a standard one and CMOS epitaxial deposition is a fast-growing part of the industry.

Epitaxial process.   A typical epitaxial process starts with a complete and rigorous cleaning of the wafer surface. Within the deposition chamber a number of steps take place to correctly deposit the film. A typical $SiCl_4$ epitaxial process is shown in Fig. 12.31. The first several steps are a gas-phase cleaning of the wafer surface in preparation for the actual deposition. Deposition follows the cleaning with a cooldown cycle at the end. During all the steps control of the temperature and gas flows is critical.

Selective epitaxial silicon

Advancement in epitaxial deposition systems has introduced the selective growth of epitaxial films. Whereas the epitaxial films for bipolar and CMOS substrates are deposited on the entire wafer, in selective growth they are grown through holes in either silicon dioxide or

Cycle	Temperature	Gas	Purpose
1	Room	N_2	Purge air from system
2	Room	H_2	Reduce any organic contaminants of wafers in system
3	(Heating)	N_2	Bring system to deposition temperature
4	Deposition Temperature	HCl	Etch wafer to prepare surface for epi deposition
5	Deposition Temperature	Source + Dopant + Carrier	Grow epitaxial film
6	(Heat Off)	N_2	Purge system of reactant gases

Figure 12.31 Typical $SiCl_4$ epitaxial deposition process.

silicon nitride films. The wafer is positioned in the reactor chamber and the epitaxial film grows directly on the silicon exposed at the bottom of the hole (Fig. 12.32). As the film grows, it takes on the crystal orientation of the underlying wafer. An advantage of such a structure is that devices formed in the surfaces of the epitaxial regions are isolated from each other by the oxide or nitride regions.

If the deposition is allowed to continue onto the isolating surface, the structure of the film switches to a polysilicon structure. Another outcome of extended deposition is that the overlaying deposited layer becomes entirely epitaxial in nature. All of these outcomes add attractive structure options for advanced device design.

Molecular beam epitaxy (MBE)

Deposition rate control, low deposition temperature, and controlled film stoichiometry are always goals in film-deposition systems. Molecular beam epitaxy (MBE) has emerged out of the laboratory to claim production status as these issues have become more important. MBE is an evaporation rather than a CVD process. The system consists of a deposition chamber (Fig. 12.33) that is maintained at a low pressure to 10^{-10} torr. Within the chamber is one or more cells (called *effusion cells*) that contain a very pure sample of the target material desired on

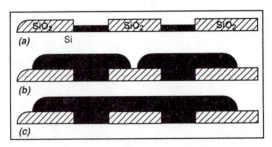

Figure 12.32 Steps in selective epitaxial growth.

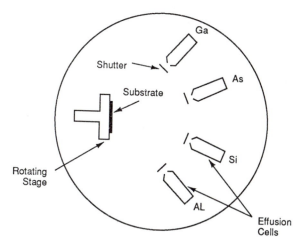

Figure 12.33 Diagram of MBE deposition system.

the wafer. An electron beam[9] is directed into the center of the target material, which it heats to the liquid state. In this state, atoms evaporate out of the material, exit the cell through an opening, and deposit on the wafers. For most applications the wafer in the chamber is heated to give additional energy to the arriving atoms. The additional energy fosters epitaxial growth and good film quality.

If the wafer surface is exposed, the depositing atoms will assume the orientation of the wafer and grow an epitaxial layer. MBE offers the intriguing option of in situ doping by the inclusion of dopant sources in the chamber. The usual silicon dopant sources are not usable in MBE systems. Solid gallium is used for P-type doping and antimony for N-type doping.[12]

The primary advantage of MBE for silicon technology is the low temperature (400 to 800°C), which minimizes autodoping and out-diffusion. Perhaps the biggest advantage of MBE is the ability to form multiple layers on the wafer surface during one process step (one pump down). This option requires the mounting of several effusion cells in the chamber and shutter arrangements to direct the evaporant beams to the wafer in the right order and for the correct time.

An advantage and disadvantage of MBE is the low film growth rate of 60 to 600 Å/min.[10] On the plus side, the films produced are very controllable. Films can be grown (and mixed) in one monolayer increments. However, most semiconductor layers do not need this level of control and quality, making the low productivity and expense of the system an expensive luxury.

A bonus possibility with MBE is the incorporation of film growth and quality-analyzing instruments in the chamber. With these instruments the process can become very controlled and produce uniform films from wafer to wafer. MBE has found production use in the fabrication of special microwave devices and for compound semiconductors such as gallium arsenide.[13]

Vapor phase epitaxy (VPE)

A VPE system[12] is a combination of a standard liquid source tube furnace and a two-zone diffusion furnace (Fig. 12.34). VPE differs from the CVD systems described in its ability to deposit compound materials. An example is the particular arrangement in Fig. 12.34 used to deposit epitaxial gallium arsenide. The creation of the GaAs layer on the wafer in the main chamber proceeds in two stages. $AsCl_3$ is bubbled into the first section of the tube where it reacts with a solid source of gallium that is sitting in a boat. The arsenic trichloride reacts with the hydrogen in the first section to form arsenic by the reaction

$$4AsCl_3 + 6H_2 \rightarrow 12HCl + As_4$$

The arsenic deposits on the gallium forming a crust. The hydrogen passing over the crust reacts in the first section to form three gases that pass into the wafer section.

$$GaAs + HCl \leftrightarrow GaCl + \tfrac{1}{2} H2 + \tfrac{1}{4} As4$$
$$\text{(solid)} \quad \text{(gas)} \quad \text{(gas)} \qquad \text{(gas)} \qquad \text{(gas)}$$

This section is at a somewhat lower temperature and the reaction proceeds in reverse, depositing GaAs on the wafers. The technique offers the advantages of clean films, since the gallium and arsenic trichloride are available in very pure forms and have higher produc-

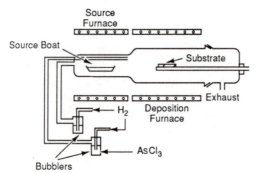

Figure 12.34 Diagram of gallium arsenide VPE deposition system.

tion rates than the MBE technique. On the downside, the film structures produced are not the quality of MBE films.

Metalorganic CVD (MOCVD)

MOCVD is one of the latest options for CVD of compound materials. Where VPE refers to a compound material deposition system, MOCVD refers to the sources used in VPE systems. There are two chemistries used, halides and metalorganic. The reactions described for the VPE deposition of gallium arsenide above is a halide process.[13] A group III halide (gallium) is formed in the hot zone and the III–IV compound is deposited in the cold zone. In the metalorganic process[14] for gallium arsenide, trimethylgallium is metered into the reaction chamber along with arsine to form gallium arsenide by the reaction

$$(CH_3)_3Ga + AsH_3 \rightarrow GaAs + 3CH_4$$

Polysilicon and amorphous silicon deposition

Until the advent of silicon-gate MOS devices (Fig. 12.35) in the mid-1970s, polysilicon layers had little or no use in device structures. The successful use of the silicon-gate device structure required reliable processes to deposit thin layers of polysilicon. Early processes involved simply placing the oxide-covered wafers in a horizontal APCVD system and letting the polysilicon deposit on the oxide. The major difference of early polysilicon depositions from epitaxial depositions was the use of silane sources. While silane is not favored for epitaxial film deposition, it is more than adequate for polysilicon depositions.

Typical polysilicon deposition processes take place in the 600 to 650°C range. The deposition may be from either 100% silane or from gas streams containing N_2 or H_2. The structure of polysilicon was previously described as a total nonarrangement of the silicon atoms. In the case of deposited polysilicon, the structure is somewhat different. During the early stages of deposition, at temperatures below 575°C,

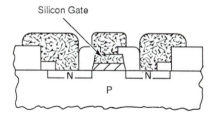

Figure 12.35 Cross section of silicon gate MOS transistor.

the structure is amorphous (no structure). The polysilicon structure formed by deposition techniques consists of small pockets (crystallites or grains) of single-crystalline silicon separated by grain boundaries. This structure is called *columnar poly*.

The importance of grain size and grain boundary consistency shows up in the electrical current flow characteristics of the films. Current resistance comes as the current crosses the grain boundaries. The larger the grain boundaries, the higher the resistance. The achievement of consistent current flow from device to device and within a device is dependent on a well-controlled polysilicon structure. One of the advantages claimed for the use of H_2 in the gas stream is the reduction of surface impurities and moisture, which in turn results in a reduced grain size. Moisture or oxygen impurities in the system cause the growth of silicon dioxide within the structure. The oxide increases the resistance of the film and its etchability in subsequent masking steps.

All of the system's usual operating parameters (temperature, silane concentration, pump speed, nitrogen flow, and other gas flows[15]) affect the deposition rate and the grain size. Often the wafers will receive a postdeposition anneal in the 600°C range to further crystallize the film. The process of recrystallization goes on whenever the wafers go through a high-temperature process. The grain size and electrical parameters of the polysilicon film on the finished device or circuit are never the same as the deposited film.

Also influencing the grain size is the presence of dopants in the gas stream. In many devices or circuits a strip of polysilicon functions as a conductor which requires doping to decrease its resistivity. Doping can be done by diffusion before or implantation after the deposition. In situ doping takes place by adding gas dopant sources in the source cabinet and metering them into the chamber. When diborane (boron source) is added there is a large increase in the deposition rate. The opposite effect takes place when phosphine (phorphorus source) or arsine (arsenic source) is the dopant gas. Undesirable effects of in situ doping are a loss of film uniformity, doping uniformity, and control of the deposition rate.

The resistivities of doped polysilicon films are less than those of equally doped epitaxial or bulk silicon. The lower resistivities are due to dopants being trapped in the grain boundaries. The development of LPCVD systems provided a more productive and lower-temperature means of depositing this layer and is the preferred method. LPCVD brings to the deposition good step coverage, a requirement since polysilicon layers are usually deposited later in the process and the surface has become varied in its topography.

SOS and SOI

These two acronyms stand for *silicon on sapphire* and *silicon on insulator*. Both refer to the deposition of silicon on a nonsemiconductor surface. The need for such structures came about from the limits placed on some MOS devices by the presence of a semiconducting substrate under the active device. These problems are resolved by forming a silicon layer on an insulating substrate. The first substrate used for this purpose was sapphire (SOS). As different substrates were investigated, the term was expanded to the more general silicon on insulator (SOI).

One technique is a direct deposition on the substrate followed by a recrystallization process (laser heating, strip heaters, oxygen implantation) to create a usable film.[16] Another approach is a selective deposition through holes in a surface oxide with an overgrowth to form the continuous film.

Insulators and dielectrics

CVD is the favored method of depositing films that will function in the device or circuit as insulators or dielectrics. The two films in widespread use are silicon dioxide and silicon nitride. In general, the two films find a multiplicity of uses in device and circuit designs. While they have processing and quality differences they must meet the same general requirements as other deposited films.

Silicon dioxide. Deposited silicon dioxide films are best known from their long-term use as a final passivation layer covering the completed wafer. In this role, they provide physical and chemical protection to the underlying circuit devices and components. Deposited silicon dioxide films used as a protective top layer are known by the proprietary terms *Vapox, Pyrox,* or *Silox.* Vapox (vapor-deposited oxide) is a term coined by Fairchild engineers. Pyrox stands for pyrolitic oxide. Silox is a registered trademark of Applied Materials, Inc. Sometimes the layer is simply called a *glass.* This protective role has expanded, and deposited silicon oxide layers are used as interdielectric layers in multimetallization schemes, as insulation between polysilicon and metallization layers, as doping barriers, as diffusion sources, and as isolation regions.

CVD-deposited silicon dioxide films vary in structure and stoichiometry from thermally grown oxides. Depending on the deposition temperature, deposited oxides will have a lower density and different mechanical properties, such as index of refraction, resistance to cracking, dielectric strength, and etch rate. These factors are highly affected by the addition of dopants to the film. In many processes the deposited film will receive a high-temperature anneal, a process called *densification*. After the densification, the deposited silicon dioxide film is close to the structure and properties of a thermal oxide.

The need for a low-temperature-deposited SiO_2 was dictated by the unacceptable alloying of aluminum and silicon at temperatures above 450°C. The early deposition process used was a horizontal conduction-heated APCVD system from silane and oxygen by the reaction

$$SiH_4 + O_2 \rightarrow SiO_2 + 2H_2$$

This process produced films that were of unacceptable quality for use in advanced device designs and on larger wafers due to the poorer film quality produced by the 450°C deposition temperature.

The development of LPCVD systems made possible higher-quality films, especially for the factors of step coverage and lower stress. LPCVD processes are the preferred deposition techniques from both quality and productivity considerations. High temperature (900°C) LPCVD of silicon dioxide is performed with a dichlorosilane reaction with nitrous oxide.

$$SiCl_2 + 2NO_2 \rightarrow SiO_2 + 2N_2 + HCl$$

In medium-temperature ranges, $Si(OC_2H_5)$ sources are used with plasma-assisted PECVD in the presence of oxygen. The source is known as tetraethyl orthosilicate or TEOS. The TEOS PECVD systems can produce quality films below 400°C.[17] The major advantages of this deposition combination are improved step coverage and stress-free films.

Another option is the reaction of silane with nitrous oxide in an argon plasma

$$SiH_4 + 4N_2O \rightarrow SiO_2 + 4N_2 + 2H_2O$$

Doped silicon dioxide. Silicon dioxide layers are doped to improve their protective characteristics and flow properties, or for use as dopant sources. The earliest dopant used with deposited oxides was phosphorus. The phosphorus source is phosphine (PH_3) gas added to the deposition gas stream. The resultant glass is called phosphorus silicate glass or PSG. Within the glass the phosphorus is in the form of phosphorus pentoxide (P_2O_5), making the glass a dual compound or, more correctly, a binary glass.

The role of the phosphorus is threefold. The added dopant increases the moisture-barrier property of the glass. Mobile ionic contaminants become attached to the phosphorus and are prevented from traveling into the wafer surface. This action is called *gettering*. The third result

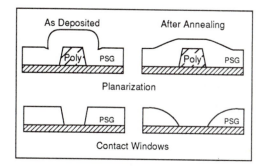

Figure 12.36 Planarization of surface by flowing glass.

is an increase of the flow characteristics (Fig. 12.36) which aid the planarization of the glass surface after a heating step in the 1000°C range. The phosphorus content is limited to about 8 weight percent. Above this level the glass becomes hydroscopic and attracts moisture. The moisture can react with the phosphorus, form phosphoric acid, and attack underlying metal lines.

Boron is often added to the glass from a diborane (B_2H_6) source. The purpose of the boron is to also aid the flow characteristics (Fig. 12.36). The resultant glass is called a borosilicate glass (BSG). The boron and phosphorus are often used together in the glass. The result is referred to as BPSG (borophosphorus silicate glass).

Silicon nitride. In most respects, silicon nitride is a superior replacement for silicon dioxide uses, especially for top layer protection. Silicon nitride is harder, which provides better scratch protection, is a better moisture and sodium barrier (without doping), has a higher dielectric strength, and resists oxidation. The latter property has led to its use in the local oxidation of silicon (LOCOS) for isolation purposes. Figure 12.37 illustrates the process, where patterned islands of silicon nitride prevent oxidation under the islands. After thermal oxidation and removal of the nitride there are wafer surface regions ready for device formation, separated by isolating regions of oxide. A disadvantage of silicon nitride is that it does not flow as easily as silicon oxide and is more difficult to etch. The etch restriction has been overcome with the development of plasma etch processes.

An early limit on the use of silicon nitride protective films was lack of a low-temperature deposition process. In APCVD systems a temperature of 700 to 900°C is required for the deposition of silicon nitride from silane or dichlorosilane (Fig. 12.38). The result is a film with the composition Si_3N_4. The reactions also take place in LPCVD reactors but at a temperature low enough for deposition over an aluminum

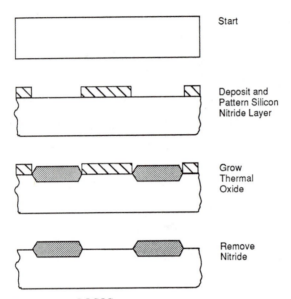

Start

Deposit and
Pattern Silicon
Nitride Layer

Grow
Thermal
Oxide

Remove
Nitride

Figure 12.37 LOCOS process.

| Silane | $3SiH_4 + 4NH_3 \rightarrow Si_3N_4 + 12H_2$ |
| Dichlorosilane | $3SiCl_2H_2 + 4NHCl + 6H_2 + 6HCl$ |

Figure 12.38 Silicon nitride deposition reactions.

metallization layer. The advent of PECVD has opened up the use of different source chemistries. One use is silane reacted with ammonia (NH_3) or nitrogen in the presence of an argon plasma.

Conductors

The traditional metal conductors are aluminum and aluminum alloys deposited by evaporation or sputtering techniques. The advent of silicon-gate MOS transistors and all their variations have added doped polysilicon as a conductor on the chip. More recently CVD refractory metals have been working their way into chip designs for a number of reasons. Those reasons are covered in the next chapter along with the CVD techniques used.

Key Concepts and Terms

Amorphous	MBE
APCVD	MOCVD
Barrel system	Pancake system
BPSG	PECVD
BSG	Polysilicon
CVD	PSG
Deposited Si_3N_4	Selective epitaxy
Deposited SiO_2	SOI
Epitaxial	SOS
Induction heating	VPE
LPCVD	

Review Questions

1. Sketch and name the major subsystems of a basic CVD system.
2. Describe the differences between APCVD, LPCVD, and PECVD.
3. Define an epitaxial film.
4. Why is the deposition temperature of a silicon dioxide passivation layer limited to 450°C?
5. List an advantage of a horizontal vertical-flow CVD reactor.
6. Write the reaction equation for the deposition of silicon from silicon tetrachloride.
7. Describe the difference between MBE, VPE, and MOCVD systems.
8. What wafer surface condition is necessary for the deposition of a polysilicon layer?
9. Describe the advantages of plasma-assisted depositions compared to APCVD systems.
10. Why, and with what, are deposited silicon dioxide layers doped?

References

1. J. Hayes and P. Van Zant, *CVD Today Seminar Manual*, Semiconductor Services, San Jose, Calif., 1985, p. 9.
2. S. Wolf and R. Tauber, *Silicon Processing for the VLSI Era*, Lattice Press, Sunset Beach, Calif., 1986, p. 165.

3. M. L. Hammond, "Epitaxial silicon reactor technology: A review," *Solid State Technology*, May 1988, p. 160.
4. M. L. Hammond, "Epitaxial silicon reactor technology: A review," *Solid State Technology*, May 1988, p. 160.
5. J. Hayes and P. Van Zant, *CVD Today Seminar Manual*, Semiconductor Services, San Jose, Calif., 1985, p. 13.
6. S. Wolf and R. Tauber, *Silicon Processing for the VLSI Era*, Lattice Press, Sunset Beach, Calif., 1986, p. 165.
7. M. L. Hammond, "Epitaxial silicon reactor technology: a review," *Solid State Technology*, May 1988, p. 159.
8. R. P. Roberge, et al., "Gaseous impurity effects in silicon epitaxy," *Semiconductor International*, Jan. 1988, p. 81.
9. S. Wolf and R. Tauber, *Silicon Processing for the VLSI Era*, Lattice Press, Sunset Beach, Calif., 1986, p. 157.
10. P. Singer, "Molecular beam epitaxy," *Semiconductor International*, Oct 1986, p. 42.
11. P. Singer, "Molecular beam epitaxy," *Semiconductor International*, Oct 1986, p. 42.
12. R. Williams, *Gallium arsenide processing techniques*, Artech House, Dedham, Mass., 1984, p. 44.
13. P. Burggraaf, "The growing importance of MOCVD," *Semiconductor International*, Nov. 1986, p. 47.
14. P. Burggraaf, "The growing importance of MOCVD," *Semiconductor International*, Nov. 1986, p. 48.
15. S. M. Sze, *VLSI Technology*, McGraw-Hill, New York, 1983, p. 119.
16. L. Jastrzebski, "Silicon CVD for SOI: principles and possible applications," *Solid State Technology*, Sept. 1984, p. 239.
17. B. L. Chin and E. P. van de Ven, "Plasma TEOS process for interlayer dielectric applications," *Solid State Technology*, Apr. 1988, p. 119.

13

Metallization

Overview

Critical to the operation of a semiconductor device or circuit is the surface metallization that connects the various parts of the devices and serves as electrical terminals for the chip. In this chapter, the materials, requirements, and methods used to create the metal patterns is presented along with other uses of metals in chip manufacturing.

Objectives

Upon completion of this chapter, you should be able to:

1. List the requirements of a material for use as a chip surface conductor.
2. Draw cross sections of single and multilayer metal schemes.
3. Describe the purpose and operation of a thin-film fuse.
4. Make a list of materials used in the metallization of semiconductor devices.
5. Describe the operation of an evaporator electron gun and its need in the evaporation of aluminum.
6. Draw and identify the parts of a vacuum evaporator.
7. Describe the principle of sputtering.
8. Draw and identify the parts of a sputtering system.
9. Describe the principle and operation of oil diffusion and cryogenic high-vacuum pumps.

Metal Film Uses

Conductors

The most common and familiar use of metal films in semiconductor technology is for surface wiring. The materials, methods, and processes of "wiring" the component parts together is generally referred to as *metallization*. The metallization process (Fig. 13.1) starts in the masking area where small holes, called *contact holes* or *contacts*, are etched through all of the surface layers, down to the active regions of the devices. Following contact masking, a thin layer (10,000 to 15,000 Å) of the conducting metal is deposited by vacuum evaporation, sputtering, or CVD techniques over the entire wafer. The unwanted portions of this layer are removed by a conventional photomasking and etch procedure or by lift-off. This step leaves the surface covered with thin lines of the metal that are called *leads, metal lines*, or *interconnects*. Generally a heat-treatment step, called *alloying*, is performed after metal patterning to ensure good electrical contact between the metal and the wafer surface.

The process just outlined is for a single-layer metal system. Increasing chip density has placed more components on the wafer surface, which in turn has *decreased* the area available for surface wiring. The answer to this dilemma has been dual- and triple-level schemes (Fig. 13.2). The multilevel schemes start with a standard metallization process which leaves the surface components partially wired together.

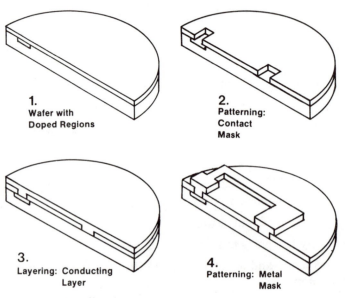

1.
Wafer with Doped Regions

2.
Patterning: Contact Mask

3.
Layering: Conducting Layer

4.
Patterning: Metal Mask

Figure 13.1 Metallization sequence.

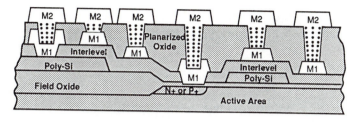

:: Via Plugs   M1 = First Metal   M2 = Second Metal

Figure 13.2 Cross section of typical planarized two-level metal VLSI structure showing range of via depths after planarization. (*Courtesy of Solid State Technology.*)

Next comes a layer of some dielectric material, called an intermetallic dielectric. This dielectric may be a deposited oxide, silicon nitride, or a polyimide film. This layer receives a masking step that etches contact holes, called *vias*, down to the first-level metal. The whole process is repeated, with the final structure having two or three levels of metal connected to each other and protected by a final passivation layer. In some schemes the via holes are filled with metals other than the main conducting material. The particular metals for via filling are discussed in the following pages.

A multilevel metal system is more costly, of lower yield, and requires greater attention to planarization of the wafer surface and intermediate layers to create good current-carrying leads.

In the role of surface conductor, a metal must meet the following criteria:

- Good electrical current-carrying capability (current density)
- Good adhesion to the top surface of the wafer (usually SiO_2)
- Ease of patterning
- Good electrical contact with the wafer material
- High purity
- Corrosion resistance
- Long-term stability
- Capable of deposition in uniform void- and hillock-free films

Fuses

The development of thin-film fuse technology allowed creation of the programmable read-only memory (PROM) circuit. The fuse allows field programming of data in the memory section of the chip. In this role the fuse is not a protective device, as in most electrical circuits, but is included specifically to be "blown" or disabled.

In the memory section of the chip, called the *array*, are a number of memory cells, each with a fuse between the cell and the main metallization system. The array is essentially a blank blackboard (Fig. 13.3). The desired information in the array is established by the number of "functioning" cells at specific locations. Cells not needed in the array are disabled by blowing the fuse. The same system is used to program logic arrays. The fuse is blown by the heat generated by a high-voltage current (blowing current) passing through the narrow neck of the fuse, heating the material to the melting point and leaving it disabled (Fig. 13.4). Once the fuse is blown the associated memory cell is permanently removed (electrically) from the circuit.

Films of nichrome or titanium-tungsten are used for fuse materials. The deposited films are very thin, in the 200 to 500-Å range. Surface conditions and deposition uniformity are very important factors influencing the "blowing current" of the fuse. The heat generated in the neck is dependent on the current level and the amount of fuse material in the neck region. A nonplanarized surface or nonuniform film can have a great effect on the total amount of fuse material in the narrow neck and change the blowing current level. A nonmetal fuse scheme employs a thin film of polysilicon or oxide in the contact hole. The fuse is blown when a high current is passed through the layer and destroys it by heating.

MOS gate electrode

In most semiconductor devices the electrical current travels from an outside source into the chip's metal system for distribution to the various components. Bipolar transistors, diodes, resistors, and fuses have

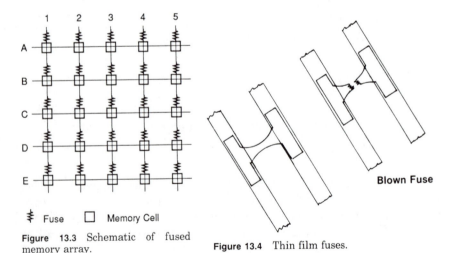

⌇ Fuse   ☐ Memory Cell

Figure 13.3 Schematic of fused memory array.

Figure 13.4 Thin film fuses.

Blown Fuse

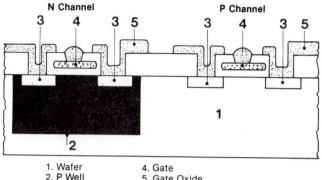

N Channel

P Channel

1. Wafer
2. P Well
3. Source/Drain
4. Gate
5. Gate Oxide
6. Metallization

Figure 13.5 CMOS transistor polysilicon gate electrode. (*From Ref. 4.*)

this outside current flowing through them. What happens to the current in the components is covered in Chap. 16.

MOS transistors have a source and drain separated by a gate region. No current travels from the metal system *into* the gate region. Instead, a voltage is applied to a metal lead over the gate region where it induces (or depletes) charges in the wafer surface under the gate. In most circuits the gate is formed from a deposited polysilicon compound (Fig. 13.5), which is doped to increase its conductivity. The doped polysilicon functions both as an MOS gate electrode and as a circuit conductor.

Intermediate barrier and protective films

Advanced circuit designs are requiring the use of metal film "sandwiches." These sandwiches of two or three different metals take advantage of some property of one of the metals. The other ones may function as an intermediate layer between the wafer and second metal. One may serve to separate two incompatible metals or separate a metal and a semiconductor. Others may serve to protect a lower-level metal from physical abuse. Such is the case in platinum-gold-molybdenum conduction systems, barrier metals in aluminum-silicon systems, and other sandwiches as explained in the sections on specific metals.

Backside plating

Gold is sometimes evaporated onto the entire back of the wafer just prior to wafer sort. The gold functions as a solder in certain packaging processes (Chap. 18).

Metal Materials

A number of metals serve various roles in semiconductor devices. Most are deposited onto the wafer from solid sources by evaporation or sputtering. To prevent contamination of the devices from mobile ionic contaminants, the solid material sources must be very pure. Aluminum sources are purchased at 5 to 6 "nines" of purity (99.999 to 99.9999 percent). Refractory metal depositions may be deposited by CVD techniques that also require very pure beginning gas sources.

Aluminum

Prior to the development of VLSI-level circuits, the primary metallization material was pure aluminum. The reasons for this primacy and aluminum's limitations are instructive to the understanding of metallization systems in general. From an electrical conduction standpoint, aluminum is less conductive than copper and gold. In fact, early device construction used some gold leads. However, gold makes a poor direct electrical contact with silicon. Solution of this problem led to the use of a platinum layer between the gold and silicon. The platinum is heated to form a platinum silicide (platinum and silicon) that provides the necessary intermediate layer. The platinum silicide also prevents gold from getting into the devices. Gold in many silicon devices causes leakage currents that destroy the device functioning. The platinum silicide also acts as an adhesion layer since gold does not readily stick to silicon dioxide. Another problem with gold is its softness; gold requires an overlayer of molybdenum for protection. Similarly, copper is not used because of its high contact resistance with silicon and its unacceptability inside silicon devices.

Aluminum emerged as the preferred metal because it avoids the problems just mentioned. It has a low-enough resistivity (2.7 $\mu\Omega \cdot$ cm),[1] and good current-carrying density. It has superior adhesion to silicon dioxide, is available in high purity, has a naturally low contact resistance with silicon, and is relatively easy to pattern with conventional photoresist processes.

Aluminum-silicon alloys

Shallow junctions in the wafer surface presented one of the first problems with the use of pure aluminum leads. The problem has to do with the need to bake aluminum-silicon interfaces to stabilize the electrical contact. This type of contact is called *ohmic* because the voltage-current characteristics behave according to Ohm's law. Unfortunately, aluminum and silicon dissolve into each other and at 577°C there exists a eutectic formation point. A eutectic formation occurs when two materials heated in contact with each other melt at temperatures much lower than their individual melting temperatures. Eutectic formations occur over a temperature range, and the aluminum-silicon eutectic starts to form at about 450°C, the temperature necessary for good electrical contact. The problem is that the alloy formation can melt down into the silicon wafer. If the surface has shallow junctions, the alloy region can extend completely through the junction, shorting it out (Fig. 13.6).

Two solutions to this problem are employed. One is a barrier metal layer that separates the aluminum and silicon and prevents the eutectic alloy from forming. The second is an alloy of aluminum with 1 to 2% silicon. During the contact heating step the aluminum alloys more with the silicon in the alloy and less with the silicon from the wafer. This process is not 100 percent effective and some alloying between the aluminum and wafer always occurs.

Aluminum-copper alloy

Aluminum suffers a problem called *electromigration*. The problem occurs when long skinny leads of aluminum are carrying high currents over long distances, as is the situation on VLSI circuits. The current sets up an electric field in the lead that decreases from the input side to the output. Also, heat generated by the flowing current sets up a thermal gradient along the lead. The aluminum in the lead becomes mobile and diffuses within itself along the direction of the two gradients. The first effect is a thinning of the lead. In the extreme the lead

Excessive Alloy

Aluminum / Silicon
melted into Wafer

Aluminum with Si

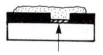

Figure 13.6 Eutectic alloying of aluminum and silicon contacts.

Barrier Metals

can become completely separated. Unfortunately, this event usually takes place after the circuit is in operation in the field causing a failure of the chip. Prevention or moderation of electromigration is achieved by depositing an alloy of aluminum and 4% copper or an alloy of aluminum and 0.1 to 0.5% titanium. Aluminum alloys containing both copper and silicon are often deposited on the wafer to resolve both alloying and electromigration problems.

Drawbacks to the use of aluminum alloys are an increased complexity for the deposition equipment and process and changed film etch rates. Another drawback is an increase in film resistivity compared to the pure aluminum. The amount of the increase varies with the alloy composition and heat treatments, but can be as much as 25 to 30 percent.[2]

Doped polysilicon

The advent of silicon-gate MOS technology made the use of deposited polysilicon lines on the chip a natural consideration for conduction leads. For use as a conductor, the polysilicon has to be doped to increase its conductivity. Generally the preferred dopant is phosphorus due to its high solid solubility in silicon. Doping is by either diffusion, ion implantation, or in situ doping during an LPCVD process. Each of the methods produces a different doping result. The differences relate to the doping temperature's effect on the grain structure. The lower the temperature, the greater the amount of dopant trapped in the polygrain structure, where it is unavailable for conduction. This is the situation with ion implantation. Diffusion doping results in the lowest film sheet resistivity. In situ CVD doping has the lowest dopant carrier mobility due to grain boundary trapping.

Doped polysilicon has the advantage of a good ohmic contact with the wafer silicon and can be oxidized to form an insulating layer. Polysilicon oxides are of a lower quality than thermal oxides grown on single-crystal silicon because of the nonuniformity of the oxide grown on the rougher polysilicon surface.

Titanium-tungsten barriers

A method of preventing the eutectic alloying of silicon and aluminum metallization is by a barrier layer. A layer of titanium-tungsten (TiW) alloy is sputter-deposited onto the wafer into the open contacts before the aluminum deposition takes place. The TiW deposited on the field oxide is removed from the surface during the aluminum etch step.

Sometimes a first layer of platinum silicide is formed on the exposed silicon before the TiW is deposited.

Refractory metals and refractory metal silicides

Although the limitations of electromigration and eutectic alloying have been made manageable by aluminum alloys and barrier metals, the issue of contact resistance may prove to be the final limit on aluminum metallization. The overall effectiveness of a metal system is governed by the resistivity, length, thickness, and total contact resistance of the metal-wafer interconnects.

Contact resistance is influenced by the materials, the substrate doping, and the contact dimensions. The smaller the contact size, the higher the resistance. Unfortunately, VLSI chips have smaller contact openings, and large gate array chip surfaces can be as much as 80 percent contact area.[3] These two factors make the contact resistance the dominant factor in VLSI metal system performance. Aluminum-silicon contact resistance, along with the alloying problem, have led to the investigation of other metals for VLSI metallization. Polysilicon has a lower contact resistance than aluminum and is in use in MOS circuits (Fig. 13.7).

Refractory metals and their silicides also offer lower contact resistance. The refractory metals of interest are titanium (Ti), tungsten (W), tantalum (Ta), and molybdenum (Mo). Their silicides form when they are alloyed on a silicon surface (WSi_2, $TaSi_2$, $MoSi_2$ and $TiSi_2$). The refractory metals were first proposed for metallization in the 1950s, but they stayed in the background due to a lack of a reliable deposition method. That situation has changed with the development of LPCVD and sputtering processes.

All modern circuit designs, especially MOS circuits, use refractory metals or their silicides as intermediate, barrier, or conducting layers.

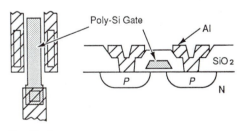

Metallization
and Metal Mask

Figure 13.7 Silicon gate electrode extended for metallization lead.

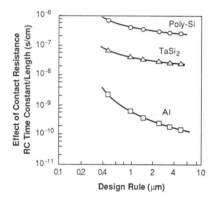

Figure 13.8 Effect of contact resistance on RC time constant.

The lower resistivities and lower contact resistances (Fig. 13.8) make them attractive for conducting films, but impurities and deposition uniformity problems make them less attractive for MOS gate electrodes. The solution to the problem has been the polyside and salicide gate structures, which are combinations of a silicon gate topped by a silicide. The details of this structure are explained in Chap. 16.

A popular use of refractory metals is the filling of via holes in multilevel metal structures. The process is called *plug filling* and the filled via is called a *plug* (refer back to Fig. 13.2). The vias are filled by either selective tungsten deposition through surface holes onto the first layer metal or by CVD techniques.[4] Of the available refractory metals, tungsten finds a lot of use as aluminum-silicon barriers, MOS gate interconnects, and for via plugs.

Deposition Methods

Metallization techniques, like other fabrication processes, have undergone improvements and evolution in response to new circuit requirements and new materials. The mainstay of metal deposition up to the mid-1970s was vacuum evaporation. Aluminum, gold, and the fuse metals were all deposited by this technique. The needs of depositing multimetal systems and alloys, along with the need for better step coverage, led to the introduction of sputtering as the standard deposition technique for VLSI circuit fabrication. Refractory metal use has added the third technique, CVD, to the arsenal of the metallization engineer.

Vacuum evaporation

Vacuum evaporation is still used for the deposition of metals on discrete devices and circuits of lower integration levels. It is also used for the deposition of gold to the back of a wafer for die adhesion into a package. Vacuum evaporation takes place inside an evacuated chamber (Fig. 13.9). The chamber can be a quartz bell jar or a stainless steel enclosure. Inside the chamber is a mechanism to evaporate the metal source, wafer holders, a shutter, thickness and rate monitors, and heaters. The chamber is connected to a vacuum pump(s) (see "Vacuum Pumps," p. 350).

Since aluminum is the most critical of the materials evaporated, we shall focus on its deposition. The vacuum is required for a number of reasons. First is a chemical consideration. If any air (oxygen) molecules were in the chamber when the high-energy aluminum atoms were coating the wafer, they would form aluminum trioxide (Al_2O_3), a dielectric which, if incorporated into the deposited film, would compromise aluminum's role as a conductor. A second requirement for vacuum deposition is uniform coating. When the pressure is sufficiently reduced, the mean free path of the coating atoms is increased to exceed the dimensions of the chamber. This situation ensures that depositing atoms will strike the wafers before hitting each other and causing nonuniform depositions. The vacuum required for successful evaporation of aluminum is from 5×10^{-5} to 1×10^{-9} torr of pressure. (See Chap. 2 for a discussion on vacuum and pressure.)

Evaporation sources. Before describing the various methods of causing a metal to evaporate, a review of basic evaporation theory is in order. Most of us are familiar with the evaporation of a liquid from a beaker. Within the liquid there is a considerable amount of atomic and/or molecular activity at the liquid surface. The energy for the ac-

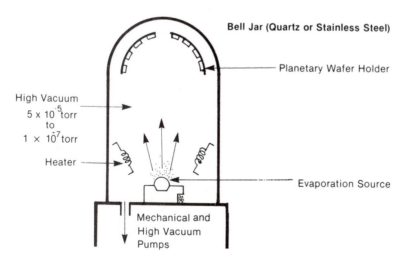

Figure 13.9 Vacuum evaporator.

tivity comes from the temperature (heat) of the liquid. In the case of water, molecules are constantly leaving and returning to the liquid. Some of the molecules have enough energy to escape the liquid completely and stay in the surrounding atmosphere. We say that the escaped molecules have evaporated into the air. Evaporation occurs because of the energy in the liquid (remember, even at room temperature materials possess energy) and the lower concentration of water molecules in the atmosphere above the liquid water. Evaporation rates can be increased by giving the water molecules more energy, that is, by heating, and/or decreasing the humidity in the air above the liquid water.

The same process of evaporation can be made to occur in solid metals. The requirement is to heat the metal to a liquid state so that the atoms or molecules evaporate into the surrounding atmosphere. Three methods are used to evaporate metals in a vacuum system. They are

- Filaments
- Electron beam
- Flash hot plate

Filament evaporation (Fig. 13.10) is the simplest of the three methods. It is used for noncritical evaporations such as backside gold layers. The material, in wire form, is either wrapped around a coiled tungsten (or other metal that is able to withstand high temperatures) wire. A high current is passed through the tungsten wire where it heats the deposition metal to a liquid and evaporates into the chamber, coating the wafers. Another version uses a flat filament with a dimple to hold pieces of the deposition material.

Filament evaporation is not very controllable due to temperature variations along the filament. For noncritical depositions the filament will be heated until all of the deposition material is gone. Additional control is added by the use of a shutter to cut off the deposition at some predetermined time. A drawback to filament evaporation is that any contaminants in the source materials or on the filament are also evaporated onto the wafers.

Alloys are difficult to deposit by this method. Each element has a

Wire Filament   Flat Filament

Figure 13.10 Evaporation filament sources.

different evaporation rate at a given temperature. When an alloy, such as nichrome, is evaporated, the nickel and the chromium each evaporate at different rates. The composition of the film on the wafer will be different than the composition in the starting material.

The need for evaporation control and low contamination led to the development of the electron beam evaporation source (Fig. 13.11) for aluminum. The system is called an *E-beam gun* or just *E-gun*. This evaporation source consists of a water-cooled copper crucible with a center cavity to hold the aluminum. At the side of the crucible is a high-temperature filament. A high current is passed through the filament, which, in turn, "boils" off electrons. The negative electrons are bent 180° by a magnet so that the electron beam strikes the center of the charge in the cavity. The high-energy electrons create a pool of liquid aluminum in the center of the charge. Aluminum evaporates from the pool into the chamber and deposits on the wafers in holders at the top of the chamber. The water cooling maintains the outer edges of the charge in a solid state, thus preventing contaminants from the copper crucible from evaporating. E-gun evaporation is relatively controlled for an elemental source such as aluminum or gold. It is less useful for the deposition of alloys due to the same limitation as with filament evaporation sources, that is, the different evaporation rates of different elements.

A more controlled evaporation of alloys is accomplished by including a second or third gun in the chamber. With multiple guns the power to each can be adjusted to achieve evaporation rates such that the alloy film on the wafer is the correct composition. Film thickness is controlled by shutters and by rate and thickness monitors. Closed automatic shutters in the path of the evaporating material allow the evaporation to reach a steady rate before depositing on the wafers. The shutter also allows rapid shutoff of the deposition. In-chamber monitors, located near or above the wafer holders, feed back informa-

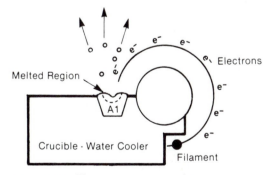

Figure 13.11 Electron gun evaporation source.

Figure 13.12   Flash evaporation source.

tion to the E-gun power supply which controls the evaporation rate. Rate control is important for consistent and uniform film structure.

The hot plate, or *flash* system (Fig. 13.12) was developed to resolve some of the problems with the evaporation of alloys with E-gun sources. This source consists of a hot plate held at a temperature well above the melting point of the particular alloy. A thin wire of the alloy is fed automatically onto the hot plate's surface. Upon contact the tip of the wire melts and the material "flashes" into a vapor and coats the wafers in the chamber. Since all of the elements are "flashed" simultaneously, the composition of the film on the wafer is very close to the composition of the wire.

A major goal of any metal deposition system is good step coverage (Fig. 13.13). This is a challenge for vacuum evaporators because the source is essentially a point source. The problem comes when material from a point source is shadowed by steps on the wafer surface. The result can be that one side of the openings in the surface oxide can be too thin or have a void.

Several methods are used to ensure good step coverage. Primary is a planetary domed wafer holder (Fig. 13.14) suspended over the source. The planetary holders are designed so that when rotated at a high speed, the wafer is positioned at many angles to the source, allowing the even deposition of the film.

Quartz heaters in the chamber aid step coverage by maintaining evaporant mobility on the wafer for a short period of time. While cooling down, the atoms "fill in" the steps by a capillary action. Typical temperatures are about 400°C and cannot exceed the 450°C limit for aluminum on silicon. The third approach to step coverage is to round the edge of the contact holes by etching (called *slope etch*) in the masking area (Fig. 13.15). The rounded corners reduce the size of the "shadow," allowing easier access into the contact area.

Sputter deposition (PVD)

Sputter deposition (sputtering) is another old process adapted to modern semiconductor needs. It is a process first formulated in 1852 by Sir William Robert Grove.[5] Sputtering is a process that (in general) can deposit any material on any substrate. It is widely used to coat cos-

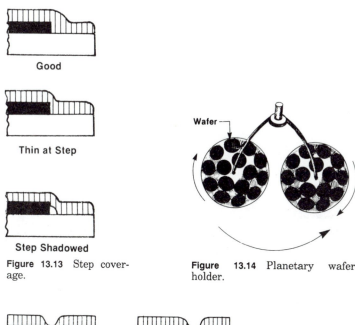

Good

Thin at Step

Step Shadowed

Figure 13.13 Step coverage.

Figure 13.14 Planetary wafer holder.

Oxide
Wafer

Figure 13.15 Slope etch step coverage.

tume jewelry and put optical coatings on lenses and glasses. Discussion of the benefits of sputtering to the semiconductor industry is best left until the principles and methods of sputtering are covered. Sputtering, like evaporation, takes place in a vacuum. However, it is a physical not a chemical process (evaporation is a chemical process), and is referred to as physical vapor deposition (PVD).

Inside the vacuum chamber is a solid slab, called a *target*, of the desired film material (Fig. 13.16). The target is electrically grounded. Argon gas is introduced into the chamber and is ionized to a positive charge. The positively charged argon atoms are attracted to the grounded target and accelerate toward it. During the acceleration they gain momentum, which is force, and strike the target. At the target a phenomenon called *momentum transfer* takes place. Just as a cue ball transfers its energy to the other balls on a pool table, causing them to scatter, the argon ions strike the slab of film material, causing its atoms to scatter (Fig. 13.17). The argon atoms "knock off" atoms and molecules from the target into the chamber. This is the sputtering activity. The sputtered atoms or molecules scatter in the chamber with some coming to rest on the wafer. A principal feature of

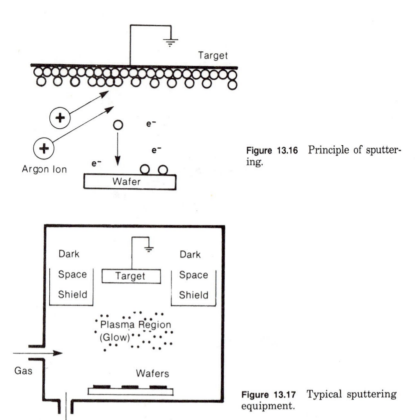

Figure 13.16 Principle of sputtering.

Figure 13.17 Typical sputtering equipment.

a sputtering process is that the target material is deposited on the wafer without chemical or compositional change.

There are several advantages of sputtering over vacuum evaporation. One is the aforementioned conservation of target material composition. A direct benefit of this feature is the deposition of alloys and dielectrics. The problem of evaporating alloys was described in the preceding section. In sputtering, an aluminum and 2% copper target material yields an aluminum and 2% copper film on the wafers.

Step coverage is also improved with sputtering. Whereas evaporation proceeds from a point source, sputtering is a planar source. There is material being sputtered from every point on the target with material arriving at the wafer holder with a wide range of angles to coat the wafer surface. Step coverage is further improved by rotating the wafer holder and by heating the wafer.

Adhesion of the sputtered film to the wafer surface is also improved over evaporation processes. The higher energy of the arriving atoms makes for a better adhesion, and the plasma environment inside the chamber has a "scrubbing" action[5] of the wafer surface that enhances adhesion. Adhesion and surface cleanliness can be increased by grounding the wafer holder and sputtering the wafer surface for a brief time prior to the deposition. In this mode the sputter system is functioning as an ion-etch (sputter-etch, reverse-sputter) machine, as described in Chap. 10.

Perhaps the greatest contribution of sputtering is the control of film characteristics available by the balancing of the sputtering parameters of pressure, deposition rate, and target material. Sandwiches of material can be sputtered in one process with multiple target arrangements.

The sputter process starts with the purchase of clean and dry argon (or neon). Extreme cleanliness is required to maintain film composition characteristics and low moisture is required to prevent unwanted oxidation of the deposited film. The chamber is loaded with the wafers and the pressure is reduced by pumps (pumped down) to the 1×10^{-9} torr range. The argon is introduced and ionized. Control of the argon amount entering the chamber is critical due to its effect of raising the pressure in the chamber. With the argon and sputtered material in the chamber the pressure rises to a level of about 10^{-3} torr. Chamber pressure is a critical parameter in the deposition rate of the system. After liberating material from the target, the argon ions, the sputtered material, gas atoms, and electrons generated by the sputtering process form a plasma region in front of the target. The plasma region is evident by its purple glow. The plasma region is separated from the target by a darkened region, known as the *dark space*.

Four sputtering methods are used in semiconductor applications. They are

- Diode [direct current (dc)]
- Diode [radio frequency (rf)]
- Triode
- Magnetron

The first two methods, called *diode sputtering*, are simple in concept. The target is connected to a negative potential with a positively charged anode present in the chamber. The negatively charged target ejects electrons, which accelerate toward the anode. Along the way they collide with the argon gas atoms, causing them to become ionized. The positively ionized argon atoms then accelerate to the target, initiating the sputtering process. The ionized argon (+) and the target (−) form a diode.

A secondary effect of the ionization process is the impact of the electrons on the gas atoms, resulting in the plasma region visible as the glowing purplish region just in front of the target. Dark spaces exist just in front and to the sides of the target. Sputtering efficiency is enhanced when the plasma is confined to the region between the target and wafers. This condition is enhanced by placing "dark space" shields to the sides of the target. The shields prevent target material from being sputtered from the sides, material that will never deposit on the wafers.

Another problem arises from the out-gassing of contamination from the chamber walls while the chamber is under vacuum. This condition is called a *virtual leak*, as opposed to an actual leak of atmosphere into the system. Besides compromising the pressure level in the chamber the contamination can be incorporated into the deposited film. This latter problem is addressed by placing a small negative bias (charge) on the wafer holder. The bias creates ions on the wafer surface and has the effect of dislodging stray out-gassed atoms from the growing film. Direct-current diode sputtering is used primarily to deposit metals.

Improved sputtering is gained by connecting the target to the negative side of a radio-frequency (rf) generator. The ionization of the gas takes place near the target surface without requiring a conductive target. Radio-frequency sputtering is necessary to sputter nonconducting materials (dielectrics) and is used also for conductors. Biasing is also used with the radio-frequency sputtering to achieve a cleaning effect at the wafer surface. Radio-frequency biasing offers the advantage of etching and cleaning of the exposed wafer surface. Etching and cleaning are achieved by putting the wafer holder at a different field potential than the argon, causing the argon atoms to impinge directly on the wafer. This procedure is called *sputter etch, reverse sputter*, or *ion milling*. The process removes contamination and a small layer from the wafer. Removal of contamination improves electrical contact between the exposed wafer regions and the film and improves adhesion of the film to the rest of the wafer surface.

In diode sputtering a number of processes occur at or near the wafer surface. Upon impact of the argon atom a number of electrons are created. These electrons cause heating of the substrate (as high as 350°C), which in turn can cause uneven film deposition. The electrons also create a radiation environment that can damage sensitive devices.

The heating produced with diode sputtering causes a serious problem with the deposition of aluminum. The heating causes residual oxygen in the target and in the chamber to combine with the aluminum to form aluminum oxide. The aluminum oxide is a dielectric and can compromise the conductive property of the deposited aluminum. More serious, a layer of aluminum oxide can form on the target surface and the impinging argon atoms do not (in diode sputtering) have enough energy to break through the layer. In effect the target becomes sealed and the sputtering stops.

Triode sputtering avoids some of the problems of diode sputtering. The electrons necessary to ionize the argon are created by a separate high-current filament. In designs where the filament is located outside the deposition chamber, the wafers are protected from radiation damage. Films deposited by a triode method are more dense.

Another problem with diode sputtering is the electrons that escape into the chamber and do not contribute to the establishment of the plasma necessary for deposition. The situation is resolved in *magnetron* sputtering systems, which use magnets behind and around the target (Fig. 13.18). The magnets capture and/or confine the electrons to the front of the target. Magnetron systems are more efficient for increased deposition rates. The resulting ion current (density of ionized argon atoms hitting the target) is increased by an order of magnitude over conventional diode sputtering systems. Another effect is a lower pressure required in the chamber, which contributes to a cleaner deposited film. Magnetron sputtering leaves a lower target temperature, which makes it a favorite for the sputtering of aluminum and aluminum alloys.

Production-level sputtering systems come in a variety of designs. Chambers are either batch systems or single-wafer in-line designs. Most production machines have load-lock capabilities. A load lock is an antechamber where a partial vacuum is created so that the deposition chamber can be maintained at vacuum. The advantage of a load lock is a higher production rate. Production machines are usually ded-

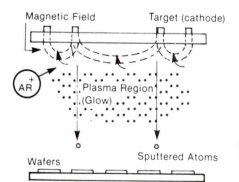

Figure 13.18 Magnetron sputtering.

icated to one or two target materials, while development machines have a wider range of capability.

CVD

Advances in CVD technology, especially LPCVD, offer the deposition department a third choice for metal depositions. CVD offers the advantages of not requiring expensive and maintenance-intensive high-vacuum pumps, conformal step coverage, and high production rates. Perhaps the most often deposited CVD refractory metal film is tungsten.

Tungsten is used in a variety of structures, including contact barriers, MOS gate interconnects, and via plugs. The filling of via holes is a key to effective multimetal systems (see Fig. 13.2). The dielectric layer is relatively thick and the via holes have to be relatively thin. These two factors make for a difficult continuous metal deposition to fill the vias without thinning of the metal in the via. Selective CVD-deposited tungsten plugs fill the entire via and present a planar surface for a subsequent conducting metal layer. For use as a barrier metal, tungsten can be deposited selectively by the silicon reduction of the gas tungsten hexafluoride (WF_6) by the reaction

$$2WF_6 + 3Si \rightarrow 2W + 3SiF_4$$

Tungsten can also be deposited selectively over aluminum and other materials from WF_6. The processes are called *substrate reduction*. Tungsten is also deposited from WF_6 by hydrogen reduction; the reaction is

$$WF_6 + 3H_2 \rightarrow W + 6HF$$

All of the depositions are performed in LPCVD systems at temperatures of about 300°C, which makes the processes compatible with aluminum metallization.

The depositions of tungsten silicide and titanium silicide proceed by the reactions

$$WF_6 + 2SiH_4 \rightarrow WSi_2 + 6HF + H_2$$

$$TiCl_4 + 2SiH_4 \rightarrow TiSi_2 + 4HCl + 2H_2$$

Vacuum Pumps

LPCVD, ion implantation, evaporation, and sputtering processes all take place in reduced-pressure chambers. The reduced-pressure atmospheres (vacuum) are clean and free of contaminating gases. In the deposition processes, the vacuum increases the mean free path of the depositing atoms and molecules, which in turn results in more uniform and controllable deposited films. LPCVD takes place in the pressure range down to 10^{-3} torr while the other processes take place at pressure ranges down to 10^{-9} torr. The first level is reached by mechanical vacuum pumps. These same pumps are used to initially reduce the pressure in the high-vacuum process chambers. In this role they are called *roughing pumps*. Additionally, mechanical vacuum pumps are used on the outlet end of high-vacuum pumping systems to assist in the removal of gas molecules from the pump to the exhaust system.

After the rough vacuum is established, a high-vacuum pump takes over to establish the final vacuum. The pumps used for this purpose are *oil diffusion, cryogenic, ion*, or *turbomolecular*. Whatever the type, all pumps are constructed of materials that will not *out-gas* into the system and compromise the vacuum. Materials used are typically type 304 stainless steel, oxygen-free high-conductivity copper (OFHC), Kovar, nickel, titanium, borosilicate glasses, ceramics, tungsten, gold, and some low-vapor-pressure elastomers.[6] Pumps used to evacuate corrosive and toxic gases or reaction by-products must have corrosion-free inside surfaces. Also, care must be taken in servicing pumps with these types of applications.

Pumps are selected and used based on a number of criteria, including

- Vacuum range required
- Gases to be pumped (lighter gases like hydrogen are more difficult to pump)
- Pumping speed
- Overall throughput
- Ability to handle impulsive loads (periodic out-gassing)
- Ability to pump corrosive gases
- Service and maintenance requirements
- Downtime
- Cost

Recall from Chap. 2 that pressure in a system results from the activity of gas atoms or molecules in an enclosure striking the chamber walls with some force. Reduction of the pressure in a system requires the removal of the gas in the chamber. This is generally accomplished by the pump establishing a lower pressure, first within itself, which allows gas material in the process chamber to flow into the pump where it is removed entirely from the system. At very low pressures there is not much material in the chamber and continued pressure reduction

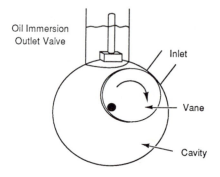

Oil Immersion
Outlet Valve

Inlet

Vane

Cavity

Figure 13.19 Mechanical rotary oil pump.

requires that the system be leak-free and not add to the pressure by its own out-gassing. Some systems require traps to prevent material from the pump from *back-streaming* into the chamber. Cold traps are explained in the section on oil diffusion pumps.

Mechanical rotary oil pumps

Mechanical oil rotary vacuum pumps trace their basic design to the 1640s when Galileo and Torricelli were investigating the theory that air had weight.[6] The pump featured an eccentric rotating vane in a cavity (Fig. 13.19). As the vane rotates, it compresses and sweeps out the gas in front of it in the cavity, simultaneously leaving a reduced-pressure region behind it. The "pushed" material exits through a valve, while another valve opens to the cavity from the chamber and allows material in the chamber to flow into the cavity. As the vane rotates, more and more material is removed from the cavity, thus reducing the pressure in the chamber.

A critical part of the pump is the efficiency of the exit valve. An ineffective valve that leaks atmosphere back into the system will limit the ultimate pressure level of the pump. This style of pump uses an oil-immersed valve that prevents back-streaming of the atmosphere.

Oil diffusion pump

The oil diffusion pump (Fig. 13.20) has been the mainstay of most semiconductor vacuum processes. The pump requires the services of an oil mechanical pump to first reduce the pressure in the chamber to the 10^{-3}-torr level. Either the same or a second mechanical pump is required at the outlet end. High vacuum, in the 10^{-8}-torr range (under production conditions), is achieved by a clever momentum transfer system. A low-vapor-pressure, hydrocarbon-based oil is heated in the base of the pump, where it rises up a structure called the *stack*. At the top of the stack is a series of downward-facing baffles. The hot oil mol-

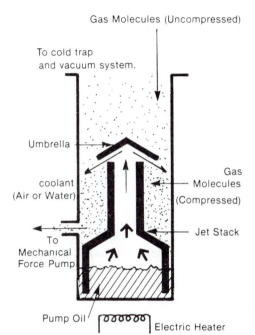

Gas Molecules (Uncompressed)

To cold trap
and vacuum system.

Umbrella

coolant
(Air or Water)

Gas
Molecules
(Compressed)

To
Mechanical
Force Pump

Jet Stack

Pump Oil

Electric Heater

Figure 13.20   Oil diffusion pump.

ecules, which have gained speed and energy from the boiling, exit the stack in a downward direction. Outside the stack they collide with gas from the chamber, causing them to be propelled toward the bottom of the pump where they are removed by a mechanical pump. The oil molecules return to the heated reservoir.

Two problems with oil diffusion pumps are the migration of oil molecules back up into the chamber and the inability of the pump to handle water vapor from the chamber. Both problems are resolved by the use of a *cold trap* between the pump and the chamber. A cold trap is similar in design to a liquid source bubbler. The fluid in the trap is liquid nitrogen, which reduces the temperature to $-196°C$. At this temperature, oil, contaminant, and water vapor molecules are frozen to the inside walls of the trap and do not add to the pressure in the system.

Cryogenic pump

Even with cold-trap technology, some processes cannot stand contamination from hydrocarbon oils such as are used in oil diffusion pumps. This situation has led to the use of cryogenic (cyro) pumps. A cryogenic pump (Fig. 13.21) uses the fact that gas molecules will "freeze" out on cold surfaces. The cold trap described in the preceding

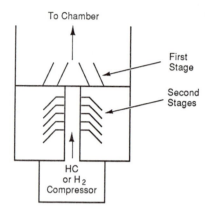

To Chamber

First
Stage

Second
Stages

HC
or H$_2$
Compressor

Figure 13.21   Cryogenic pump.

section and the frost that collects on the insides of refrigerators are examples of cryogenic activity.

Cryogenic pumps are designed with a central finned stack. At the low temperatures, gas from the chamber collects on the vanes, removing material from the system, which, in turn, reduces the pressure. The central stack, called an *expander*, is cooled as a compressor releases the liquid helium or nitrogen into it from the bottom. Cooling is by a phenomenon called adiabatic expansion. This is the same phenomenon that causes a pressurized can to cool when the nozzle is opened and the gas expands rapidly into the lower-pressure atmosphere. The top of the expander is at a higher temperature than the bottom, resulting in different gas molecules freezing out at different levels.

Cyropumps operate without cold traps or mechanical roughing pumps. Due to their action of trapping gas material on the fins, they are of a type known as capture pumps rather than the displacement-type pumps that move the chamber material to the atmosphere. The total capture nature of the pump and its oilless operation drastically reduce the possibility of contamination. However, when the pump is brought to room temperature, by mistake or for maintenance, the frozen gases are released and care must be exercised to vent any toxic or flammable gases trapped on the vanes. The buildup of gases also affects the pumping speed. The pump's speed must be monitored and the system cleaned when the speed falls. Cryopumps are simpler to operate and maintain, since no cold trap liquids or messy oils are needed. Additionally, cryopumps can handle bursts of out-gassing from the process chamber and feature a fast pumping speed.

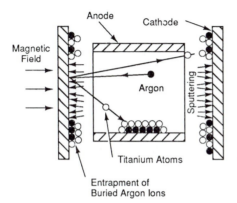

Figure 13.22 Principle of ion vacuum pump.

Ion pumps

Another capture type pump is the ion pump, also known as a *sputter ion pump* or a *getter pump* (Fig. 13.22). An ion pump operates in a manner similar to an ionization section in an ion implanter or sputter machine; only in this application, the atoms and molecules come from the chamber. A portion of those that drift into the ionization chamber are ionized to a positive charge by bombardment with electrons and attracted to a titanium cathode (negative potential). Upon collision with the titanium, some of the titanium is sputtered away and travels into the pump. The titanium atoms are chemically active enough to combine with other gases in the pump, which also accumulate on the pump walls. Again, material is removed from the chamber which reduces the pressure. Ion pumps are capable of pressures down to 10^{-11} torr which is the ultra-high vacuum range.

Turbomolecular pumps

Turbomolecular pumps are similar in design to a jet turbine engine. A series of blades (Fig. 13.23) with openings are mounted and rotated at very high speeds (24,000 to 36,00 rpm[7]) on a central shaft. Gas molecules from the chamber encounter the first blade and gain momentum from the collision with the rotating blade. The momentum direction is downward to the next blade, where the same thing happens. The net result is a removal of gas from the chamber. The use of a momentum transfer makes the pumping principle the same as an oil diffusion pump.

Major advantages of turbomolecular pumps are no back-streaming

To Vacuum Chamber

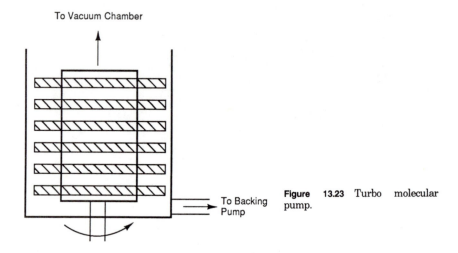

To Backing
Pump

Figure   13.23 Turbo   molecular
pump.

from oils, no need to recharge, high reliability, and pressure reduction
into the high vacuum range. Drawbacks are a slower pumping speed
compared to oil diffusion and cryogenic pumps and vibration and wear
due to the high rotational speeds.

Figure 13.24 is an overview of the metals used, their uses and deposition methods.

Key Concepts and Terms

Barrier metals   Fuses

Conducting metals   Ion pumps

Contact resistance   Mechanical oil rotary pumps

Cryogenic pumps   Oil diffusion pumps

CVD metal deposition   Refractory metal

Doped polysilicon   Refractory metal silicides

E-gun evaporation source   Sputter deposition (PVD)

Electromigration   Turbomolecular pumps

Filament evaporation source   Vacuum evaporation

Metal	Use			Deposition Method		
	Conductor	Fuse	Other	Evaporation	Sputtering	CVD
Aluminum	X			X	X	
Aluminum/Silicon	X			X	X	
Aluminum/Copper	X			X	X	
Aluminum/Copper/Silicon	X			X	X	
Gold/Molybdenum/Platinum	X			X	X	
Titanium/Platinum/Gold	X			X	X	
Titanium/Tungsten		X	Barrier		X	
Nichrome		X		X		
Gold			Die Attach/Lifetime Control	X		
Doped Silicon	X					X
Molybdenum	X				X	X
Platinum Silicide	X		Barrier		X	X
Titanium Silicide	X				X	X
Tungsten Silicide	X				X	X

Figure 13.24 Metallization overview.

Review Questions

1. Discuss the reasons aluminum metallization is superior to gold and copper.

2. Explain electromigration and how it is controlled.

3. Explain aluminum and silicon eutectic alloying and two processes used to prevent it.

4. Why are refractory metal and their silicides favored for VLSI circuit metallization?

5. What is a silicide and how is it formed?

6. Draw a diagram, label it, and use it to describe the evaporation of aluminum from an electron gun.

7. Why would there be two or more electron guns in an evaporator?

8. Draw a diagram of and label the parts of a sputter machine.

9. What are the advantages of a sputter process compared to an evaporator process?

10. List the four types of high-vacuum pumps used in semiconductor processes and their principles of operation.

References

1. S. Wolf and R. Tauber, *Silicon Processing for the VLSI Era*, Lattice Press, Sunset Beach, Calif., 1986, p. 332.
2. S. M. Sze, *VLSI Technology*, McGraw-Hill, New York, 1983, p. 347
3. Y. Pauleau, "Interconnect materials for VLSI circuits," *Solid State Technology*, Feb. 1987, p. 61.
4. D. M. Brown, "CMOS contacts and interconnects," *Semiconductor International*, p. 110ff.
5. A. J. Aronson, "Fundamentals of sputtering," *Microelectronic Manufacturing and Testing*, Jan. 1987, p. 22.
6. J. Ballingall, "State-of the-art vacuum technology," *Microelectronic Manufacturing and Testing*, Oct. 1987, p. 1
7. S. Wolf and R. Tauber, *Silicon Processing for the VLSI Era*, Lattice Press, Sunset Beach, Calif., 1986, p. 95.

14

Wafer Test and Evaluation

Overview

The wafer-fabrication process requires a high degree of precision. Because one mistake can render the wafer completely useless, it is imperative to identify out-of-spec or potentially low-yield wafers and process problems as soon as they arise. Throughout the process a variety of tests and measurements are made to judge wafer and process quality.

Objectives

Upon completion of this chapter you should be able to:

1. Explain the difference between resistance, resistivity, and sheet resistance.

2. Draw a sketch of the parts and current flow in a four-point probe.

3. Compare the principles and uses of color interference, fringe counting, spectrophotometers, ellipsometers, and stylus for film thickness measurements.

4. Compare the principles and uses of groove and stain, SEM, and spreading resistance for junction depth measurements.

5. List the methods and advantages of microscope and SEM inspection of wafer surfaces.

6. Draw sketches of diodes in forward and reverse bias and their companion current-voltage curves.

7. Explain the effect of surface current leakage on a junction performance characteristic.

8. Draw sketches of a bipolar and MOS transistor in operation and their companion current-voltage characteristics.

9. List the process steps for a capacitance-voltage measurement and the principle of contamination detection.

10. Explain the method used for pinhole detection.

Introduction

Characterization of the process and the circuit parameters is required for production-line control and product stability. Good characterization can warn of a process about to go out of control and device characterization is essential to analyze circuit performance and conformance to customer specifications. Consequently, every significant process step is followed by an evaluation of the process results. The parameters measured and the tests used have been identified in previous chapters. Here, we explain the basic theory, applicability, range of sensitivity, and frequency of measurement of those tests.

The term *test wafer* refers to blank wafers or wafer pieces that are included on the process wafer holders. Many of the tests are destructive and cannot be performed on the device wafers which have complete chips.

Resistance and Resistivity

The object of the fabrication process is to form, in and on the wafer surface, solid-state electrical components (transistors, diodes, capacitors, and resistors) that are wired together to form the circuit. Each of the components must meet individual electrical performance specifications if the entire circuit is to function as a whole. Throughout the process, electrical measurements are performed to judge the process and the electrical device performance.

Wafers are purchased and quality assurance tested to a specific resistivity specification before being put in the line. During the process additional measurements are made to ensure that the doped and doped deposited layers meet electrical design specifications.

Resistivity measurements

The addition of dopants to the wafer, both during crystal growth and during the doping processes, alters the electrical characteristics of the wafer. The altered parameter is its resistivity, which is a measure of a material's specific "resistance" to the flow of electrons (Fig. 14.1).

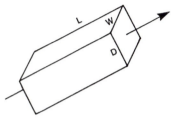

$$R = \rho\,\frac{L}{A} = \rho\,\frac{L}{W{\times}D}$$

Figure 14.1 Relationship of resistance to resistivity and dimensions.

Whereas the resistivity of a given material is a constant, the resistance of a specific volume of the same material is a function of its dimensions and resistivity. This relationship parallels that of density and weight. For example, the density of steel is a constant, whereas the weight of a particular piece depends on its size.

The units of resistance R are ohms (Ω) and the units of resistivity (ρ) are ohm-centimeters ($\Omega \cdot$ cm). Since adding dopants to a wafer will alter its resistivity, measurement of resistivity is actually an indirect measure of the amount of dopants added.

Four-point probe

The parameters of resistance, voltage, and current are governed by Ohm's law. The three parameters are related mathematically in the following way:

$$R = V/I = (\rho)L/A = (\rho)L/(W \times D)$$

where R = resistance
  V = voltage
  I = current
  ρ = resistivity of sample
  L = length of sample
  A = cross-sectional area of sample
  W = width of sample
  D = depth of sample

Theoretically, the resistivity of a wafer can be measured with a multimeter (Fig. 14.2) by measuring the voltage at a constant current through a sample of known dimensions and calculating the resistivity. However, the resistance between the probes and the wafer material is too great to accurately measure the resistivity of semiconductors with their relatively low quantity of dopants.

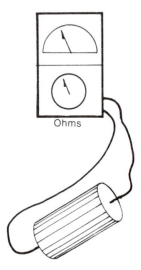

Figure 14.2   Multimeter.

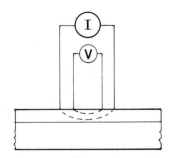

Figure 14.3 Four-point probe measurement of a thin layer.

The four-point probe is the instrument used to measure resistivity on wafers and crystals. It is an instrument with four, thin, in-line probes, connected to a power supply and voltmeter. A four-point probe consists of four thin metal probes arrayed in a line. The two outside probes are connected to a power supply and the inside probes are connected to a voltage meter. During operation, the current passes between the outer probes and the voltage drop (change) is measured between the inner probes (Fig. 14.3). The relationship of the current and voltage values is dependent on the resistance of the space between the probes and the resistivity of the material. The four-point probe cancels out the effects of probe-wafer contact resistance on the measurement.

Using a four-point probe, the voltage and current are related to the resistivity by the relationship:

$$\rho = 2\pi s V/I$$

where s is the distance between probes when s is less than the wafer diameter and less than the film thickness

Sheet resistance

The four-point probe measurement just described is used to measure the resistivity of wafers and crystals. It is also used to measure the resistivity of thin layers of dopants added into the wafer surface by the dopant processes. When a four-point probe measurement is made on a

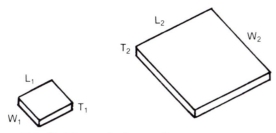

Figure 14.4 Resistance of a "square."

thin layer of added dopants, the current is confined in the layer (Fig. 14.3). A thin layer is defined as a layer thinner than the probe spacing (distance between probes).

The electrical quantity measured on a thin layer is called sheet resistance R_s. This quantity has the units of ohms/square ($\Omega/\square$). The concept of ohms per square can be understood by considering the resistance of two squares of the same thin material of equal thickness (Fig. 14.4). Since the resistivity ρ is the same for each piece and $T_1 = T_2$, the sheet resistance is the same for each piece. Or, the resistance of the thin sheet is a constant for any square of the same material.

The formula relating sheet resistance to the voltage and current is

$$R_s = 4.53 \ V/I$$

where 4.53 is a constant that arises from the probe spacing. Some companies elect to drop the constant 4.53 from the formula and just measure the V/I of a wafer as in Fig. 14.5.

Four-point probe thickness measurement

The thickness of uniform conducting layers on an insulating layer can be determined using a four-point probe. For thin films the formula is

$$T = \rho_s/R$$

where T = layer thickness
ρ_s = resistivity
R_s = sheet resistance

Since the resistivity is a constant for pure materials such as aluminum (Fig. 14.5), the sheet resistance measurement is actually a measurement of the film thickness. This formula does not calculate the

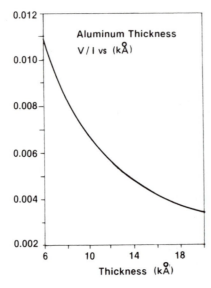

Figure 14.5 Voltage/current (*V/I*) versus thickness of aluminum.

thickness of a doped layer since the dopants are not evenly distributed throughout the layer.

Concentration profile

The distribution of dopant atoms in the wafer is a major influence on the electrical operation of a device. The distribution (or dopant concentration profile) is determined by the spreading resistance technique. After doping, a test wafer sample is prepared by the bevel technique (see p. 513). After the junction is exposed by the beveling, a series of electrical two-point probe measurements are made sequentially down the bevel (Fig. 14.6). At each point, the vertical drop of the probes is recorded and a resistance measurement made. The resistance value at each point changes with the change in dopants at each level.

A computer is used to perform calculations that relate the depth and resistance values to the dopant concentration at each level. The computer uses the data to construct a dopant concentration profile for the sample. This measurement is usually made periodically off-line or when electrical device performance indicates that the dopant distribution may have changed.

Another method is to incrementally remove thin layers from the wafer surface with the anodic oxidation technique. After the oxide is grown, it is removed by etching and a four-point probe measurement is made on the new surface. The distance down into the wafer is a

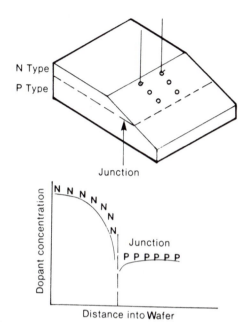

N Type

P Type

Junction

Dopant concentration

N N N N
N
N
N
N
N

Junction
P P P P P P

Distance into Wafer

Figure 14.6 Spreading resistance.

function of the oxide thickness and is related to the dopant concentration four-point probe sheet resistance values by a computer program.

Layer Thickness Measurements

Color

Both silicon dioxide and silicon nitride layers exhibit different colors on the wafer. We know that while silicon dioxide is transparent (glass is silicon dioxide), an oxidized wafer has a color. The color seen is actually the result of an interference phenomenon, the same phenomenon that creates the colors of rainbows.

The silicon dioxide layer on a silicon wafer is actually a thin transparent film on a reflecting substrate. Some of the light rays impinging on the wafer surface reflect off the oxide surface while others pass through the transparent oxide and reflect off the mirrored wafer surface (Fig. 14.7). When the light rays exit the film, they combine with the surface-reflected ray, giving the surface an appearance of having a color. This phenomenon is the reason oxidized wafers change color as the angle of viewing is changed.

The exact color is a function of three factors. One, which is a property of the transparent film material, is the *index of refraction*. A second factor is the *viewing angle*. The third factor is the *thickness* of the

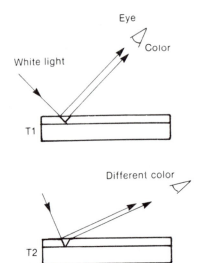

Figure 14.7 White light interference.

film. The color of a thin transparent film becomes an indication of the thickness when the nature of the viewing light is specified (i.e., daylight, fluorescent), along with the viewing angle. The classic color versus thickness chart (Fig. 14.8) is a regular feature at oxidation and diffusion stations. Color alone is not an exact indication of thickness because of the consequences of the interference phenomenon.

As the film gets thicker, the colors change in a specific sequence and then repeat themselves. Each repetition of the color is called an *order*. To determine the exact film thickness, a knowledge of the color order is necessary. A principal use of color charts is for process control.

Each oxidation or silicon nitride process is set up to produce a specified thickness. Naturally the thickness will vary from run to run. Operators quickly become sensitive to the wafer color. When a variation occurs, a quick check of the chart will indicate if the film thickness is out of specification. Rarely is a process so far off that the film thickness is a whole order (same color, different thickness) out of specification. The accuracy of color chart thickness determination is limited to the accurate perception of the colors (what exactly is red-orange?). A typical chart is accurate to ±300 angstroms (Å).

Fringes

When the order is not known, a fringe-counting technique can be used. When a test wafer edge is dipped in hydrofluoric acid for a few seconds the acid quickly eats through the oxide at an angle, leaving the oxide exposed to view (Fig. 14.9). When the wafer is viewed in

Film thickness, μm	Color* and Comments
Order I	
0.050	Tan
0.075	Brown
0.100	Dark Violet to Red Violet
0.125	Royal Blue
0.150	Light Blue to Metallic Blue
0.175	Metallic to Very Light Yellow
0.200	Light Gold or Yellow-Slightly Metallic
0.225	Gold with Slight Yellow Orange
0.250	Orange to Melon
0.275	Red Violet
0.300	Blue to Violet Blue
0.310	Blue
0.325	Blue to Blue Green
0.345	Light Green
0.50	Green to Yellow Green
Order II	
0.365	Yellow Green
0.375	Green Yellow
0.390	Yellow
0.412	Light Orange
0.426	Carnation Pink
0.443	Violet Red
0.465	Red Violet
0.476	Violet
0.480	Blue Violet
0.493	Blue
Order III	
0.502	Blue Green
0.520	Green (Broad)

*When viewed perpendicularly in daylight fluorescent light

Figure 14.8 Silicon dioxide thickness color chart. [*From IBM J. Res. Develop., 8, 43 (1964).*]

white light, colored fringes are formed between the wafer surface and the top of the film. Thickness determination is made by first determining the order of the film thickness. It is easy to see the repeated sequence of colors. If three blue-red fringes exist, the thickness corresponds to the surface color in the third order on the color-thickness chart.

A more accurate fringe-counting method uses monochromatic light as the viewing light. Monochromatic light consists of one color (wavelength), whereas white light is polychromatic (many wavelengths). The sample is prepared in the same way as it is for color fringe count-

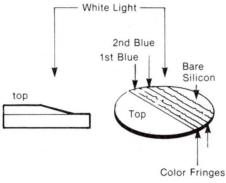

Figure 14.9   Color fringes.

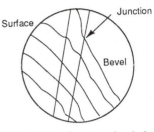

Figure 14.10 Junction depth determination by monochromatic fringes.

ing. However, in a microscope eyepiece, using monochromatic light, the fringes appear as alternating, evenly spaced black and white stripes (Fig. 14.10) with each fringe separation representing a specific vertical distance. Film thickness is determined by counting the number of fringes and multiplying by a correction factor. The correction factor is determined by the wavelength of the monochromatic light used. For sodium light the wavelength is 5890 Å.

Spectrophotometers

Film-thickness interference measurement techniques can be automated. To understand the method, let's review the interference effects. Light is actually a form of energy. The interference phenomenon can also be described in terms of energy. White light is really a bundle of rays (different colors), each with different energies. When the rays interfere through the transparent film, the result is a ray of one color, one wavelength, and one energy level. It is our eyes that interpret the energy as a color.

In a spectrophotometer, which is an automatic interference instrument, a photocell takes the place of the human eye. Monochromatic light in the ultraviolet range is reflected off the sample and analyzed by the photocell. To ensure accuracy, readings are made at different conditions. The conditions are changed by either using another monochromatic light (to change wavelength) or changing the angle of the wafer to the beam. Spectrophotometers specifically designed for use in semiconductor technology have onboard computers to alter the measurement conditions and calculate the film thickness.

Spectrophotometers are also used to measure silicon film thickness. Since silicon is opaque to ultraviolet light, an infrared source is used. The accuracy of spectrophotometer-type instruments stems from multiple measurements and changed measurement conditions. Spectrophotometers lose their accuracy for film thicknesses in the first order (below 1100 Å). For thinner transparent films, ellipsometers are the preferred measuring instrument.

Ellipsometers

Ellipsometers are film-thickness instruments that use a laser light source and operate on a different principle than a spectrophotometer. The laser light source is polarized. Polarization is the creation of a wave with all the rays traveling in only one plane. Polarization can be imagined by considering looking into the beam of a flashlight. In an ordinary beam light rays come to your eyes in many planes, like an arrow with many feathers. A polarized beam has all of the light in only one plane, or an arrow with only one feather (Fig. 14.11).

In the ellipsometer the polarized beam is directed to the oxide-covered wafer at an angle. The beam enters the transparent film and reflects off the reflective wafer surface. During its passage through the film the angle of the beam plane is rotated. The amount of rotation of the beam is a function of the thickness and index of refraction of the film. A detector in the instrument measures the amount of rotation and an onboard computer calculates the thickness and index of refraction.

Ellipsometers are used to measure thin oxides (50 to 1200 Å) and the index of refraction of the film. Their accuracy in this range is unequaled by any of the other techniques.

Stylus

Some thin films such as aluminum cannot be measured by optical techniques. And in the case of aluminum and other very thin conductive films, the four-point probe thickness measurement is not suffi-

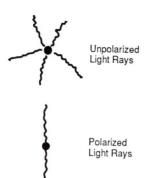

Unpolarized
Light Rays

Polarized
Light Rays

Figure 14.11 Unpolarized and polarized light.

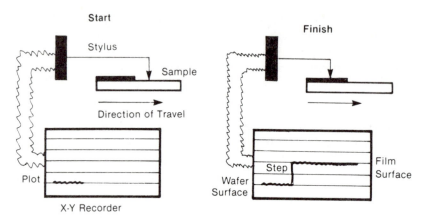

Figure 14.12 Step height measurement.

ciently accurate. In these situations, a mechanical moving-stylus apparatus is used (Fig. 14.12). The method requires that a portion of the film be removed, creating a step on a test-wafer surface. This is normally done by a masking and etching step. The prepared sample is mounted and leveled on the pivoting stylus instrument stage.

After leveling, the measuring stylus is lowered gently onto one of the surfaces. The measurement is made as the stage is slowly moved under the stylus. As the stylus goes over the step, its physical position is changed. The stylus is linked to an inductor that generates an electrical signal in response to the stylus vertical position. This signal is amplified and fed into an x-y recorder.

While the leveled wafer is moving under the stylus, it does not move in the vertical direction and no change in signal is produced. The trace on the x-y plotter is a straight line. When the stylus reaches the surface step, it changes position, causing a change in the signal output. This change is evidenced by a change of pen position on the x-y chart trace. The change in position is relative to the step height, which is read directly from the calibrated x-y chart.

Junction Depth

A critical device parameter is the junction depth of the various doped regions. This parameter is measured after each of the doping steps. The measurement methods described are all performed off-line; that is, the test wafers or device wafers have to be taken to a measurement station or laboratory for the measurement.

Groove and stain

Junction depths are measured by the groove (or bevel) and stain technique. Grooving or beveling is a mechanical method of exposing the

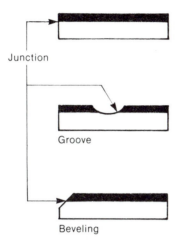

Figure 14.13 Exposure of junction by groove or beveling.

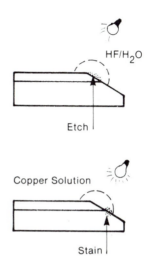

Figure 14.14 Etching or staining of junction.

junction for viewing and measurement from the horizontal plane (Fig. 14.13). The extreme shallow depth of the junction requires either grooving or beveling of the wafer to expose the junction.

The junction itself is not visible to the naked eye. Two techniques, called *junction delineation*, are available to make it visible. Both techniques utilize the electrical differences between N-type and P-type regions. The first technique, the etch technique, starts with the placement of a drop of hydrofluoric acid and water mixture over the junction (Fig. 14.14). A heat lamp is directed onto the exposed junction. The heat and light cause holes or electrons to flow in each region. As a result of the flowing current, the etch rate of the HF-H_2O mixture is higher on the N-type region, making it appear darker.

The second delineation technique is electrolytic staining. A mixture containing copper is dropped on the exposed junction. Again, the heat lamp is directed onto the junction. In effect a battery is formed, with the poles of the junction being the poles of the battery and the copper solution being the electrolytic connection. The current flowing in the liquid drop causes the copper to plate out on the N-type region side of the junction.

The final step, after exposure and delineation of the junction, is depth measurement. An interference method is used, in which a monochromatic light is directed through a thin piece of glass positioned over the groove or bevel (Fig. 14.15).

The fringes are normally created through an attachment to a microscope. The fringes are observed through it. The junction depth is read by counting the number of superimposed fringes from the surface

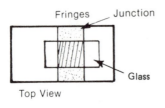

Top View

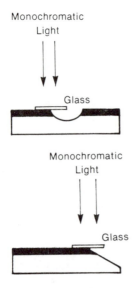

Figure 14.15   Creation of fringes.

down to the junction. For a given wavelength, each fringe represents a specific depth. A common light source is sodium. The junction depth (X_j) is calculated from the wavelength of the light and the number of fringes to the junction. Accurate fringe counting is important. Usually a photograph of the fringes superimposed over the junction is made through the microscope, and the fringes are counted from it (Fig. 14.16).

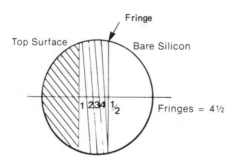

Figure 14.16   Fringe counting.

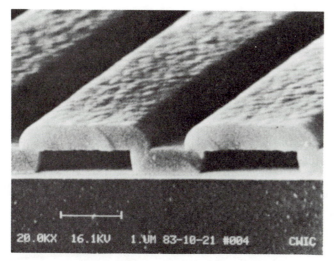

Figure 14.17 SEM declination of device cross section.

Scanning electron microscope (SEM) thickness measurement

The SEM technique (described below) can also be used to measure junction depths and film thickness. The wafer is scribed and broken with the break position over the junction. The exposed wafer junction is delineated by one of the methods described.

The exposed cross section is positioned to the SEM beam at right angles to the wafer surface and a photograph is taken. The depth is determined from the photograph and the scale factor of the SEM (Fig. 14.17).

Spreading resistance

Spreading resistance is a technique also used for measuring the junction depth (see p. 364). As the probes pass through the junction, they sense the change in conductivity type (N or P). This information, when plotted on the profile curve, also gives the junction depth (Fig. 14.6).

Contamination and Defect Detection

Detection of contamination and visual defects is essential to high yields and process control. Particulate contamination is detected primarily by visual techniques including high-intensity lights, microscopes, SEMs, and automatic machines. Chemical contamination is detected and identified by both Auger and ESCA (see p. 377) techniques. Mobile ionic contamination in the wafer is detected by capacitance-voltage plotting and by interpretation of the transistor and diode electrical tests. Many other sophisticated techniques can be used to make these inspections. The ones described here are those employed in a typical wafer-fabrication line.

1× Visual inspection

The first line of defense is to look at the wafers, which is a magnifying power of 1, or, in microscope terminology, a 1× power. Operators quickly become sensitive to the way "normal" wafers look. Even minor changes in the surface appearance are picked up by the experienced eye.

1× Collimated light

The resolving power of the naked eye (1×) can be assisted by using a high-intensity white light, like the beam of light from a slide projector (Fig. 14.18). Particulate contamination is highlighted in the light beam when the wafer surface is viewed at an angle. The effect is similar to the highlighting of dust in the air by light streaming through a window.

1× Ultraviolet

In actuality, the eye cannot see ultraviolet light, but ultraviolet light from a mercury-vapor lamp emits blue, green, and even some red light. Since the ultraviolet is harmful to the retina, a filter is frequently placed over the light source to block out the ultraviolet. The primary benefit of the ultraviolet lights used in fabrication areas is that they are very bright, which means the intensity of the scattered light is greater, therefore increasing the detection of surface contamination.

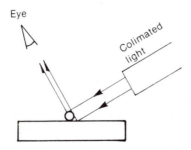

Figure 14.18 Collimated light inspection.

Light-field microscope.   The metallurgical microscope is the workhorse tool of surface inspection. The term *metallurgical* differentiates it from the standard microscopes found in biology labs. A biological microscope illuminates transparent samples by shining the light up through the sample. In a metallurgical microscope, the light is passed down to the nontransparent sample through the microscope objective (Fig. 14.19). The light reflects off the sample surface and is transmitted back up through the optics to the eyepieces. With white light illumination, the picture in the field of view exhibits the surface colors, which helps identify particular components on the wafer. Use of filters will change the surface colors.

A typical fabrication microscope is fitted with 10× or 15× eyepieces and a range of objectives from 10 to 100×. Increasing the total viewing power (eyepiece power times the objective power) reduces the field of view. This reduction requires more inspection time for the operator to look at the required sample inspection area on the wafer. The consequence is a slower inspection process. A trade-off power level, when inspecting LSI and VLSI devices, is 200 to 300× magnification.

The industry typically uses a microscope inspection procedure requiring the operator to look at three to seven specific locations on the wafer. This procedure is easily automated with motorized stages. Most of the automated microscope inspection stations feature automatic wafer placement on the stage, automatic focusing, and automatic binning of the completed wafers. At the touch of a button, the operator can direct each wafer to a boat for passed wafers or into one of several boats for different reject categories. Obviously, a microscope inspection procedure is used to judge surface and layer quality and (in masking) pattern alignment.

Dark-field inspection.   Dark-field illumination is achieved by fitting a metallurgical microscope with a special objective (Fig. 14.19). In this

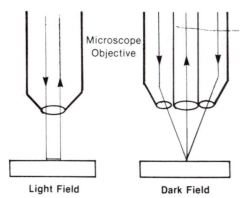

Microscope
Objective

Light Field   **Dark Field**

Figure 14.19 Light and dark field inspection.

objective, the light is directed to the wafer surface through the outside of the objective body. It impinges on the surface at an angle and passes up through the center of the objective. The effect on the "picture" in the eyepieces is to render all flat surfaces black. Any surface irregularities, such as a step or pieces of contamination, appear as bright lines. Dark-field illumination is more sensitive than light-field to any surface irregularity. It has the drawback of limiting the operator's ability to discern the nature of the surface irregularity. A passable surface dimple may look the same as a rejectable piece of contamination. In practice, dark-field illumination is rarely used in on-line inspections but is reserved for diagnostic purposes.

Advanced microscope techniques. Optical technology is capable of providing many evaluation techniques beyond simple light- and dark-field viewing such as phase contrast and fluorescence microscopes. Each allows the viewer to determine additional visual information about the surface. Phase contrast brings out surface irregularities in the vertical plane and fluorescence-illuminated microscopes use ultraviolet illumination sources. In the ultraviolet light, organic residues (photoresist, cleaning chemicals) not easily visible in white light are brought into view. In practice, these instruments are confined to quality assurance laboratories. Their use and interpretation generally require technicians trained beyond the level of production operators.

Scanning electron microscope (SEM). Conventional optical microscopes are limited in their ability to provide accurate information about the wafer surface. First, their resolving power is limited by their optical light source. The ability of a viewing system to distinguish detail is related to the wavelength of the light (radiation used). The shorter the wavelength, the smaller the detail that can be seen.

Depth of field is another viewing factor. It relates to the ability of the system to keep two planes in focus simultaneously. A conventional photograph with the subject in focus and the background out of focus has a background beyond the depth of field limit of the camera. In a microscope, the depth of field decreases as the power (magnification) of the system is increased. If the power is increased to see the surface "closer," the operator may not be able to see the top and bottom surfaces in focus. Constant refocusing results in loss of information and a slower inspection time.

Magnification is the third limiting factor of optical microscopes. An optical system with white light illumination is limited to about $1000 \times$ magnification with conventional objectives. The oil immersion

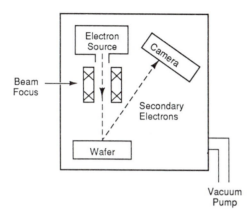

Figure 14.20 SEM analysis.

technique pushes the limit up, but it is unacceptable because it is too slow, too messy, and a possible source of contamination to the wafer.

All three limitations are overcome by using a scanning electron microscope. The microscope varies from an optical one in many aspects. The "illumination" source is an electron beam scanned over the wafer or device surface. The impinging electrons cause electrons on the surface to be ejected. These secondary electrons are collected and translated into a picture of the surface (Fig. 14.20) on either a screen or a photograph.

SEM requires that the wafer and beam be in a vacuum. The electron beam has a much smaller wavelength than white light and allows the resolution of surface detail down to submicrometer levels. Depth of field problems do not exist; every plane on the surface is in focus. Magnification is similarly very high, with a practical upper limit of $50,000\times$. A tilting wafer holder in an SEM allows the viewing of the surface at angles, which enhances the three-dimension perspective (Fig. 14.17). Surface details and features can be viewed at advantageous angles.

Some materials such as photoresist do not give off secondary electrons in response to E-beam bombardment. To inspect a photoresist layer in a SEM the resist layer is covered with a thin layer of evaporated gold. The gold layer conforms to the topography of the photoresist layer. Under E-beam bombardment the gold gives off secondary electrons, thus resulting in a SEM picture that is an accurate reproduction of the underlying photoresist layer.

Auger analysis

In an SEM, a range (spectrum) of secondary electrons is released by the impinging electron beam. One portion of this spectrum is the electrons that are released from the top several nanometers of the surface.

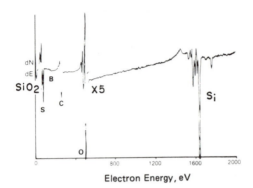

Figure 14.21   Typical auger trace.

These electrons, known as *Auger electrons*, have energies characteristic of the element that emits them. Thus sodium and chlorine each give off different Auger electrons.

The collection and interpretation of Auger electrons allows the identification of the surface materials, including contamination. In operation the E-beam is scanned across the wafer. The ejected Auger electrons are analyzed for their energies (wavelengths) and printed out on an *x-y* plotter (Fig. 14.21). Energy peaks at specific wavelengths indicate the presence of specific elements on the surface.

Scanning Auger microanalysis (SAM) is limited to the identification of elements. This technique cannot identify the chemical state of the element, what it may be combined with, or its quantity on the surface. Salt (NaCl) contamination is identified only as the presence of sodium and chlorine on the surface.

Electron spectroscope for chemical analysis (ESCA)

Solving surface contamination problems often requires a knowledge of the chemical state of the contaminant. An Auger detection of chlorine on the surface does not reveal whether the chlorine is present as hydrochloric acid or a trichlorobenzene. Knowledge of the form of the chlorine expedites locating and eliminating the process source of contamination.

The electron spectroscope for chemical analysis (ESCA) is an instrument used to determine surface chemistry. The instrument works on principles similar to the Auger technique. However, x rays instead of electrons are used for the bombarding radiation. Under bombardment, the surface gives off photoelectrons. Analysis of the photoelectron information leads to the determination of the chemical formula of the contamination. Unfortunately the ESCA x-ray beam is wider than most integrated circuit features. The beam diameter limits the technique to a macro surface analysis. By contrast, an Auger electron beam can zero in on specific bits of contamination.

Laser scanning

The detection of ever smaller sized particles on the wafer surface has led to the use of laser beams as the detection illumination. Two advantages accrue from the use of lasers. First is the obvious ability to detect smaller particle sizes (Fig. 14.22) due to the high intensity of the reflected light, which makes very small surface particles detectable (helium and neon lasers are the usual sources).

The second advantage is automation. Laser inspection equipment is easy to automate so that the inspection is automated from cassette to cassette and the quantity and size of the contamination on the surface can be determined. The information is used to produce a map of the wafer surface showing the size, location, and density of surface contamination. This last factor is a big help in process control. Laser inspection of masks has become a standard production technique.

Image recognition

Advances in image processing have allowed automatic defect and pattern distortion detection instruments. The machines mate image processing and computer technology. They have scanners with laser or optical light sources that move over the wafer surface. In one version, the computer is preprogrammed with the design pattern for the circuit. The scanner looks for added or missing parts from the expected pattern. Their location is recorded and surface maps printed out. Engineers can go back to the wafer or mask and determine the exact nature of the problem.

Another system compares adjacent die on the wafer or mask. A die is scanned and the pattern recorded in the computer memory. A second die is also scanned and any deviations between the two are recorded. This system will not detect any repeated pattern defect that occurs on every die but will pick up random defects that have a very low probability of occurring in exactly the same spot on two adjacent die.

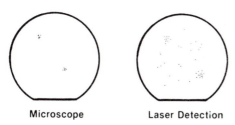

Microscope   Laser Detection

Figure 14.22 Laser detection of defects.

Cold plate

Wafer-surface stains from solvents and water are often difficult to detect using conventional microscope techniques. An alternative method to detect them is a cold-plate method. This technique is also called the *freeze-plate method*. The apparatus consists of a wafer vacuum chuck that is cooled by circulating cold water, circulating freon, or a thermal electric device. The wafer is placed on the chuck and rapidly cooled to below room temperature (Fig. 14.23). The rapid cooling causes water in the air to condense on the wafer surface. Actually, the condensation occurs first on hydrophilic regions such as stains. Observation of the wafer surface, just as the condensation is forming, reveals any surface stains or other anomalies.

This technique is also used prior to metallization deposition to determine if the contact holes are free of an oxide layer. In this version of the test, the contact holes are observed through a microscope as the condensation begins. If the contact holes are free of oxide, the condensation will form on the surrounding field oxide before forming in the contact holes.

In both uses of the test, the observation must be made immediately as the condensation begins. After a minute or so the entire wafer surface becomes covered with condensation and any information about the surface is obliterated.

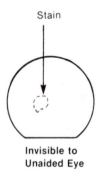

Stain

Invisible to
Unaided Eye

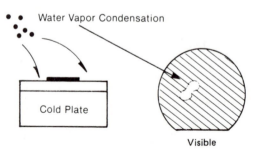

Water Vapor Condensation

Cold Plate

Visible

Figure 14.23   Cold plate.

Critical Dimensions Measurement

The exact dimensions required of each component in the circuit are controlled and influenced by *all* processes. In the doping area, the depth of each doped layer is controlled by the process conditions. Likewise, CVD processes are set to produce the required layer thicknesses. The horizontal surface dimensions are produced in the photomasking area. As part of that process the critical dimensions are measured at both develop inspection and final inspection with microscopes. Two measurement techniques are used.

Filar measuring eyepiece

The filar measuring eyepiece is a dimension-measuring instrument that is fitted to a microscope. The eyepiece features a movable hairline that requires calibration to an outside standard, usually a stage micrometer. This instrument is a glass slide that has an accurate grid etched in the center (Fig. 14.24). For calibration a segment of the grid is measured through the filar microscope. A calculation is made to "correct" the filar measurements to the actual dimensions on the stage micrometer.

In manual operation, the pattern to be measured is focused in the field of view (Fig. 14.25). The eyepiece control is rotated to position the hairline perpendicular to the measuring location path. Once oriented, the hairline is moved to the starting point of the pattern to be measured, and the value on the micrometer barrel is noted. The hairline is

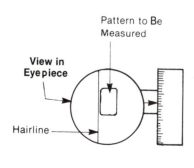

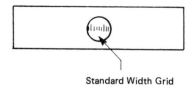

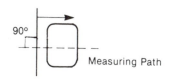

Figure 14.24 Manual filar measurement.

Figure 14.25 Manual filar width measurement.

then moved in a smooth continuous motion to the other side of the pattern. The ending value is also noted.

The actual width is calculated by subtracting the starting value from the ending value and multiplying the resulting number by the previously determined correction factor. This system is highly accurate and is used by the National Bureau of Standards. Accurate measurements require good operator techniques.

Filar systems are easily automated. The hairline movement mechanism can be motorized and the correction factor programmed into an onboard computer, resulting in a direct digital readout of the actual width. Operator fatigue is minimized by a video monitor rather than requiring the operator to view the wafer through a high-power microscope.

Image shearing

An image-shearing attachment on a microscope is another method of critical-dimension measurement. In the manual version, the operator rotates the pattern to be measured to a vertical position in the field of view. A control on the unit allows the operator to optically separate (shear) the pattern into two images. To start the measurement, the two images are butted against each other (Fig. 14.26). The operator notes the value of this position on the shearing control. In step two, the shearing control is rotated until the images are rejoined into one. Again the ending value on the shearing control is noted. The difference between the starting and ending values is the width of the pattern in "image shearing units" multiplied by a previously determined correction factor. The correction factor is determined by use of a stage micrometer, exactly like the calibration of a filar eyepiece.

The procedure just described is called *single-image shearing* because the pattern is sheared once. Shearing the image twice is another technique. This is accomplished by following the same procedure used in single shearing but continuing the shear until the image is again doubled (Fig. 14.27). The pattern is actually sheared twice and the calculated width must be divided by 2. This procedure can be automated for production work. The images are displayed on a cathode-ray tube and the shearing motion motorized. And like the filar automated measure-

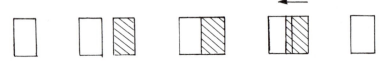

Figure 14.26   Single-image shearing.

Figure 14.27   Double-image shearing.

ment, a correction factor is programmed into the unit. The width of the pattern is read directly from the unit.

Reflectance

Both the filar and image-shearing techniques require some operator decision, which can be a source of error. A third type of dimension-measurement instrument is based on reflectance. Like the other two, the operator locates the pattern to be measured on a monitor. One edge of the pattern is positioned under a marker on the screen. The measurement is made automatically as a laser beam is swept along the direction of measurement and the reflection energy of the beam recorded (Fig. 14.28). When the beam comes to the edge of the pattern, it steps up (or down) to a new surface. The new surface causes a different reflectance of the beam which is recorded by the detector. The width of the pattern is the difference between the starting point and the point where there is a change in reflectance. This value is read out automatically. Operator decision is limited to determining the starting point, resulting in a more constant accuracy.

The limitation of this method is with patterns that have a number of steps in them, such as metal steps. The instrument is designed to read to the first change in reflectance it senses. Another limitation is in measuring pattern edges that are sloped. The reflecting beams change gradually rather than abruptly, making a width determination difficult.

Some dimension-measuring microscopes use transmitted light for the measurement of mask and reticle critical dimensions. A light beam is passed from the bottom through the mask. The pattern to be measured is moved in front of the beam and a detector above the mask

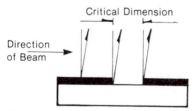

Figure 14.28   Reflectance CD measurement.

Method	Visual contamination	Surface Defects	Alignment	Contamination Elements	Contamination Compounds	C.D.'s
1x Incident White Light	X	X				
1x Incident U.V. Light	X	X				
Microscope- Light field	X	X	X			
Dark field	X	X	X			
S.E.M.	X	X	X			X
Auger				X		
E.S.C.A.					X	
Filar						X
Image Shearing						X
Reflectance						X

Figure 14.29   Overview of surface inspection techniques.

senses when the light is blocked by the opaque pattern. The horizontal distance is measured until the light again is detected by the sensor. The distance between the two is translated into the pattern dimension. A summary of the various surface-inspection techniques and their use in a fabrication process is tabulated in Fig. 14.29.

Device Electrical Measurements

During the process, it is necessary to make measurements of the actual device parameters. These measurements are usually made on special devices in the test die. At the end of the process, the actual devices are more fully characterized. This operation has various names, including electrical test, E-test, and pre-sort. (Readers unfamiliar with semiconductor device operation should read Chap. 16.)

A great deal of information about the process is available from these electrical tests. In this section we explain the basic tests and several of the more obvious and frequent device failures. A complete device and process troubleshooting guide is beyond the scope of this text.

Equipment

The basic equipment required to perform device electrical tests are a probe machine with the capability of positioning needlelike probes on the devices, a switch box to apply the correct voltage, current, and polarities to the device, and a curve tracer to display the results (Fig. 14.30). A curve tracer is a special-design oscilloscope set up to display

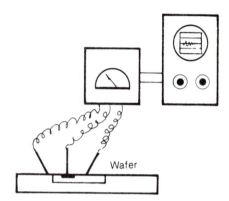

Figure 14.30 Device measurement equipment.

Wafer

voltage on the horizontal scale and current on the vertical scale.

In advanced systems, the probe station may be automated to sequentially test several die. The switching can also be automated to allow the equipment to perform a series of tests in a predetermined sequence. In some systems the results of a specific test may be displayed on a digital display rather than an oscilloscope screen. Automated systems also include hard-copy printouts of the test results as well as analysis of the data. The individual tests will be explained, as performed on manual equipment. The shape and relationship of the traces displayed on the oscilloscope are very helpful for understanding device operation.

All the device measurements are made in basically the same way. A voltage is applied to the component contact probe and the resultant current flowing between the contacts is measured. The results are displayed on the oscilloscope screen. The shape of the trace is governed by the dimensions of the device, the doping, and the presence of junctions. The current is modified by the resistance of the component or the presence of junctions. Properly working components will display known traces. The fundamental relationship is that of Ohm's law:

$$R = V/I$$

Resistors

Resistor measurements are made by contacting each end of the resistor and applying a voltage. The current passing through the resistor is a result of the resistance of the material between the probes. The oscilloscope allows the variation of the applied voltage, from zero to higher values. On the screen the voltage value is displayed on the x axis, with current measured on the y axis (Fig. 14.31).

The value of the resistance is calculated by dividing one of the voltage values by the corresponding current value. One might ask, "Why

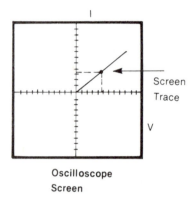

Oscilloscope
Screen

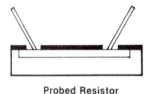

Probed Resistor

Figure 14.31   Resistor.

not determine the resistance by simply measuring the voltage and current values with a meter?" In other words, why display the values on the screen? The answer lies in the quality of information gained from the trace. A resistor's V/I relationship should be a linear one (straight line). Any deviation from linearity could indicate a process problem, such as high contact resistance or leaking junction.

Diodes

Diodes function as switches in a circuit. This means that a diode can pass current in one direction (forward bias) and not in the other (reverse bias). Checking diode operation in either direction requires probing the diode with the proper polarities, as shown.

As the voltage is increased in the forward direction, current immediately starts flowing across the junction and out of the diode (Fig. 14.32). The initial resistance to that flow comes from contact resistance and a small resistance at the junction. After the resistances are overcome there occurs a "full" flow of current through the diode. A diode is designed to have this condition occur at some minimum voltage value. If the diode forward voltage value exceeds the design value, it is out of specification. This voltage is called the *forward voltage.*

In the reverse direction, the diode is designed to block current flow as long as the voltage stays below a specified value. In the reverse condition a small current, called a *leakage current,* always flows across

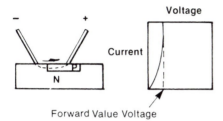

Figure 14.32   Diode forward bias.

the junction (Fig. 14.33). Eventually the applied voltage can reach a value that causes a breakdown of the junction, allowing "full" current flow. This *breakdown voltage* is a design value of the diode. Circuits are designed to operate at a voltage level below the designed breakdown voltage of the diode. If improper processing results in mistakes or excessive contamination, the diode will *break down* at a lower voltage.

Junction breakdown is normally a temporary condition. Exceeding the breakdown voltage does not permanently damage the junction, unless the applied voltage is extremely high. In that case the diode (junction) can sustain physical damage from a high current flow.

A second value determined during this test is the leakage current or current at breakdown. A small amount normally occurs as illustrated above. Contamination and/or improper processing can result in additional leakage current.

Trace 1 in Fig. 14.34 shows a diode with a small amount of leakage. The amount of current increases as the voltage is increased. Eventually, the breakdown voltage is reached and the diode becomes fully conducting. In trace 2, gross leakage is demonstrated. The junction leaks current with every increase in voltage, and the problem is so se-

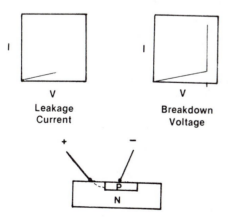

Figure 14.33   Diode reverse bias measurement.

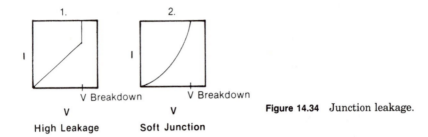

Figure 14.34 Junction leakage.

vere that a breakdown level is never reached and the diode never operates as a current block.

Oxide rupture (BV$_{ox}$ or rupture voltage)

An electrical measurement, BV$_{ox}$, is used as a measure of oxide quality. The test structure used is the same as for capacitance-voltage (C/V) analysis. But in this case, the voltage is continually increased until the oxide is physically destroyed and current flows freely from the aluminum dot to the silicon. The maximum voltage that the oxide can withstand before breakdown is a function of its thickness, structural quality, and purity.

Bipolar transistors

Bipolar transistors, as explained in Chap. 16, are three-region, two-junction devices. Electrically, they can be thought of as two diodes back to back. Many tests are performed to characterize bipolar transistors. The individual junctions are characterized separately and the whole transistor operation measured. The parts (or junctions) are probed for forward and reverse characteristics. The breakdown voltage (BV) tests are followed by the probing of the entire transistor.

Individual junction probe measurements are designated by the letters BV followed by lowercase letters that indicate the particular junction being probed. BV$_{cbo}$ indicates the breakdown voltage measured between the collector and base regions. The "o" indicates that the emitter region is "open"—it has no voltage applied to it. BV$_{ceo}$ is the breakdown measured between the collector and emitter.

The forward voltages of the two bipolar structure junctions are also measured. V_{be} is the forward voltage of the emitter-base junction and V_{bc} is the forward voltage of the base-collector junction. BV$_{iso}$ probes the collector-isolation junction for leakage currents.

The principal electrical measurement of a bipolar transistor is the beta (Fig. 14.35) measurement. This is a measurement of the amplification characteristic of the transistor. In Chap. 2 transistor operation

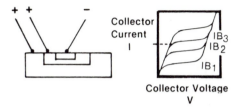

Figure 14.35   NPN transistor beta measurement.

was compared with the flow of water through a valve. In a bipolar transistor, the current flows from the emitter to the collector, through the base. The base current is varied to change the resistance in the base region. The amount of current flowing out of the collector (from the emitter to base) is regulated by the base resistance.

The amplification of the transistor is defined as the collector current divided by the base current. This number is known as *beta*. Thus a beta of 10 means that a 1-mA (milliampere) base current will give rise to a 10-mA collector current. The beta of a transistor is determined by junction depths, junction separations (base width), doping levels, concentration profiles, and a host of other process and design factors. Measurement of beta is done by a variation of the BV_{ceo} measurement. A BV_{ceo} measurement at a specific base current is performed. In this mode, the emitter base junction is forward-biased.

The collector characteristic of the transistor is displayed on the oscilloscope screen. The almost horizontal lines represent increasing base current values (IB_1, IB_2, etc.). With every increase in base current, a corresponding increase in collector current occurs. Calculation of the beta value takes place from the data displayed on the screen. Collector current is determined from the vertical axis (dotted line). The base current is calculated by multiplying the number of horizontal lines (steps) by the scale value for each step (from the oscilloscope).

MOS transistors

MOS circuits are also made up of resistors, diodes, capacitors, and transistors. The first three are measured by the same methods used to measure bipolar circuit components. Like the bipolar transistor, the MOS transistor is composed of three regions, in this case called the source, gate, and drain (Fig. 14.36). Measurement of this type of transistor consists of determining the reverse and forward values of the source and drain junctions. The functioning of the gate is determined by the threshold voltage test.

An MOS transistor has the source region forward-biased. Because of the high resistivity of the gate region, the forward current does not

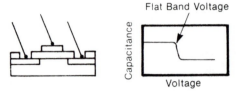

Figure 14.36 Threshold voltage measurement.

reach the drain. A voltage applied to the gate at a specified level (threshold) will cause enough charges to appear in the gate region to form a conducting channel that allows the source current to reach the drain region. Every MOS transistor is designed to operate at a specific threshold voltage. This value is measured using the capacitance-voltage technique. The gate voltage is continuously increased, while the capacitance of the gate structure is monitored.

A capacitor is a storage device. Initially, during the voltage-increase portion of the measurement, the capacitance does not change. At the threshold voltage level, the inversion layer forms and acts like a capacitor. Since two in-series capacitors have a combined lower capacitance than the sum of the two, the result is a drop to a combined lower capacitance. MOS transistors also exhibit amplification characteristics. The gain is defined as the source-drain current divided by the gate current. The source-drain characteristic for various gate currents is shown (Fig. 14.37).

Capacitance-voltage plotting

A variation of the threshold voltage test is used to test for the presence of mobile ionic contamination in the oxide. The test is performed on specially prepared test wafers. A thin oxide is grown on a "clean" silicon wafer. After oxide growth, aluminum dots are formed on the wafer by evaporation through a mask (Fig. 14.38). Dot evaporation is usually followed by an alloy step to ensure good electrical contact between the aluminum and oxide.

$$\text{Gain} = \frac{\Delta I_{DS}}{\Delta V_{GS}}$$

Figure 14.37 Gain characteristic of an MOS transistor.

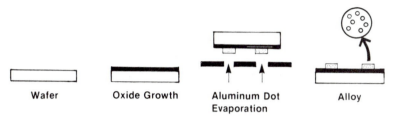

Figure 14.38 Preparation of *C/V* test wafer.

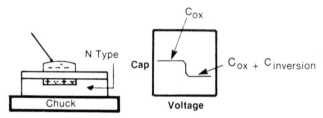

Figure 14.39 *C/V* plotting—first plot.

The "dotted" wafer is placed on a chuck and a probe is placed on the aluminum dot. This structure is actually an MOS capacitor. A voltage is applied to the dot and gradually increased as the capacitance of the structure is simultaneously measured. The results are printed out on an *x-y* plotter with capacitance on the *y* axis and voltage on the *x* axis (Fig. 14.39).

At a voltage level known as the *threshold voltage* (or inversion voltage) charge starts to build up at the silicon surface. The charges "invert" the conductivity type from N-type to P-type. The inverted layer has a capacitance of its own. Electrically, the structure now has two capacitors in series. The total capacitance value of the two is less than the sum of the two by the relationship

$$1/C_{total} = 1/C_{oxide\ layer} + 1/C_{inversion\ layer}$$

The trace on the *x-y* plotter drops vertically to the new capacitance level.

The second step in the process is to force the mobile positive ions in the oxide to the SiO_2-silicon interface. This is done by simultaneously heating the wafer to the 200 to 300°C level and placing a positive 50-V bias on the structure (Fig. 14.40). The elevated temperature increases the mobility of the ions and the positive bias "repels" them to the oxide-silicon interface.

The last step in the process is a repetition of the initial *C/V* plot. However, as the voltage increases, inversion does not start at the same level as in the initial test (Fig. 14.41). The positive charges at

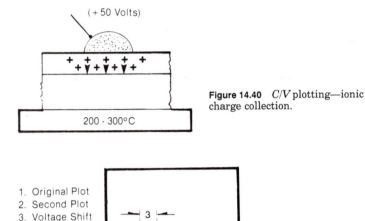

Figure 14.40 *C/V* plotting—ionic charge collection.

1. Original Plot
2. Second Plot
3. Voltage Shift
   or Drift

Figure 14.41 *C/V* replot.

the interface require additional negative voltage to "neutralize" them before inversion can happen. The result is a *C/V* plot identical to the original but displaced to the right. The additional voltage required to complete the plot is known as the *drift* or *shift*.

The amount of the shift is proportional to the amount of mobile ionic contamination in the oxide, the oxide thickness, and the wafer doping. *C/V* analysis cannot distinguish the element (Na, K, Fe, etc.) that was in the oxide, only the amount. Neither can this test determine where the contamination came from. It may have come from the wafer surface, any cleaning step, the oxidation tube, the evaporation process, the alloy tube, or any other process(es) the wafer has been through.

C/V analysis is usually a part of the evaluation of any process changes that may contaminate a wafer, such as a new cleaning process. To make the evaluation, *C/V* wafers are divided into two groups. One group receives normal processing as detailed above. The second group goes through the proposed cleaning process, usually between the oxidation and aluminum steps of the test. The drift on the standardly processed wafers is compared with the drift of the experimental group. An increased drift on the experimental wafers would indicate that the proposed cleaning process actually contaminated the wafers with mobile ionic contamination.

Acceptable *C/V* drifts vary between 0.1 and 0.5 V, depending on the sensitivity of the device being made. *C/V* analysis has become a standard test in a fabrication area. The test is made after any process change, equipment maintenance, or cleaning that could have the potential of contaminating the wafers. The C/V plot provides a wealth of other information, such as flat-band voltage and surface states, to the process engineer, in addition to the voltage drift.

Pinhole Counting

A photoresist film will contain circular discontinuities too small to be resolved at normal magnifications. To test for pinholes the resist is spun on an oxidized wafer. The determination of which "dots" are actually pinholes is made by processing the wafers through a photomasking step. The resist-coated wafer is either flood exposed or exposed through a blank mask. After exposure and development only the actual pinholes are left in the resist film.

These pinholes are transferred into the oxide layer at an oxide etch step. One more step is necessary to differentiate the real pinholes from oxide film surface irregularities. The step is a silicon etch. The etch penetrates through the pinhole in the oxide layer and etches the silicon underneath (Fig. 14.42). The hole in the silicon is visible with a microscope as a pattern around the pinhole. The shape of the pattern is either a triangle for ⟨111⟩ material or a square for ⟨100⟩ wafers.

Normally a grid is inserted in the microscope eyepiece and the pinholes within the grid are counted. A typical specification for a VLSI resist is less than 1 pinhole per square centimeter. Figure 14.43 is a summary of the four-point probe measurements.

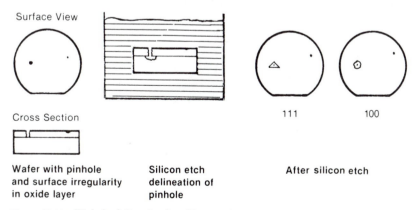

Surface View

Cross Section

Wafer with pinhole
and surface irregularity
in oxide layer

Silicon etch
delineation of
pinhole

After silicon etch

111   100

Figure 14.42   Pinhole delination by silicon etching.

QUANTITY MEASURED	FORMULA	CONDITIONS
Resistivity	$\varrho = 2\pi SV/I$	Layer thickness greater than probe spacing.
Sheet Resistance	$R = \pi V_{In} = 4.53\,V/I$	1. Layer thickness much smaller than probe spacing 2. Sample area larger than probe spacing.
Thickness of Conductors	$T = \varrho_s/R$	1. Layer thickness less than probe spacing 2. Uniform resistivity

Figure 14.43 Summary of four-point probe measurements.

Key Concepts and Terms

Contamination detection	Junction depths
Critical dimensions	Light-field microscope
C/V measurements	Measurements
Dark-field microscope	Resistance/resistivity
Device	SEM
Ellipsometer	Spectrophotometer
Four-point probe	Transistor measurement
Fringe counting	Visual inspections
Groove and stain	

Review Questions

1. Changing the size of a resistor will change which of the following?
   a. Resistance
   b. Resistivity
   c. Both
2. In a four-point probe the current is flowing in the inside probes. (true/false)
3. The color of an oxide film on silicon is an indication of its exact thickness. (true/false)
4. Which is the most accurate determination of a thin oxide layer (less than 1,000 Å)?
   a. Color
   b. Fringes
   c. Ellipsometer
   d. Spectrophotometer

5. Explain why an oxidized silicon wafer changes color when rotated.
6. Name a method to delineate an N-P junction for the groove and stain technique.
7. A collimated light is used to detect sodium contamination. (true/false)
8. An SEM uses which magnifying power as compared with a light-field microscope?

   a. Higher
   b. Lower
   c. The same

9. Name two methods of measuring CDs.
10. If an oxide layer is contaminated with sodium, what is the effect on the C/V drift?

    a. C/V drift will be higher
    b. C/V drift will be lower
    c. C/V drift will remain the same

15

Manufacturing Technology

Overview

Now into its fifth decade, the semiconductor industry has advanced to the manufacturing stage. While product and process innovations are still very much an industry focus, more emphasis is being placed on the "business" factors. These include automation, cost control, computer-automated manufacturing, computer-integrated manufacturing, and statistical process control.

Objectives

Upon completion of this chapter you should be able to:

1. List the major cost factors that influence fabrication costs.
2. Define and discuss the principle of payback.
3. List the advantages of statistical process control.
4. Identify the parts and use of a control chart.
5. List and discuss the different levels of automation.
6. List the factors that enter into an evaluation of a particular piece of equipment.
7. Define WIP, JIT, and their importance to production costs.
8. Sketch a "product manager"-oriented organizational chart.
9. Define the terms CIM and CAM and their use in a manufacturing setting.

Fabrication and Factory Economics
Overview

The semiconductor industry started supplying commercial products in the late 1940s. The manufacturing lines were little more than laboratories, and the workers were primarily trained technologists. By the 1970s, total factory sales of integrated circuits alone were some one billion dollars per year. The manufacturing scene had changed to clean rooms with highly specialized equipment attended by skilled production workers. A fabrication area of 2000 to 3000 ft^2 could be built for 2 to 3 million dollars.

The VLSI era dramatically changed the look and cost of wafer-fabrication areas. The size stayed about the same but the cost had shot up to some 100 million dollars. Most of the increase has been in the expense of building class 10 and 1 clean rooms and newer, even more specialized and automated equipment and process control systems.

Along with these increases have come increases in materials (wafer costs have grown faster than their area), labor, and chemicals. The newer processes require more steps which require more inventory expense. All these cost increases have come as individual chip sale prices continue to erode through improved productivity and competition. To stay profitable, chip manufacturers have to continually improve their efficiency, yields, and control costs at all levels. These requirements have driven the introduction of new systems and skills to the semiconductor plant.

Throughout all the changes in products, processes, facilities, and management systems, the overall financial measure of a fabrication area has remained the same: it is *the cost per functioning die shipped out of fabrication*. When extended to a complete merchant facility with assembly capabilities, the measure becomes *the cost per die shipped*. The remainder of this chapter examines the various factors and considerations that go into reducing the fabrication cost per die produced.

Wafer-Fabrication Costs

A number of factors contribute to the cost of producing a functioning die (Fig. 15.1). They are:

- Overhead—administration and facilities
- Materials—direct and indirect
- Equipment
- Labor
- Yield

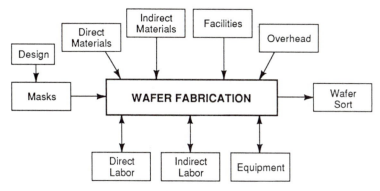

Figure 15.1 Fabrication cost factors.

The facility, overhead, equipment, and labor costs are usually considered fixed. They exist as a cost burden whether or not any die are produced. Over the long run, direct and some indirect labor become a variable cost as people are hired and fired to produce changing production levels. The variable costs are primarily the materials that fluctuate with the volume of wafers processed.

Overhead

Overhead costs are all those incurred by the administrative and executive staff plus the cost of providing and maintaining a facility. A curious fact of company growth is that after a certain level the number of administrative personnel grows faster than the manufacturing workers. The driving factor behind this is the growing importance of information in a growing company and the economy in general. As companies grow, more information is generated internally and more information must be handled from customers and suppliers. To be effective, the information must be available to an ever-growing staff. The two needs result in more and more staff processing "information" rather than product. Currently some 50 percent of the work force of the industrialized economies is involved in information processing.[1] A primary overhead expense is the design activity. With expensive CAD systems and a large professional design team, the cost of circuit design is considerable. Design cost goes up for ASIC fabrications where a large number of individual circuits are designed and produced with a fast turnaround requirement.

The facility cost and maintenance of the facility is a major contributing cost. A fabrication area occupies only about 20 percent of the total facility area, yet requires the majority of the expense. Air conditioning, chemical storage and delivery, and the cost of the clean room are all major expenses. Fabrication clean-room costs for a VLSI facility are as high as several thousand dollars per square foot.

Materials

Manufacturing materials are divided into direct and indirect. Direct materials are those that go directly in or on the chip. This includes the wafer materials, the added layer materials, and the packaging material costs. Indirect materials are the masks and reticles, chemicals, stationery supplies, and other materials that support the process but do not enter into the product.

Equipment

This cost is the equipment used directly in the fabrication of the devices and wafers. It shows up in the cost calculations as fixed overhead or as depreciation. Depreciation is the loss of value of the machines as they wear out or become obsolete.

Labor

Labor also has a direct and indirect component. Direct labor takes in those workers actually handling the wafers and equipment. Indirect labor is the support personnel such as supervisors, engineers, facility technicians, and office workers.

Yield

The overall fabrication yield (see Chap. 6) determines how the various costs affect the final die cost. If the die yield is low, the cost per die goes up. Not only are the fixed costs distributed over fewer die but the variable costs go up as more materials are required to get out the die. When the yield is calculated into the cost, the term used is *yielded die cost*. The cost of producing the wafers without considering die yield is the *unyielded die cost*.

Die costs are a function of the wafer size, die size, and wafer sort yield. A $300 wafer-manufacturing cost for a 6-in-diameter wafer with 300 die will translate into an unyielded die cost of $1 each. If the die sort yield is 50 percent, the die cost rises to $2 each ($300 wafer cost divided by 150 functioning die).

Cost factor distribution

Typical contributions of each of the factors to the overall unyielded die cost are shown in the following table.[2] The percentages will vary with the wafer and feature size, degree of automation, and number of process steps.

Factor	Contribution, %
Overhead and utilities	8
Materials (total)	44
Equipment depreciation	25
Labor	
Direct	17
Indirect	6

Actual wafer and die costs

In any discussion about wafer cost factors, one is naturally curious about the actual costs. This number changes from line to line, with the complexity of the devices and with the position of the product in the maturity cycle (Fig. 15.2). The more mature the product and process, the higher the yields and the lower the equipment depreciation factor. Newer products suffer from operator learning curves, equipment shakedown times, and development of new processes. The reader should realize that the costs given are circa 1989 and are going out of date as they are written. Wafer-fabrication costs are about $100 for discrete device products and simple integrated circuits. VLSI wafers cost in the $200 to $500 range to process to wafer sort.

Figures of merit

Aside from the financial cost figures, fabrication-area efficiencies are judged by several figures of merit. One is the number of wafers out per

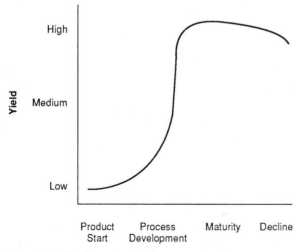

Figure 15.2 Wafer production lost versus product maturity.

operator. This number, for a reasonably defined process, is 16. A second figure is the number of wafers out per square foot of fabrication area. This number is about two wafers out per month per square foot of fabrication.

Yield Improvements

While it is true that more attention is being given to traditional business factors, no less attention is paid to process and wafer yields. The effect of a yield improvement can be significant in terms of dollars. Consider the yield figures in Fig. 15.3. Line 2 shows that a wafer-fabrication yield improvement of five percentage points increases the overall yield from 38 to 40.4 percent, a 1.2 percent increase. If the fabrication area starts 10,000 wafers per month and there are 350 die per wafer and the selling price is $5 per die, the increased revenue is

10,000 wafer starts × 350 die per wafer × 0.012 (1.2%) × $5

$$= \$210,000 \text{ per month}$$

Whether or not this amount of increased income is significant depends on the cost of effecting the improvement.

Statistical Process Control (SPC)

This text for the most part has presented the process technology with few mathematical formulas. The next few sections on statistical process control will break the general approach. SPC has emerged as a powerful and needed tool to maintain process control and improve yields. Unfortunately SPC is based on statistics, which is applied with the language of mathematics. While using illustrative charts and graphs, no formulas will be presented. Instead, there will be an attempt to explain the basis and the application of the most used statistical tools.

The first question to address is, "What is the object of process control?" The answer is simple: to produce a product that falls within its

		Yield			
Process	Change	Fab	Sort	Assembly	Overall
Base Line		0.80	0.50	0.95	0.38 (38%)
	Fab to 0.81	0.81	0.50	0.95	0.384 (38.4%)
	Fab to 0.85	0.85	0.50	0.95	0.404 (40.4%)

Figure 15.3   Yield impact of fabrication yield improvement.

Reading no.	Run A	Run B
1.	25	28
2.	24	25
3.	26	23
4.	23	26
5.	25	25
6.	26	22
7.	25	23
8.	24	25
9.	25	27
10.	26	25
Average	24.9	24.9
Range	26 − 23 = 3	28 − 22 = 6

Figure 15.4   Sheet resistance reading ($\Omega/\square$).

design and operational specifications. The simple approach to achieving that goal is to develop a set of processes that produce the in-spec product, document them, and perform them the same way every time. The problem is that the outputs of the individual processes are the result of a number of operating parameters and that the precision asked of the equipment and processes is very high.

Processes do not run in control without monitoring and adjustments. Process control provides the tools for the monitoring so that the operating personnel can make the necessary changes and keep the process in control. Process control tools vary from simple to very complex. The simplest (and most familiar) is the calculation of the average of a group of numbers (called a population). We also know that the extremes of a population (range), along with the average, give us an idea of the distribution of the data within the population. For example, the two groups in columns A and B (Fig. 15.4) have the same average. If the numbers were the sheet resistance of wafers from furnace A and B, we could easily come to the decision that furnace A is producing a tighter distribution and a more consistent product and is under more control.

This fact is illustrated when the data is plotted (Fig. 15.5). The plots are called *histograms* and visually display data distributions that a simple average calculation will not reveal. Average calculations and histograms are statistics in action. Histograms are usually the first step in determining whether or not a process is in control.

Their power comes from a mathematical distribution known as the gaussian distribution. Named after the famed mathematician Gauss, its origin is interesting. Gauss set out to reconcile the different star positions reported by different astronomers. His approach was to make all the necessary corrections to the observations, taking into ac-

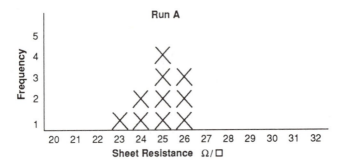

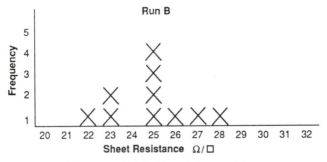

Figure 15.5   Frequency distribution of sheer resistances.

count the time of the year and from where on the earth the observations were made. He expected that when all the corrections were made there would be agreement from all the position calculations on the position of a particular star. After all, reason dictates that a star can occupy only one position at a time, and we should be able to determine that position.

However, the final data did not confirm his hypothesis. After all corrections were made, there were still a number of locations for each star. Fortunately Gauss did not scrap the project but went on to establish the basis for the field of statistics and distribution probabilities. If the math-aversion readers will hang on, they will find that the concept of probability is not so difficult. The way Gauss analyzed the data was to plot the various locations for a star. He calculated the center point (average) and drew a circle that encompassed the star location the farthest from the center. He reasoned that there was a 100 percent probability (a probability of 1) that the real star location was within that circle. He also reasoned that the probability of finding the star within the confines of smaller circles was less than 1. In fact, the smaller the circle the lower the probability that the star was within those boundaries.

Further analysis of the data revealed what has come to be known as the gaussian distribution. It is a mathematically constant distribution of data that results from certain conditions. A good example is the height of blades of grass in a lawn. If all the blade heights are measured and plotted on a histogram, the distribution will be the familiar bell-shaped curve, also called a *normal curve* (Fig. 15.6). From a probability consideration, it predicts that there is a higher probability that any given blade of grass will have a height closer to the average (center values) and that there is a lower probability that any given blade will be very short or very tall. The mathematical conditions that result in the distribution include human heights, IQ distributions, and (in most cases) semiconductor process parameter distributions, such as sheet resistance.

A first step in process control is to make a histogram of the particular process parameter and determine if the distribution is a normal distribution. If it is not, the chances are good that there is something wrong in the process. If the distribution is a normal one, the next step is to compare the range of the distribution with the design limits for the particular parameter (Fig. 15.6). This comparison is made to determine if the natural process distribution limits fall within the design limits. If they do not, the process must be fixed or some percentage of the parameter readings (and the wafers) will always be out of specification.

Another useful statistical tool is the paredo chart. This is a form of the histogram but with the x axis divided into unrelated sections rather than a continuum of some parameter such as sheet resistances. Defect inspection results are a good candidate for paredo charts. On the x axis is a list of the defect categories. A mark or vertical column

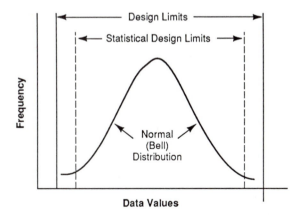

Figure 15.6 Comparison of normal distribution to design limits.

over each defect type indicates the frequency of occurrence of each of the defect types. The operator or process engineer can see from the chart which defects are occurring most often and what processes must be improved.

So far the statistical methods explained are the after-the-fact history of a process. A most powerful statistic of real-time process control is the X-R control chart (Fig. 15.7). The chart is constructed in two parts with the y axis representing the parameter values. The top graph has a horizontal line for the historic average of the parameter. On each side of the average are control limit lines that are calculated from the historic data.[3] A control limit represents the limits that the individual data values will range between when the process is in control. Also on the chart are the process or design limits which represent the extremes the individual data points may have, before being rejected. The bottom graph is constructed by calculating and plotting the amount that each data point varies from the average. When plotted, these values give further visual evidence of the amount of control in the process.

The value of the X-R bar control charts is their predictive powers. A process in control will produce data points that tend to vary in a regular pattern about the average (top of Fig. 15.7). The mathematics of a controlled process predicts this regular fluctuation. It also predicts *when* a process is going out of control *before the data points exceed the control limits* (bottom of Fig. 15.7). The data points in part B have shifted to the top of the control limit range. This is a unnatural pat-

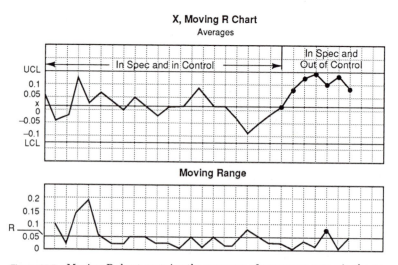

Figure 15.7   Moving R chart contains the averages of measurements x in the upper plot. The lower plot shows the moving range R which is descriptive of process stability.

tern for an in-control process. When this situation occurs, the production operators, who maintain the charts as the data is produced, alert the proper personnel so the process can be brought back into control *before the data points exceed the control or design limits* and wafers have to be scrapped. A number of more sophisticated controls used in processing are beyond the scope of this text.

Another powerful statistical tool is multivariable experiment analysis. Most measured quality control parameters (sheet resistance, line width, junction depth, etc.) are influenced by a number of variables in the process. Line width, for example, varies with the resist solution, film thickness, exposure radiation time and intensity, baking temperatures, and etch factors. Any one or all can contribute to an out-of-spec condition. The tool of multivariable experimentation allows the engineer to run tests that identify the contribution(s) of each of the individual variables.

Equipment

The previous section looked at various statistical process control techniques used at the shop floor level. In this section we examine some equipment issues, also from the shop floor perspective. The issue of factory-level integration of the control of the process, equipment factors, and material control is addressed in the section on CIM and CAM.

Over the years semiconductor processing has changed from a laboratory activity to full manufacturing using sophisticated and dedicated equipment. The development of semiconductor-specific equipment has allowed the advances into the VLSI era and the spread of the technology throughout the world. Up to the early 1970s, most of the equipment was basic and manual with process effectiveness being in the way in which the equipment was run, the process "tricks" of the trade. With the advent of more automatic equipment the processes became part of the machines and therefore available to anyone purchasing them.

IBM has pioneered the use of the generic term *tool* for all process equipment. As the tools have become more sophisticated, the prices have risen to the point where they are the major expense in any new facility. VLSI research placed the tab for a 1980s VLSI facility photolithographic equipment at about 1 million dollars.[4] They estimated the tab for a 1990s fabrication to be around 42 million dollars!

Equipment factors

A number of considerations (Fig. 15.8) go into the selection of a particular vendor's pieces of equipment. They can be roughly divided into

Technical/Process	Economic/Manufacturing
Capability	Cost
Repeatability	Operating Expense
Flexibility	Uptime
Upgrade Potential	Upgrade Costs
Ease of Operation	Ease of Operation
Automation Level	Automation Level

Figure 15.8   Equipment purchase considerations.

two broad categories, technical and economic. Keep in mind that in order for a machine to be a workhorse contributor to the bottom line it must be well thought out in *all* categories and that the overall machine performance is the result of all of them.

First and foremost is the ability of the machine to meet the design requirements of the device (*process capability*). If the machine cannot routinely produce the right product, it is of no use. Second, the machine must do its job repeatably. Successful semiconductor processing is a volume business and with more emphasis on the equipment repeatability is a basic requirement. Flexibility relates to the ease with which a machine can be switched to run a variety of products and processes. Upgradability is the ability of the machine to handle future process requirements. This factor is very important for fabrication lines doing ASIC work where staying abreast of the latest product technology is vital. In fabrication areas dedicated to long production runs of the same (or similar) products, the ability to upgrade is less important. In general, the lifetime of a particular product is around 7 to 10 years but the technology of leading-edge products advances every 3 to 5 years.

On the economic side, cost is a major factor. Cost includes the initial purchase price, the cost of operation and repair, spare parts, and operator training. Under the cost of operation and repair are some important factors. Lumped together as downtime factors, they are *scheduled maintenance frequency and time, mean time to failure* (MTF), and *mean time to repair* (MTR). Given the expense of most machines, having backups is a costly luxury. It falls on the vendor to provide a machine that runs for long periods of time (MTF), can be repaired quickly (MTR) when failure occurs, and does not need frequent and lengthy routine maintenance. Cost also includes the price of materials needed for the machine process. Machines that waste materials also waste money.

Ease of operation shows up as both a technical and an economic factor. On the technical side, a machine difficult to operate will result in product that is variable and/or out-of-spec. On the economic side, difficult operation of the machine will result in excessive downtime and/or scrapped product. Safety is an important factor, especially for processes that use toxic or hazardous materials.

Vendor support has emerged as a critical factor. The advancement of process technology and equipment sophistication has forced close cooperation and alliances between the chip manufacturers and their suppliers. The cost of machine development requires input from the using process engineers, and the detailed nature of the machines requires on-time backup from the vendors. Management guru Tom Peters sees modern corporations as having "fuzzy edges." By this he means that more and more a corporation must rely on its vendors and customers for information and codevelopment of needed equipment and materials. This is certainly true of the semiconductor industry and is the model practiced by the very successful Japanese semiconductor industry.

Automation

The last but certainly not the least equipment factor is the degree of automation. Automation is a term bandied about as a cure-all for many manufacturing woes, and it *is* a powerful factor. Since 1940, automation of oil refineries has reduced the number of workers by a factor of 5.[5] Automation of semiconductor equipment has been in stages, and any discussion of a particular process must be based on the particular automation level of that process.

Process automation. The first level of automation is of the process itself. Most semiconductor equipment, by definition, automates the process, such as photoresist spinners that automatically dispense the primer and resist at the correct speeds and for the correct time. Automatic gas-flow controllers are also examples of process automation. Process automation brings consistency to the process and the product by reducing reliance on operator skills that vary from operator to operator and with the human factors of training, personal skill, morale, and fatigue.

The instruction to the machine received from an onboard computer which contains the proper process parameters is a *recipe*. The recipe is loaded into the machine by the operator or from a central host computer.

Wafer-loading automation. The next level of automation is the loading and unloading of the wafers. The industry has settled on the wafer cassette as the primary wafer holder and transfer vehicle. Cassettes are placed on the machine by various mechanisms. Elevators and/or wafer extractors or robots feed the wafers into the particular process chamber, spin chuck, etc. In some processes, such as process tubes, the entire cassette is placed in the process chamber. This level of automation is referred to as the "one button" operation. With one button the operator activates the loading system and the wafers are processed and returned to the cassette. At the end of the cycle, the machine sounds an alarm or light and the operator removes the cassette.

Some machines have buffer storage systems that maximize the machine efficiency by always having fresh wafers available for processing. The operator places the cassette(s) on the machine loader and pushes a start button, after which the machine takes over the processing.

Wafer-delivery automation. The third level of automation is when wafers are automatically brought to and loaded on and removed from the machines. Early delivery systems used traveling robotic carts that duplicated human delivery (Fig. 15.9). Called *automated guided vehicles* (AGVs), the carts travel along the aisles and dispense wafer cassettes when the machines need them. This approach has the advantage of being retrofitable to fabrication lines where the equipment is lined up in rows.

Another approach is the use of an overhead rail.[5] The wafer cassettes arrive at the process tool area (Fig. 15.10) where a secondary system removes them from the overhead rail and places them in the

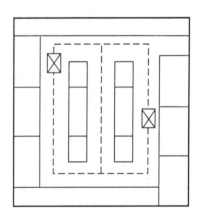

AGV

AGV Path

Figure 15.9 Automated guided vehicle wafer delivery system.

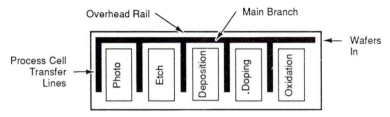

Figure 15.10 Overhead rail wafer delivery layout.

tool buffer. This system requires that the process machines be grouped in bays rather than in the traditional linear layout.

Some fabrication areas use centrally located robots (Fig. 15.11) that feed cassettes to several process machines. This delivery system also uses a bay system of machine placement. Bay layouts that group like-type machines offer the production advantage of maximum usage when one machine is inoperable for repair or maintenance.

Control-system automation. The industry is just embarking on the last level of total automation, the closed-loop feedback system. These are machines that have, or are connected to, automatic inspection or measuring subsystems. The subsystem measures important parameters in real time (as they are happening), compares them with a standard, and feeds the information back to built-in mechanisms which make the necessary process parameter changes. This final stage of automation is difficult to achieve for many processes. It requires measuring

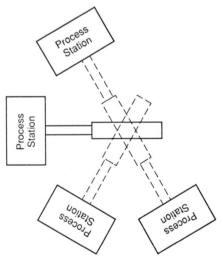

Figure 15.11 Robot wafer delivery in process island.

sensors that operate in hostile environments, such as heated-tube furnaces and sputtering chambers. One area of progress is in the area of mask making where recognition systems are advanced enough to compare the manufactured plates with the design criteria. The development of closed-loop process and machine control is necessary before the goal of "lights out, peopleless fabrication" is possible.

Inventory Control

A critical issue in fabrication cost control and yield is inventory levels and control. As the number of processing steps has increased so has the length of the processing time and the number of wafers in process (WIP). The problem is that the company pays for the wafers when purchased and doesn't receive payment until the finished devices are shipped. This period can vary from 2 to 8 months for a production line making similar circuits. Fabrication lines doing ASIC circuits have an even heavier burden when many different circuit types are going through the system. To get an idea of the burden, consider a CMOS-type process with 50 major steps and 4 substeps each for a total of 200 processes. The high cost of the equipment generally requires some buffer inventory to ensure that the machines are operating at maximum efficiency. If each buffer has 4 cassettes of 25 wafers each, the total inventory of WIP is 40,000 wafers. At a cost of $50 per wafer (large diameter) the total inventory burden becomes 2 million dollars.

Excessive WIP affects productivity by having the capability of hiding process and equipment problems.[7] With a lot of inventory at the stations, wafers can keep flowing out the back end while parts of the process are shut down. With a lower WIP these problems are readily apparent and force the fixing of problems. WIP also influences the overall fabrication yield. The collective experience of the the industry is that the longer wafers are in the process the lower their wafer sort yield.

Just-in-time inventory control (JIT)

Just-in-time inventory control is a philosophy based on the objective of "Make only what is required, only as required."[7] The system is simple in concept. All buffer inventories are reduced to an absolute minimum, from the storeroom to the machine buffers. To work effectively, excellent vendor relationships must be established. The incoming materials have to be of the highest quality as JIT leaves little cushion for extensive incoming inspections and returns. Second, the vendor is asked to maintain ready-to-ship materials at their facility. In effect, the vendors are asked to hold the inventory previously held by the chip manufacturer, a situation that they are not always happy about. The chip manufacturers have to be very efficient in assessing their quantity, quality, and delivery needs, and must have a system that expedites material to the proper process tool and detects quality problems quickly.

JIT also has applications in the process flow. Some companies will only commit wafers to a process if the work can flow unimpeded through all of the substeps. This system gives higher yields and, if on a properly balanced line, increases throughput time, even though parts of the line are idle. This system is called *demand-pull*. Wafers "upstream" are worked on only when there is demand from downstream process stations about to run dry of wafers.

An effective JIT procedure can reduce the number of operators in the fabrication area since a good proportion of their time is spent sorting, staging, and delivering work to the process tools. Fabrication-area layouts can change from the traditional linear arrangement of process stations. A linear layout benefits a line making only a few different product types. Wafers are introduced into the front of the line and physically move on a first in–first out (FIFO) basis. Problems arise when particular lots of wafers must be moved quickly. Usually tagged as *hot lots*, they are given priority processing at each station. Besides the control problems associated with keeping track of the lots, hot lots have the effect of making regular product sit in a queue waiting for the hot lots to clear. This layout is particularly cumbersome for ASIC lines where shipping dates and product types have constantly changing priorities.

JIT-CAM systems offer the advantage of knowing where all the lots are in the production cycle. The computer can stage the work to minimize disruption and keep a steady flow of product. Further, JIT-CAM systems teamed with automatic delivery systems can be more efficient when the process stations are grouped rather than strung out in a line. This concept, called *work cells*, is more efficient, especially when a cell can run a number of different products. When a machine is inoperable in a linear layout, the line comes to a halt and inventory builds up in front of the machine that is down.

Factory-Level Automation

The advent of higher levels of process and tool automation and JIT requires higher levels of centralized control and information sharing. Most companies have computer-based management information systems (MIS) handling the paperwork and details of employment and finances. These systems are being expanded to the entire manufacturing environment in a process called *computer-integrated manufacturing* (CIM).

CIM is the computerization of all plant operations and the integration of those operations into one computer design, control, and distribution system. The processes involved in CIM are all related and interdependent, as illustrated in Fig. 15.12. The major activities of CIM are business functions, product design (mask and circuit), manufacturing planning (inventory, shop floor priorities, etc.), manufacturing control, and the fabrication processes. A complete CIM system is interactive at all levels. This means that each of the five functional areas inputs data to the system in real time and that the information is available to all who need it.

Computer-aided design (CAD) and computer-aided manufacturing (CAM) are two subsystems within the CIM system. The role of CAD has been discussed in mask and circuit design. CAM is the part of the system that does the planning and control of the manufacturing operation. A CAM system includes a computer network and the automated process tools and material delivery systems.

In concept a CIM system kicks into operation when a customer order is received. The computer logs the order and initiates the CAD system to start the design (if it is a custom order). It also triggers the ordering of needed materials in the right quantities and schedules their delivery times. This subsystem goes by the name *computer-aided process planning* (CAPP). Through the CAM program the individual process recipes are downloaded to the individual equipment comput-

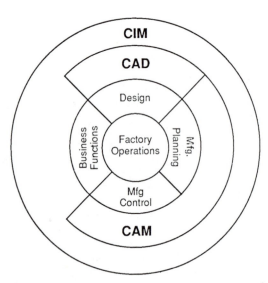

Figure 15.12 Provinces of factory computer control systems.

ers. Once processing begins, the CAM system controls the WIP and makes necessary priority decisions to meet shipping schedules. It also keeps track of equipment performance and schedules repairs and maintenance.

An important feature of the CAM system is yield monitoring and reporting. Important measuring systems are connected directly to the factory computer. If a poor yield problem appears, the system reports it to the engineering and facility staff and (if necessary) will reorder materials to make up the losses. At the end of the production run the system calculates costs and yields and schedules shipment to the customer.

Other packages included in CIM systems are facilities monitoring, process modeling, and security systems. Facility monitoring might include power consumption and environmental factors inside the plant. Some facility CAM systems include monitoring the levels of liquids and gases in storage with automatic alerting and/or reorder. Monitoring and precise control of these factors can save appreciable amounts of money. Process modeling is a system of testing a particular design against the known process variations. The computer can run many variations simulating the changes expected in fabrication. Good modeling can identify weak points in the design or process before wafers are committed to the line. Security monitoring may include the entering and exiting of employees (and/or intruders) and the securing of expensive finished product or materials. A security system may also include fire and other hazards control.

Line Organization

Most fabrication areas are organized around the product-line concept. In this concept fabrication areas are built to accommodate products with similar processing needs. Thus there are bipolar lines, CMOS lines, etc. This arrangement makes for more efficient processing since most of the machines are in use most of the time and the staff can gain experience in processing a few products.

The staff of these lines are also fairly self-contained. The primary responsibility falls to the fabrication or product manager (Fig. 15.13). Reporting to this individual are an engineering supervisor, a production manager (or general supervisor), a design department, and the equipment maintenance group. The production manager is responsible for producing the finished wafers to specification, to cost, and to schedule. The engineering group is responsible for the developing of high-yield processes, documentation of the processes, and the daily sustaining of the line process. Both the production and engineering staffs are divided into groups focusing on a particular part of the pro-

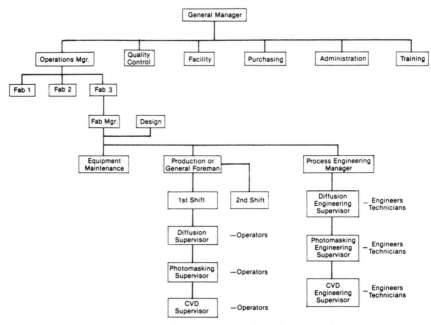

Figure 15.13 Typical semiconductor product line table of organization.

cess. This organization has the virtue of high focus on the fabrication area's primary goal of producing chips at a profitable level.

As the processes become more automated and arranged in process cells small group organizational teams and responsibilities are emerging. A cell is attended by the operator(s), the equipment technicians, and the process engineer(s). These small groups make floor-level decisions with the information provided by the CIM system. However, few companies have formalized this arrangement with an organization structure, and the teams tend to exist as cross-department cooperatives.

Key Concepts and Terms

Automation levels	Inventory control
CAD	Just-in-time (JIT)
CAM	MIS
CAPP	Normal distribution
CIM	Process limits
Control chart	Product-line organization
Data averages	Statistical process control
Design limits	Wafers out per operator
Fabrication cost factors	Wafers out per square foot
Histogram	WIP

Review Questions

1. List the major factors determining the manufacturing cost of a wafer.

2. Rank the factors according to their level of influence.

3. Describe how WIP impacts profit.

4. Discuss how statistical process control can enhance wafer yields.

5. List the four levels of automation in a fabrication process.

6. Describe the philosophy and operation of a JIT inventory system.

7. What is the difference between CIM and CAM?

8. What is a process cell and how does it compare with a traditional line layout?

9. List the production, process, and yield advantages of a totally automated fabrication area.

10. Draw an organizational chart of a product-line-managed fabrication area.

References

1. S.M. Sze, *VLSI Technology*, McGraw-Hill, New York, 1983, p. 5.
2. J. G. Harper, L.G. Bailey, "Flexible Material Handling Automation in Wafer Fabrication," *Solid State Technology*, July 1984, p. 94.
3. D. M. Campbell, Z. Ardehale, "Process Control for Semiconductor Manufacturing," *Semiconductor International*, June 1984, p. 127.
4. K. Levy, "Productivity and Process Feedback," *Solid State Technology*, July 1984, p. 103.
5. S. Shinoda, "Total Automation in Wafer Fabrication," *Semiconductor International*, September 1986, p. 87.
6. L. G. Cory, "Just-in-Time Approach to IC Fabrication," *Solid State Technology*, May 1986, p. 178.
7. K. Levy, "Productivity and Process Feedback," *Solid State Technology*, July 1984, p. 177.

Semiconductor Devices and Integrated Circuit Formation

Overview

Integrated circuits are composed of individual conductors, fuses, resistors, capacitors, diodes, and transistors. The formation of these components in the wafer surface and the added techniques and structures required for formation of the major integrated circuits are presented in this chapter.

Objectives

Upon completion of this chapter you should be able to:

1. Sketch and identify the structural parts of the individual components of an integrated circuit.
2. Explain the role and different isolation structures used for integrated circuits.
3. Sketch and identify the operation of a bipolar and MOS transistor.
4. List the types and advantages of the different MOS gate structures.
5. Sketch and identify the parts of a Bi-MOS circuit.

Semiconductor-Device Formation

The previous chapters have focused on the individual processes used to make semiconductor devices (also referred to as components or circuit components) and integrated circuits. In this chapter the actual formation of the devices and circuits is explained. It is assumed that the reader has already read about (or is familiar with) the processes and has a good understanding of the electrical performance of the individual components as explained in Chap. 14.

There are literally thousands of different semiconductor device structures. They have been developed to achieve specific performances, either as discrete components or in integrated circuits. In this chapter the basic structures are explained along with the major variations used for specific applications. The circuits components are:

- Resistors
- Capacitors
- Diodes
- Fuses
- Conductors

Resistors

Resistors have the effect of limiting current flow. This is accomplished by the use of dielectric materials or high-resistivity portions of a semiconductor wafer surface. In semiconductor technology resistors are formed from isolated sections of the wafer surface, doped regions, and deposited thin films.

The value of a resistor (in ohms) is a function of the resistivity of the resistor and its dimensions (Fig. 16.1). The relationship is

$$R = \rho L / A$$

where ρ = resistivity
L = length of resistive region
A = cross-sectional area of the resistive region

The area (A) becomes $W \times D$, where W = width of the resistor and D = depth of the resistive region. If the resistor is formed from a doped region, the length and width are the surface pattern openings and the depth is the junction depth.

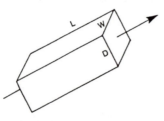

$$R = \rho \left| \frac{L}{A} \right. = \rho \left| \frac{L}{W \times D} \right.$$

Figure 16.1 Relationship of resistance to resistivity and dimensions.

It is important to master the resistor-rho-dimensional relationship since it applies to every region in a device structure. It governs the flow of current through any region. A conductor is simply a resistor with a low resistance. Conversely a resistor is a conductor with a low current flow. The conceptual importance of Ohm's law is that the electrical resistance of any region in the device or circuit is altered by any change in dimensions or change in the doping level.

Doped resistors. Most of the resistors in integrated circuits are formed by a sequence of an oxidation, masking, and doping operation (Fig. 16.2). A pattern is opened in the surface oxide. Typical resistor shapes are dumbbells (Fig. 16.3) with the square ends serving as contact regions and the long skinny region in between serving the resistor function. The resistance of this region is calculated from the sheet resistance of the region and the number of squares (□'s) contained in the region. The number of squares is calculated by dividing the length by the width.

After doping and a subsequent reoxidation, contact holes are etched in the square ends to contact the resistor into the circuit. A resistor is a two-contact, no-junction device. The term no-junction means that the current flows between the contacts without crossing an N-P or P-N junction.

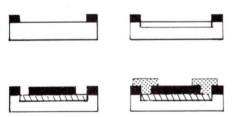

Figure 16.2   Diffused resistor formation.

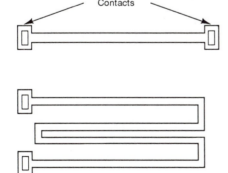

Contacts

Figure 16.3   Resistor shapes.

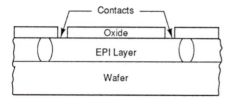

Figure 16.4 Epitaxial layer resistor.

Resistors doped by ion implantation have more controlled values than those in diffused regions. Doped resistors can be formed during any of the doping steps performed during the fabrication process. The mask for the doping step has patterns for the transistor part, such as a bipolar base or MOS source and drain, and a separate resistor pattern. The transistor part and the resistor are doped simultaneously and the resistor has the same doping parameters (sheet resistance, depth, and dopant quantity) as the transistor part.

EPI resistors. A resistor can be formed by isolating a section of an epitaxial region (Fig. 16.4). After surface oxidation and contact hole masking what is left is a three-dimensional region with resistive properties.

Pinch resistors. Ohm's law shows that the cross-sectional area of the resistor is a factor in its value (Fig. 16.5). One way to reduce the cross-sectional area (and increase the resistance) is to dope the resistor region and then do another doping of the opposite conductivity type. This occurs in bipolar processing when a resistor region is formed during the P-type base doping with a "pinched" cross section formed from a subsequent N-type region formed along with the emitter doping.

Thin film resistors. Doped resistors don't always have the resistance control needed in some circuits and are poor performers in radiation environments. Radiation, such as found in space, generates unwanted holes and electrons that allow the current to leak across the confining junction. Resistors formed from deposited thin films of metal do not have this radiation problem.

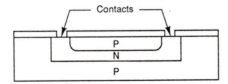

Figure 16.5 N-type resistor "pinched" with P-type doped region.

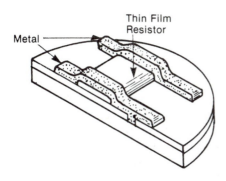

Figure 16.6 Formation of a thin film resistor.

The resistors (Fig. 16.6) are formed by either film deposit and masking sequences or by lift-off techniques. After resistor formation of the wafer surface, it is "wired" into the circuit by contact between the resistor ends and the leads of the conducting metal. Nichrome, titanium, and tungsten are typical resistor metals.

Capacitors

Oxide-silicon capacitors. Silicon planar technology is based on a silicon wafer with a grown silicon dioxide layer on top. The conducting metal leads lie on top of the oxide. In the chapter on oxidation it was explained that oxide layers are normally thick enough to act as a dielectric between the metal and semiconductor. In MOS gate structures thin layers of oxide are grown specifically to cause charge accumulation in the gate region (see MOS operation).

When the oxide thickness is about 1500 Å[1] it can be part of a capacitor. A capacitor is a device that stores charge. A battery is a capacitor. An oxide-silicon capacitor (Fig. 16.7) is a sandwich made up of the underlying silicon, the oxide, and a top layer of conducting metal. Also required is a metal contact to the wafer surface. When a voltage is applied to the metal a charge builds up in the wafer layer under the oxide. The amount of charge is a function of the oxide thickness, the dielectric constant of the oxide, and the area, as defined in the metal top plate. Capacitors of this structure are called parallel-plate, monolithic, or MOS (after the *m*etal *o*xide *s*emiconductor materials of the sandwich).

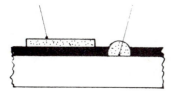

Figure 16.7 Monolithic capacitor.

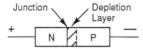

Figure 16.8 Depletion layer junction capacitor.

Junction capacitors. A capacitor is formed at every junction in the devices. When a voltage is applied across any junction, carriers on each side of it move away from the junction, leaving a depleted region (Fig. 16.8). This depleted region acts as a capacitor in the device or circuit. The value of this junction capacitance must be taken into account when the circuit is designed. Some circuits actually use junction capacitors as part of the circuit design. In some circuits the natural junction capacitance has the effect of slowing up the circuit operation. This is due to the time required to "fill up," or charge, the depleted region before current flows. A finite time is also required for the various junction capacitors to discharge. Both of these times affect circuit switching and operational speeds.

Trench capacitors. Preservation of wafer surface area is always a design criterion. One of the problems with oxide-metal capacitors is their relatively large area. Trench (or buried) capacitors solve the problem by creating a capacitor in a trench etched vertically into the wafer surface (Fig. 16.9). The trenches are etched either isotropically with wet techniques or anisotropically with dry etch techniques. The trench sidewalls are oxidized (the dielectric material) and the center of the trench filled with deposited polysilicon. The final structure is "wired" from the surface, with the silicon and polysilicon serving as the two electrodes with the silicon dioxide dielectric between them.

Diodes

Doped diodes. A diode is a two-region (two-contact) device separated by a junction. A diode either allows current to pass easily or acts as a

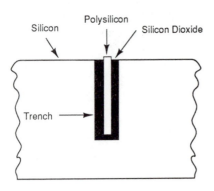

Figure 16.9 Trenching capacitor.

current block. Which function it performs is determined by the voltage polarity *biasing* of the current relative to the region polarities (Fig. 16.10). When the current voltage is the same as in the diode region the diode is in forward bias and the current flows easily. When the diode is reverse-biased the current is blocked. A reverse-biased diode can be forced into a conducting state by raising the current voltage until the junction goes into *breakdown*. This condition is temporary; when the voltage is reduced the diode once again becomes a blocking device.

Diodes are used in circuits to steer the current around the circuit. By proper choice of the circuit current polarities and the correct diode polarities the current is allowed to pass into some branches and is blocked out of others. A planar diode is from a doped region and two contacts on either side of the junction where it intersects the surface (Fig. 16.11). Diodes are usually formed along with transistor doping steps. Thus in bipolar circuits there are base-collector diodes and

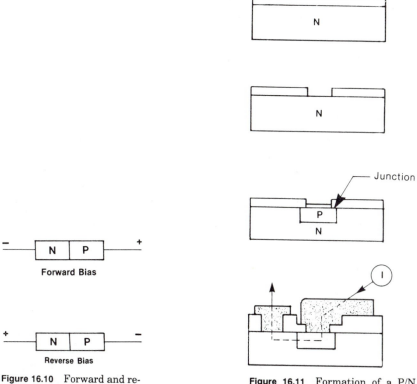

Figure 16.10 Forward and reverse biasing.

Figure 16.11 Formation of a P/N planar diode.

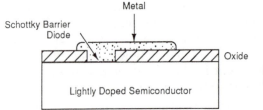

Figure 16.12 Schottky barrier diode.

emitter-base diodes. In MOS circuits most of the diodes are formed with the source-drain doping step.

Schottky barrier diodes. In 1938 (ten years before the invention of the transistor) W. Schottky[2] discovered that whenever a metal is in contact with a lightly doped semiconductor a diode is formed (Fig. 16.12). This diode has a faster forward time (it responds faster) and operates with a lower voltage than a doped silicon junction diode. Metal contacts to highly doped regions (greater than 5×10^{17} atoms per cubic centimeter) are regular ohmic contacts. This is the situation for the majority of contacts in a silicon circuit. This Schottky diode effect is taken advantage of in some NPN bipolar transistors. The structure and effect are explained in the section on bipolar transistors.

Bipolar transistors

Transistor operational analogy. A transistor is a three-contact, three-part, two-junction device that performs as a switch or an amplifier. An often used analogy to explain the role of the parts and the operation of a transistor is the water flow system in Fig. 16.13. The flowing water represents current flow. In this system one part is the source of the water (the tank), the valve controls the flow, and the bucket collects the water. The system can be operated as a switch simply by turning the valve on and off. It can even be imagined in an amplifier role. Consider the valve as a high-mechanical advantage miniature water wheel activated by a small external stream of water. A small trickle onto the valve wheel could open the valve to allow a large flow through the system. If the whole system was enclosed so that an observer saw only the trickle going in and a large flow coming out, they might conclude that the system was *amplifying* the water trickle.

Solid-state bipolar transistor. The same basic parts and functions are present in solid-state transistors. A bipolar transistor is shown in two forms in Fig. 16.14. The current flows from the emitter region (tank) through the base (valve) into the collector (bucket). When the base is

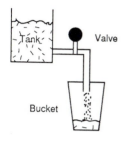

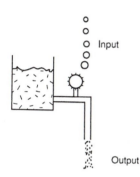

Figure 16.13 Water analogy of transistor operation.

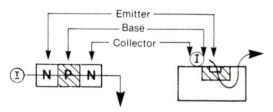

Figure 16.14 Bipolar transistor operation.

turned off there is no current flow. When it is on the current flows. The base is turned on and off by a small current applied to it. The size of the base current regulates the larger amount of current flowing through the transistor (called the *collector current*). There is an amplification of the base current to the collector current. The base current in effect changes the resistance of the base region. In fact the term *transistor* comes from an early term for a bipolar transistor: *transfer resistor*. During operation both positive and negative currents flow in the base. This fact led to the term *bipolar*, literally, two polarities.

The amplification property, called gain or beta, is numerically the result of dividing the collector current by the base current. For efficiency the emitter region has a higher doping concentration than the base, and the base doping is higher than the collector. A typical doping concentration versus distance plot is shown in Fig. 16.15.

Most bipolar circuits are designed with NPN transistors. NPN represents the respective conductivity types of the emitter, base, and collector. Some applications require PNP transistors, with many of them being formed laterally (Fig. 16.16). NPN transistors are more efficient because of the ease (higher mobility) of electrons in the N-type regions.

Bipolar transistors are favored for their fast switching speed. The speed is governed by a number of factors, of which the most important is the base width. Applying common sense, it stands to reason that the shorter the distance an electron or hole has to travel the less time it will take. Bipolar transistors can switch on and off in as little as a billionth of a second. To achieve this speed the transistor is maintained at an "on" state. This means that the base always has power applied, which is a downside of bipolar-based circuits. Another penalty for this necessary condition is a buildup of heat in the transistor. This heat will eventually affect circuit operation and is the reason for the cooling fans and air-conditioned rooms of earlier bipolar-based computers.

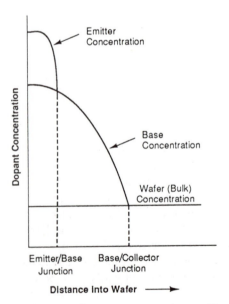

Figure 16.15 Dopant concentration profile of bipolar transistor.

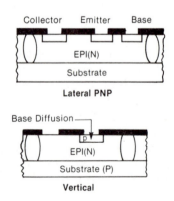

Figure 16.16 Lateral and vertical PNP transistors.

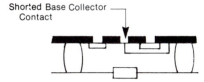

Shorted Base Collector
Contact

Figure 16.17   Schottky barrier diode.

Schottky barrier bipolar transistors.   The Schottky barrier diode principle mentioned in a previous section is put to use in some bipolar transistors (Fig. 16.17). The construction requires that the base contact be extended into the collector region. When covered with metal a Schottky diode is formed between the base and collector, which results in a faster-responding transistor. In a circuit the time required to turn the transistor on and off (switching speed) is critical when millions of transistors are operating.

Field-effect transistors

Metal-gate MOSFET.   As early as 1948 William Shockley noted another type of transistor operation, a field-effect-actuated transistor current flow. That effect was developed into the field-effect transistor, with the MOS structure now the most popular design. An MOS transistor, like a bipolar (Fig. 16.18), has three regions, three contacts, and two junctions but in a different structure. There is a similar analogy to the water system described previously. Current travels from the source region (tank), through the gate (valve), and into the drain (bucket) before exiting the device.

It is the gate that operates by a *field effect*. The MOS structure shown in Fig. 16.18 is a metal-gate type. The MOS stands for the materials of the gate. Electrically a gate structure is a capacitor and operates the same as a capacitor. When a voltage (the gate voltage) is applied to the gate through the gate metal an effect takes place in the surface of the semiconductor. The effect is either a buildup of charge

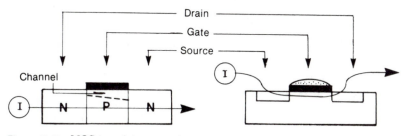

Figure 16.18   MOS transistor operation.

or a depletion of charges. Which event occurs depends on the doping conductivity type in the wafer under the gate and the polarity of the gate voltage.

The buildup or depletion of charge creates a channel under the gate which connects the source and drain. The surface of the semiconductor is said to be *inverted*. In operation the source is biased with a voltage and the drain is grounded relative to the source. In this condition a current starts to flow as the connecting channel is formed. The source and drain are essentially shorted together. Applying more voltage to the gate increases the size of the channel, allowing more current to flow through the transistor. By controlling the gate voltage an MOS transistor can be used as a switch (on/off) or as an amplifier. However, MOS transistors are voltage amplifiers, unlike the current amplification of bipolar transistors.

If the source and drain are N type formed in a P-type wafer the channel must be of N type for conduction to occur. This type of MOS transistor is called N channel. MOS transistors with P-type sources and drains are P channel. Most high-performance MOS circuits are built around N-channel transistors due to the higher mobility of electrons in silicon. The mobility makes N-channel transistors faster and they consume less power than P-channel circuits. They often are referred to as NMOS transistors. Figure 16.19 shows the steps in the formation of an N-channel metal-gate MOS transistor.

Silicon gate MOS.   A certain amount of voltage must be applied to the gate metal before the channel forms. This voltage is called the *threshold voltage* or V_t. The value of the threshold voltage is an important and critical circuit parameter. Higher V_t means more power supplies and slower circuits.

A primary parameter that determines the threshold voltage is the *work function* between the gate material and the doping level in the semiconductor. The work function can be thought of as a kind of electrical compatibility. The lower the work function, the lower the threshold voltage, the lower the power required to run the circuit, etc.

Deposited doped polysilicon has a lower work function than aluminum as an MOS gate material and has become the standard gate electrode material for MOS transistors. The formation of the transistor is shown in Fig. 16.20. The polysilicon is heavily doped N type to reduce its resistance. Thus doped it serves as the gate electrode and as a circuit conduction line. A polysilicon gate can withstand subsequent high-temperature processing without degradation.

An additional benefit of the silicon-gate process is the self-aligned gate. In the metal-gate process sequence a hole for the gate oxide must be patterned between the source and drain. To ensure that the source

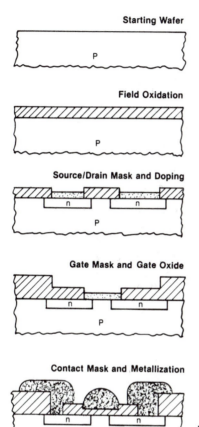

Starting Wafer

Field Oxidation

Source/Drain Mask and Doping

Gate Mask and Gate Oxide

Contact Mask and Metallization

Figure 16.19 N-channel metal gate MOS process.

Not Shown: Passivation Layer

and drain are bridged by the gate, overlap for alignment tolerances must be allowed. This results in some overlap of the gate into the source and/or drain. The overlap becomes a region of unwanted capacitance. In the silicon-gate process the gate is formed first and acts as a mask to locate the source and drain. Thus, whatever the gate placement, the source and drain *self-align* to it.

Other factors, besides the gate metal, affecting the gate threshold voltage and operation are:

- Gate oxide thickness
- Source-drain separation (channel length)
- Gate doping level
- Sidewall capacitance of the doped source and drain regions

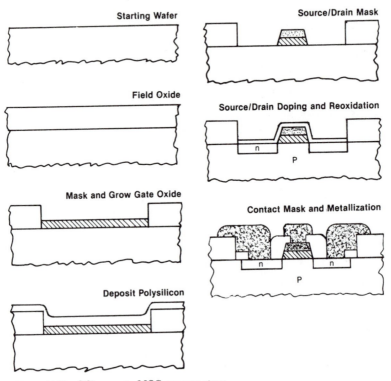

Figure 16.20 Silicon gate MOS process steps.

The thinner the gate oxide the faster the device and the lower the threshold voltage. Gate oxide thicknesses of advanced devices are in the 100 to 300 Å range. Channel lengths also affect the speed. Channel lengths have continually shrunk to the submicrometer range. In self-aligned structures the channel length is established by the gate width. The gate doping level influences the threshold voltage by modifying the work function difference between the gate metal and surface. Ion implantation, with its ability to dope through thin oxides, is often used to set the gate doping level. Sidewall capacitance of the doped source and drain also serves to slow up the operation of the device, as enough charge must be built up to overcome the junction capacitance.

Polyside-gate MOS. The discussion of MOS transistors so far has concentrated on the gate metal. Equally important is the gate oxide. MOS development was impeded by the inability of the industry to grow noncontaminated and thin oxides in the 1960s. Contamination, espe-

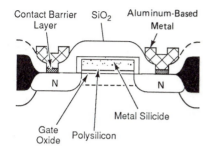

Contact Barrier   SiO₂   Aluminum-Based
Layer   Metal

Metal Silicide

Gate
Oxide   Polysilicon

Figure 16.21   Polycide gate structure.

cially the mobile ionic variety, interferes with the field effect, making very unreliable gates. In fact, clean gate oxides and the oxide-silicon interface are so well understood and effective that no gate oxide replacement is under consideration.

Consequently the quest for a higher-efficiency gate has led to the polycide structure (Fig. 16.21). The gate sandwich retains the thin oxide on the wafer surface topped by a layer of polysilicon. The polysilicon provides a low work function (and lower threshold) gate and the reliable polysilicon-oxide interface is preserved. The new layer is a refractory metal silicide on top of the polysilicon. The silicide makes a low contact resistance with the polysilicon (as compared with aluminum) and reduces the overall sheet resistance of the polyside sandwich.

Salicide-gate MOS.   The self-aligned process that uses the polycide-gate structure is called a *salicide* gate. Its formation is illustrated in Fig. 16.22. The process combines the best features of a polysilicon gate with self-alignment. The source and drain are lightly diffused around the polysilicon gate. Then a layer of silicon dioxide is deposited and anisotropically etched to form spacers on the side of the gate. These spacers act as ion implantation masks for a subsequent heavier doping of the source and drain. The more lightly doped "finger" under the gate is called a lightly doped drain extension (LDD) and is necessary for channel lengths less than 2 μm.[3] After ion implantation the refractory metal is deposited, and the silicide is formed by reaction with the underlying polysilicon layer by an alloy step. The final step is the removal of the unreacted refractory metal from the wafer surface.

V-groove MOS (VMOS).   VMOS stands for *V-groove MOS* technology. Invented in the late 1970s, this MOS transistor structure offered the promise of higher-density circuits since the transistors were formed on the sidewalls of a V-groove etched into the surface (Fig. 16.23). The formation of the transistors on the sidewalls eliminated device prob-

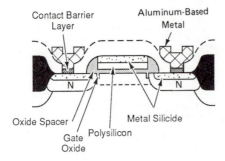

Figure 16.22   Salicide-gate structure.

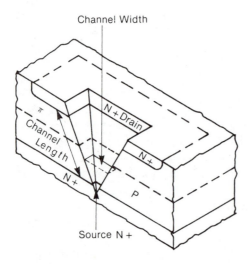

Figure 16.23   VMOS structure.

lems associated with surface damage and contamination. VMOS is a difficult structure to fabricate owing to the inconsistencies in patterning the components on the nonplanar groove sides. While not finding a home as an integrated circuit technique, VMOS has proved a good structure for discrete power MOS transistors.

Diffused MOS (DMOS).   DMOS refers to a diffused MOS structure (Fig. 16.24). The channel length is established by two diffusions through the same opening. As the second diffusion is taking place the first moves laterally to the sides. The second diffusion functions as the source and the bulk semiconducting material of the wafer functions as the drain. The difference between the two diffusion widths is the channel length of the transistor. DMOS structures feature narrow and well-controlled gate widths.

Memory MOS (MMOS).   MMOS is a structure that provides a more or less permanent storage of the charge in the gate region. The storage is

Diffusion #1

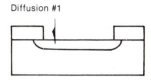

Diffusion #2   Source   Gate

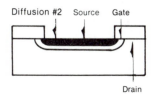

Figure 16.24   DMOS structure.

Drain

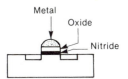

Metal

Oxide

Nitride

Figure 16.25   MMOS structure.

provided by a thin layer of silicon nitride between the wafer and the gate oxide (Fig. 16.25). When the gate is charged to store data the silicon nitride layer traps and retains it. This type of transistor is used in nonvolatile circuits where protection against memory loss is important (see Chap. 17).

Junction field-effect transistors (JFET).   A junction field-effect transistor (Fig. 16.26) is similar in construction to a MOSFET but has a junction formed under the gate. During operation the current flows *under* the diffused region from the source to the drain. As the gate voltage is increased a region depleted of charge (the depletion region) spreads under the junction toward the N-type and P-type interface. The depleted region will not support current flow and has the effect of restricting the current flow as it increases in depth.

A JFET operates opposite from an MOS transistor. In the MOS version, increasing the gate voltage *increases* the current flow. In a JFET, increasing the gate voltage *decreases* the current flow. JFETs are a standard gallium arsenide device. The transistor is formed in an N-type GaAs layer that is on top of a semi-insulating GaAs wafer (Fig. 16.27).

Metal semiconductor field-effect transistor (MESFET).   The MESFET is the basic GaAs transistor structure (Fig. 16.28). The MESFET oper-

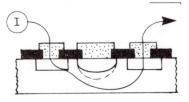

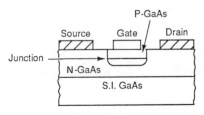

Figure 16.26 Junction field-effect transistor.

Figure 16.27 GaAs JFET.

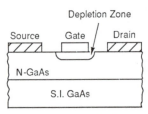

Figure 16.28 GaAs MESFET.

ates in the same manner as the JFET but the gate metal is deposited directly onto the N-type GaAs layer.

Fuses and conductors

Some memory and logic circuits require thin film or oxide fuses. The formation of these components and surface wiring were detailed in Chap. 14. In very dense circuits precious surface area is preserved by the use of subsurface *underpass conductors*. These are created from heavily doped regions formed under surface wiring leads.

Integrated-Circuit Formation

Integrated circuits contain all the components described in the previous sections. The components are formed in specific sequences with the process flows designed around the transistor in the circuit. The process designer will attempt to form as many component parts as possible with each doping step.

Circuits are designated by the transistor type. A bipolar circuit means that the circuitry is based on bipolar transistors. MOS circuits are based on one of the MOS transistor structures. For the first thirty years of the semiconductor industry the bipolar transistor and bipolar circuits were the structures of choice. Bipolar transistors had fast speed (switching times), control of leakage currents, and a long history of process development. These qualities fit nicely into the logic, amplifying, and switching circuits which were the first offerings of the industry. These circuits handled the computational requirements of the growing computer industry. The internal memory functions of the early computers were handled by core memories. These memories were limited in capacity and slow. Much of the information needed was stored outside the computer on tape, disks, or punch cards. While bipolar memory circuits were available, the earlier circuits could not compete economically with core memories.

MOS transistor circuits held the promise of fast, economical, solid-state memory, but earlier metal-gate MOS circuits suffered from high leakage currents and poor parameter control. Even so, the built-in advantages of MOS transistors drove the development of MOS memory circuits. The advantages are smaller dimensions which allow denser circuits and relatively faster switching speeds. Yields tend to be higher on smaller-dimensioned circuits since a given defect density will affect fewer transistors and components.

Perhaps the biggest density factor advantage of MOS components is the smaller area required for isolation of adjacent components. The various isolation schemes used are discussed in the next sections. Another advantage is low-power operation. First, MOS transistors sit in the circuit in the "off" mode, not soaking up power or generating heat like bipolar transistors, which must be turned on to be in a "ready" state. Second, MOS transistors, being a voltage-controlled device, require a lower power to operate. CMOS circuits (see p. 441) are an integrated circuit design that reduces power requirements even further.

An initial advantage of MOS circuits was fewer processing steps and smaller die sizes, which made for lower processing costs and higher yields. These advantages have disappeared as MOS circuits have evolved to VLSI size with the additional steps required to fabricate CMOS circuits. In general, the faster switching speed of bipolar circuits has made them favored for logic circuits. MOS circuits, with their smaller component dimensions and lower power requirements, have been incorporated into memory circuits. By the 1980s these traditional uses blurred, with bipolar memories and MOS logic circuits readily available. These topics are addressed further in Chap. 17.

Bipolar circuit formation

Junction isolation. The bipolar transistor structure and basic performance have been illustrated. Early discrete bipolar transistors were formed from double-diffused (base and emitter) structures. These structures are unworkable in an integrated circuit. If two transistors are fabricated next to each other they will not work because electrically they would share a common collector (Fig. 16.29). The need to fabricate bipolar transistors electrically *isolated* from each other led to the epitaxial layer bipolar structure (Fig. 16.30).

The process starts with a P-type wafer into which an N-type diffu-

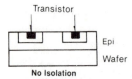

Figure 16.29 Adjacent bipolar transistors with common collectors.

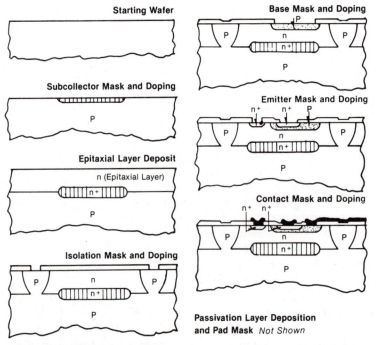

Figure 16.30 NPN bipolar process.

sion is made (the flow diagram does not show the oxidation and masking steps required to create the diffused layer). After the diffusion step an N-type epitaxial layer is deposited, leaving the N-type diffused region "buried" under the epitaxial layer. The N-type region is known as a *buried layer* or as the subcollector of the transistor. Its function is to provide a lower-resistance path for the collector current as it flows out of the base region on its way to the surface collector contact.

After the deposition of the epitaxial layer it is oxidized and a hole opened up on each side of the buried layer. A P-type doping step is performed deep enough to reach the P-type wafer surface. The doping step divides (*isolates*) the epitaxial layer into N-type islands, each surrounded on the sides (the doped regions) and the bottom (the P-type

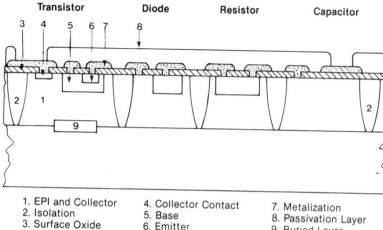

Figure 16.31 Bipolar structures.

1. EPI and Collector   4. Collector Contact   7. Metalization
2. Isolation   5. Base   8. Passivation Layer
3. Surface Oxide   6. Emitter   9. Buried Layer

wafer) by P-type doped regions. Components formed on the surface of each of the islands are electrically isolated from each other (Fig. 16.31). The electrical isolation occurs because the N-P junction is "wired" into the circuit to function in the reverse bias mode; that is, no current crosses the junction. This scheme is called *junction isolation* or *doped junction isolation*.

Dielectric isolation. In high-radiation environments, such as outer space or in the vicinity of atomic weapons, doped junctions produce holes and electrons that compromise the junction functions. Besides causing circuit component failure the radiation swamps out the isolation protection of the doped regions. Dielectric isolation schemes provide the necessary electrical isolation and radiation protection.

The process (Fig. 16.32) starts with the etching of pockets or trenches in a wafer surface. The etching can be either an isotropic wet etch or an anisotropic dry etch. Isotropic etch profiles follow the orientation structure of the wafers. Dry etch processes allow the shaping of the trenches. One goal of the etch step is to minimize the area of the pocket on the wafer surface. Wide pockets limit the packing density of the circuit.

After etch the pocket sides are oxidized and backfilled with deposited polysilicon. Next the wafer is turned over and the silicon of the wafer is lapped until the oxide layer is reached. These steps leave a wafer surface containing oxide-isolated pockets of the original single silicon material. The circuit components are fabricated in the silicon pockets, with each pocket being isolated on three sides by the layer of

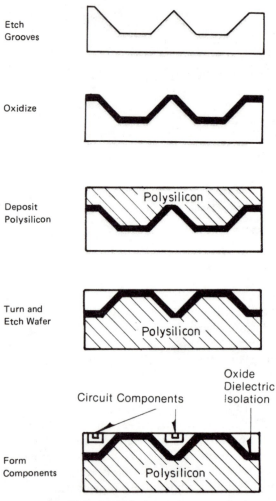

Figure 16.32 Dielectric isolation.

silicon dioxide. The dielectric property of the silicon dioxide prevents leakage currents in both normal and radiation environments.

Collector contact.   Note that in the cross section (Fig. 16.31) of a circuit integrated bipolar transistor there is a doped region under the collector contact. This doped region is put into the surface along with the emitter N-type doping. The emitter is usually designated N^+ to indicate that it is highly doped. The N^+ region under the collector contact is present to create a lower resistance between aluminum metallization and the silicon of the collector.

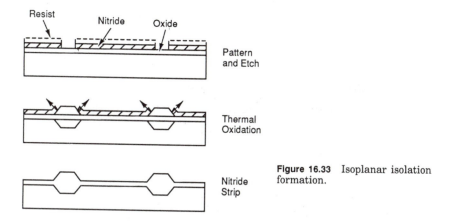

Resist   Nitride   Oxide

Pattern
and Etch

Thermal
Oxidation

Nitride
Strip

Figure 16.33   Isoplanar isolation
formation.

MOS Integrated circuit formation

Isoplanar isolation. MOS transistors function with a current flow
along the surface between the source and drain. When two MOS transistors are formed next to each other and at some distance apart there
are no shared parts, as with bipolar transistors. However, with
densely packed circuits there is a need for isolated adjacent components from surface leakage currents. The scheme used is called
isoplanar isolation.

The isolation scheme (Fig. 16.33) starts with depositing a layer of
silicon nitride over a layer of silicon dioxide in the regions where the
components are to be formed. Next the exposed surface is oxidized.
The oxide grows down into the wafer surface in a hexagonal shape.
When the silicon nitride is removed, the surface is left with flat regions separated by isolating sections of silicon dioxide.

Trench isolation. Trench isolation is also used for MOS circuits
(Fig. 16.34). The procedure is the same as forming trench capacitors.

CMOS. Complementary MOS (CMOS) is an MOS circuit formed with
both N-channel and P-channel transistors. CMOS has become the
standard circuit for many applications. It is the CMOS circuit that has
made possible digital watches and hand-held calculators. It allows circuits on one chip that would require several chips using N-channel
and P-channel only circuits. CMOS circuits also use lower amounts of
power than comparable circuits.

CMOS structures (Fig. 16.35) are formed by first fabricating an N-

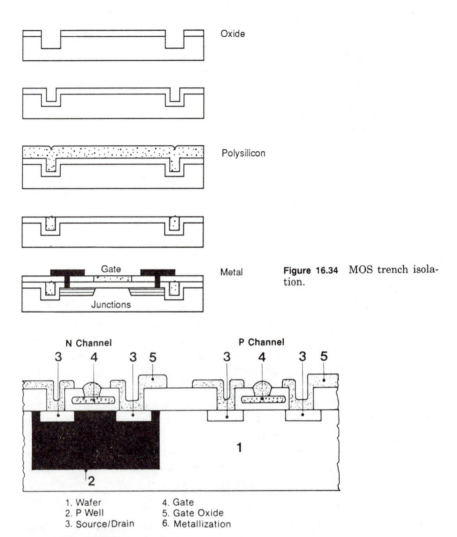

Oxide

Polysilicon

Metal   **Figure 16.34** MOS trench isolation.

Gate

Junctions

N Channel   P Channel

3   4   3   5   3   4   3   5

1

2

1. Wafer   4. Gate
2. P Well   5. Gate Oxide
3. Source/Drain   6. Metallization

Figure 16.35 CMOS structure.

channel MOS transistor in a deep P-type well formed in the wafer surface. After N-channel transistor formation a P-channel transistor is fabricated. The transistor structures are silicon gate or other advanced structures. CMOS processing uses the most advanced techniques since smaller, more densely packed, and higher-quality components all increase the advantages inherent in the CMOS design.

Bi-MOS

The unique advantages of bipolar and CMOS transistors and their respective circuits come together in Bi-MOS circuits. The circuits (Fig.

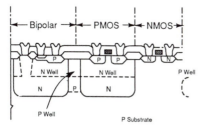

Figure 16.36 Bi-MOS structure.

16.36) have in them bipolar, P-channel, and N-channel transistors along with memory cells (see Chap. 17). The low-power advantage of CMOS is used in logic and memory sections of the circuit while the high-speed performance of bipolar circuitry is used for signal processing.[4] These circuits represent great challenges to the processing area with their large die size, small component size, and large number of processing steps. Figure 16.37 lists the number and type of implants required for a typical Bi-MOS structure.

Superconductors

Much interest has been generated by developments in super-conductivity materials. Superconductivity is a phenomenon that occurs in certain materials when they are cooled close to absolute zero ($-273°C$). In ordinary metals current is passed along by the flow of electrons. In a rest (nonconducting) state the electrons exist in orbits around the nuclei of the atoms. To become conducting they must gain energy to overcome the internal resistance of the particular material. The energy must be continuously supplied to maintain the current flow.

In a superconducting material electrons exist in a "conducting state" and can support an electrical current with little or no additional energy input. The prospect of a resistanceless material has the potential of revolutionizing electronic devices.

Semiconductor researchers have investigated the superconducting effect for years. In 1962, B. D. Josephson described the effect (named after him): when a thin oxide separates two superconducting materials, electrons will pass through the oxide with zero resistance. The structure is called a *Josephson junction*. This effect (called *tunneling*) is very complicated and requires quantum physics concepts to understand (well beyond the scope of this text!). The result is that the oxide has the functional aspects of a gate and Josephson junctions can perform basic switching, logic, and memory functions.[5] While IBM and

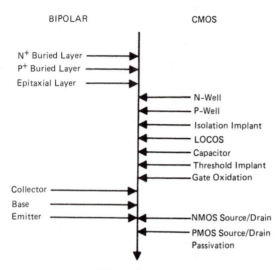

Figure 16.37   Bi-MOS implant steps.

Bell Labs (now AT&T) have development programs in superconductivity devices, no practical devices are currently available.

Physical Semiconductor Devices

Semiconductor junctions have the property that almost anything done
to them produces a change in their functioning. The anything can be a
physical strain, exposure to light, radiation, or heating. A device that
takes advantage of a junction reaction to physical pressure is the
strain gauge. The gauge is made by back-etching through the wafer
until only a very thin membrane is left. A junction and supporting circuitry are formed in the membrane. When the membrane is deflected
by some force, as in a weight scale or pressurized gas line, the semiconductor circuit produces an output proportional to the deflection.
The output of the circuit is correlated with the amount of pressure on
the membrane and displayed on the appropriate meter.

Light-Emitting Diodes

One effect when certain compound junctions are reversed-biased is the
production of photons. Photons are a form of radiation that humans
see as light. The devices are called *light-emitting diodes* (LED). These
are the displays that are used in consumer electronics equipment and
automobile displays.

The devices (Fig. 16.38) are made on gallium-arsenic-phosphide
(GaAsP) wafers that are covered with thousands of diodes, wired so
that they can be turned on or off individually. Groups of diodes are
turned on in groups to form letters and numbers. GaAsP material pro-

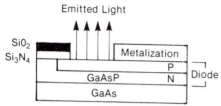

Figure 16.38   LED structure.

duces the familiar red displays, with other colors being produced when dopants are added to the wafer.

Solar Cells

Not only can semiconductor junctions emit light, they respond to it. This property is taken advantage of in the solar cell (Fig. 16.39). The cell is composed of diodes formed in a thin layer of semiconducting material such as amorphous silicon. When sunlight strikes the junction region, a current passes across it. The current is captured by onboard circuitry.

Temperature Sensing

Semiconductor junctions are temperature-sensitive. Heating a device will produce more current across the junction. This effect is taken advantage of in a variety of devices such as solid-state medical thermometers and industrial control units.

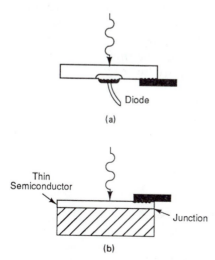

Figure 16.39 Light-sensitive semiconductor structures. (a) Photodiode; (b) solar cell.

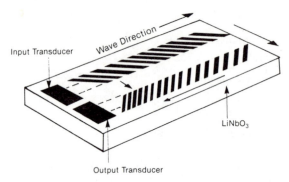

Figure 16.40   Acoustic wave device.

Acoustic Wave Devices

Acoustic wave devices (Fig. 16.40) are nonsilicon solid-state components used in microwave communication systems. They serve the
function of converting microwaves into electrical impulses. A compound material such as $Be_{12}GeO_{20}$ has the property of reacting to the
wave and setting up an electrical response in a solid-state circuit
formed on the chip by normal semiconductor processes.

Bubble Memories

Bubble memories are nonsemiconductor, solid-state devices used for
memory storage (Fig. 16.41). The material is an iron alloy deposited
on a substrate. The alloy is capable of forming tiny magnetic *bubbles*
in response to an electrical pulse from a semiconductor-type metal system formed on the surface. The information to be stored is coded in
long lines (strings) in the alloy surface as magnetic bubbles that are
oriented either north or south (up or down). The information is re-
trieved by moving the strings of bubbles under a surface sensor that
records the polarity of the bubble and sends the information to other
circuitry that reconstructs it into the required word form.

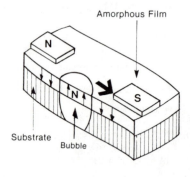

Figure 16.41   Bubble memory device.

Key Concepts and Terms

Acoustic wave	Isoplanar isolation
Bipolar transistors	MESFET
Bubble memory	Oxide capacitors
CMOS LED	Polycide gate
Doped diodes	Salicide gate
Doped resistors	Schottky barrier diodes
Epitaxial resistors	Strain gauge
FETMOS transistors	Thin film resistors
JFET transistors	VMOS
Junction capacitors	DMOS
	MMOS
Junction isolation	
Dielectric isolation	

Review Questions

1. Sketch and name the three parts and functions of a bipolar and MOS transistor.

2. Explain the function of a diode, resistor, and capacitor.

3. What is the operational advantage of a Schottky barrier diode?

4. Sketch and name the layers in a metal, polyside, and salicide MOS gate structure.

5. What is the structural difference between FETMOS, JFET, and MESFET gate structures?

6. Describe why component isolation is required in integrated circuits.

7. What are the differences and advantages of junction and dielectric isolation schemes?

8. Sketch the process steps used to create an isoplanar isolation scheme.

9. Name three ways that a semiconductor junction in operation can be influenced.

10. How does a CMOS structure differ from an NMOS structure?

References

1. H.R. Camenzind, *Electronic Integrated Systems Design*, Van Nostrand Reinhold, Princeton, N.J., p. 85.
2. H.R. Camenzind, *Electronic Integrated Systems Design*, Van Nostrand Reinhold, Princeton, N.J., p. 141.
3. Y. Pauleau, "Interconnect Materials for VLSI Circuits," *Solid State Technology*, April 1987, p. 157.
4. C.B. Yarling, "M.I. Current, Ion Implantation for the Challenges of ULSI and 200 mm Wafer Production," *Microelectronic Manufacturing and Testing*, March 1988, p. 15.
5. P.W. Anderson, "*Electronic and Superconductors*," in E. Ante'bi (ed.), *The Electronic Epoch*, Van Nostrand Reinhold, Princeton, N.J., p. 66.

Integrated Circuit Types

Overview

The solid-state semiconductor components are wired into integrated circuits. Those circuits perform many different functions in electronic instruments and machines. In this chapter the major circuit types and their functions are explained.

Objectives

Upon completion of this chapter you should be able to:

1. Explain the concept of binary numbering.
2. List the three major integrated circuit functions.
3. Compare the basis of analog and digital logic circuits.
4. List the user and production advantages of logic gate arrays and PAL circuits.
5. Explain the two major memory circuit types.
6. Make a list of the four nonvolatile memory circuits.
7. Compare the performance and cost factors of DRAM and SRAM memory circuits.

Introduction

Over 60 percent of the units produced by the semiconductor industry are in the form of integrated circuits. The number of individual circuits that can be created using the technology and processes described

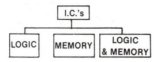

Figure 17.1   IC circuit functions.

in this text is endless. A circuit catalog from a major integrated circuit (IC) producer like National Semiconductor or Motorola is as large as a New York City phone book. IBM estimates that in the 1990s their internal circuit catalog will list over 50,000 separate circuits!

Becoming familiar with integrated circuits is not as awesome a task as the high numbers might imply. The burden is eased by the fact that most circuits fall into three functional categories: logic, memory, and hybrid (Fig. 17.1). Additionally, within each circuit family there are a few basic designs and functions.

The major functional circuit categories and their circuit designs are explained in this chapter. In the last section, we look at the future in IC circuitry from the perspective of the industry today. What the circuits will actually be like in 2010 can only be imagined, just as in 1950, no one predicted the megabit RAM or the microprocessor.

Circuit Basics

The question of how any particular integrated circuit actually works is the subject of other texts, but all circuits are based on the processing of data in binary notation. The binary system is a way of representing any number with just two digits—a zero and a one. It is actually an accounting system that keeps track of the place and value of the components of a number. Numbers can be expressed as the sums of numbers. For example:

$1 = 1 + 0$

$3 = 2 + 1$

$7 = 4 + 2 + 1$

$10 = 8 + 2$

Another way to express numbers is as powers of their factors. Yet another way is to express numbers as powers of their roots.

The basis of binary notation is the powers of the number 2. In Fig. 17.2 are the numbers 1, 2, 4, 8 expressed as powers of 2. In effect those numbers can be *represented* by the exponent of 2. Thus 1 becomes 0, 2 becomes 1, 4 becomes 2, and 8 becomes 3.

A number expressed in binary notation is first expressed as the sum

$$1 = 2^0$$
$$2 = 2^1$$
$$4 = 2^2$$
$$8 = 2^3$$

Figure 17.2 Powers of 2.

Standard	32	16	8	4	2	1
Power of 2	2^5	2^4	2^3	2^2	2^1	2^0

Number Standard = Binary Number						
1	0	0	0	0	0	1
7	0	0	0	1	1	1
18	0	1	0	0	1	0
33	1	0	0	0	0	1

Figure 17.3 Binary notation.

of numbers that are factors of 2. The number 25 can be expressed as the sum $16(2^4) + 8(2^3) + 1(2^0)$. In the number 25 there is one 2^4, one 2^3, zero 2^2, and one 2^0. Or the number 25 can be represented by the code 1101, each of the digits representing the presence or absence of a particular power of 2. The chart in Fig. 17.3 lists some numbers in binary notation.

Translating numbers into binary notation is easily accomplished by establishing a grid with each column representing a power of 2. The actual number is represented by a string of zeros or ones that indicate the presence or absence of the particular powers of 2 that make up that number.

Binary notation has been known for centuries. Buckminster Fuller, in his book *Synergistics*, has an amusing account of the use of binary coding by the ancient Phoenicians to keep track of cargo amounts. He claims that the Phoenician sailors were considered stupid because they could not count in the system of the day, when actually they were accurately keeping track of large amounts of cargo with only two "numbers"; they were counting in binary notation.

With binary notation only two numbers are necessary, a one or a zero. In the discussion above, binary notation was represented by the numbers zero and one. In the physical world, binary numbers can be represented by any system that has two conditions. Figure 17.4 shows several different ways to code the number 7. The last row could represent binary coding by the off-on states of a transistor or memory cell.

Inside a circuit the numbers are coded, stored, and manipulated in their binary code. The numbers can be stored and manipulated because capacitors can be charged to have a charge or not have a charge and transistors can be either on or off in the correct sequence. The smallest piece of information in a circuit is called a "binary digit" or "bit."

The binary coding system is simple. The problem of how the coded numbers could be added, subtracted, and multiplied was solved by

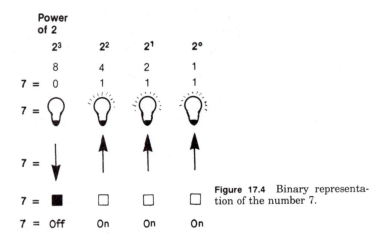

**Power
of 2**

2^3	2^2	2^1	2^0
8	4	2	1

7 =   0   1   1   1

7 =

7 =

7 =

7 = Off   On   On   On

Figure 17.4 Binary representation of the number 7.

George Boole, a nineteenth-century mathematician, He developed a logic system capable of handling numbers in binary notation. Until the development of computer logic, his Boolean logic (or Boolean algebra) was an academic curiosity.

Chips and computers are designed to handle a specific size of binary number or word. An 8-bit machine manipulates numbers with 8 binary bits at a time. A 32-bit machine can handle a number composed of 32 binary bits. The more bits a machine can handle at one time the faster and more powerfully it processes data. Every 8 bits is known as a byte. Thus, an 8-megabyte storage capacity can hold 8 million bits of information.

Integrated-Circuit Types

A solid-state integrated circuit is comprised of a number of separate functional areas. Each chip, regardless of the circuit function, has an input and encode section where the incoming signals are "coded" into a form that the circuit can understand. The majority of the circuit area contains the circuitry required to perform the circuit function, either memory or logic. After the data is manipulated by the circuit it goes to a decode section where it is changed back into a form that is usable by the machine's output mechanism. The circuit's output section actually sends the data to the outside world.

Although this is an overly simplified explanation of a circuit, it illustrates the fact that the interior of a chip is composed of definite separate functional areas. In many circuits these areas perform the same functions as the main parts of a computer. Circuit types fall into three broad categories: logic, memory, and logic and memory (microprocessors).

Logic circuits perform a specified logical operation on the incoming data. For example, pushing the plus (+) key on a calculator instructs the logic portion of the chip to add the numbers presented to it. An onboard automobile computer goes through a logical operation to direct the signal from a sensor indicating an open door to light up the correct warning light on the dashboard.

Memory circuits are designed to store and give back data in the same form in which it is entered. Pushing the (pi) key on a calculator activates the memory part of the circuit where the value of pi (3.14) is stored. The value 3.14 is displayed on the screen. Every time that key is activated, that value is displayed.

A third type of circuit function combines both logic and memory in a circuit called a microprocessor. In 1972, Intel Corporation introduced the first practical microprocessor. It was the microprocessor that allowed the design of powerful personal computers, digital watches, and one-chip calculators and the transfer of so many business machines to solid-state electronics, from phone systems to vending machines.

Microprocessors can be programmed to perform many different circuit functions. To accomplish this, they contain logic and memory circuitry as well as the necessary encode, decode, input, and put sections. The microprocessor has also made possible the one-chip electronic calculator, the digital watch, and the personal computer.

The microprocessor has been dubbed "a computer on a chip." While it contains all the functional areas of a computer, it is not truly a complete computer. Even simple computers require vast amounts of memory capacity which microprocessors do not have. Within many personal computers, microprocessors function as the central processing unit (CPU). Additional memory chips have to be included in order to make the computer of practical use.

Actually, every integrated circuit contains both logic capability and memory sections. For example, the logic circuitry of a calculator must have certain constants stored in a memory section in order to perform calculations. And memory circuits must have some logic functions to direct the flow of electrons and holes to the right parts of the circuit for storage.

Logic circuits

Analog-digital logic circuits. Logic circuits fall into two main categories: analog and digital (Fig. 17.5). Analog logic circuits were the earliest logic circuits developed. An analog circuit has an output that is

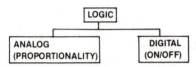

Figure 17.5   Logic circuit types.

proportional to the input. Digital circuits, on the other hand, feature a predetermined output in response to a variety of inputs. A wall light dimmer is an analog device. Turning the control varies the voltage to the dimmer, which in turn varies the brightness of the light. A standard on-off light switch is a digital device. Only two brightness conditions are possible: on or off. Most audio circuits are of the analog type. Changing the level setting of the volume produces a proportional change in the sound coming out of the speaker.

Analog logic circuit types.   Analog circuits were the first type designed in integrated form. The home computer hobby kits of the 1950s were the analog type. These simple circuits were based on Ohm's law $(R = V/I)$. The circuit contains a resistance meter and a means for generating a current and measuring a voltage. The three quantities are related by Ohm's law. Any other three variables similarly related can be represented by the resistance, voltage, and current. Varying one changes the other two. The circuit thus becomes a computer for solving any equation of the form $A = B/C$.

The accuracy of analog circuits is dependent on the precision of the relationship between the input and the output. In the simple computer illustrated, accuracy is dependent on the precision of the components in the circuit, the clarity of the meters for setting the input and reading the output, and the immunity of the circuit to outside "noise." Unless the circuit contains a section to regulate incoming voltage levels, a change in the line voltage would alter the output, and hence the accuracy.

Both simple and complex analog circuits are vulnerable to variations in the incoming signal and to internal noise. Analog circuits are also dependent on precise control of the resistor values. Unfortunately, diffused resistors cannot be fabricated with a resistance variation from design value better than 3 to 5 percent, which is unacceptable for many applications. Greater resistance precision is gained by the use of matched resistor pairs, in which the effective resistance in the circuit is the difference between two resistors. This difference can be more tightly controlled than one resistor alone.

Ion implantation also provides the analog circuit designer with a tool for producing resistors with a higher degree of control. Many analog circuits feature thin film resistors to achieve the required precision. The growth and popularity of digital circuits is based on their ability to produce a set output. If a 5-V signal is absolutely required to operate a device, this can be achieved by designing a digital circuit that produces a 5-V signal every time, regardless of the input variation or internal electrical noise.

Digital circuits, however, do not respond as fast as linear circuits. The term used in electronics is *real-time response*. In some applications, such as airplane controls, real-time response is mandatory. Recent development in digital circuit speed is speeding the encroachment of digital circuitry into this traditional use of analog circuits. A major advantage of digital circuits over analog circuits is in general-purpose computers. Analog circuits are more difficult to design to respond to a general range of problems. All modern general-purpose computers are based on digital circuits.

The most popular use for analog circuits is in amplifiers. They are designed in a variety of configurations, for many different applications. All have the same basic principle—the incoming signal or pulse is amplified. Audio circuits require amplification of a weak signal from the record tone arm or other input in order to produce the level required to operate a speaker.

The real-time aspect of analog circuits also makes them real-world circuits. Wherever there is a real-world measurement such as temperature or movement, analog circuits are used. Even when the majority of the circuitry in a system is digital, analog circuits are often part of the interface with the outside world.

Most analog amplifier circuits are of the differential operational type. These circuits produce an output voltage amplified from and proportional to the difference of two input signals. Bipolar technology is favored for these circuits because bipolar circuits are modular electric current devices and are better suited to the applications required of analog circuits.

The output signal of an analog device can have a "one-to-one" relationship with the input signal. These circuits are called *linear*. If the input is changed, the output changes in a linear manner. So many analog circuits are of linear design that the two terms are often used interchangeably. However, there are nonlinear circuits, such as those featuring a logarithmic relationship between the input and output.

Logic circuitry is built around the logic gate. A gate controls and directs passage through a barrier. The size and design of a gate influence the amount of passage allowed. For example, a room with many "in" doors and only one "out" door is a gate. Many people can enter the room but their exit is restricted because only one door is provided. This example gate can also be operated in reverse, allowing people to enter through only one door and leave through many.

Electronic logic gates perform similar functions but with electrical signals. In a circuit they perform the necessary logic operation by the dictates of Boolean logic. A discussion of their incorporation in logic design is beyond the scope of this text. The point for this text is that gates, both analog and digital, are formed in a logic circuit by wiring together various components. The circuits are designated by the first letters of the components in the gate. The major bipolar analog logic circuit designations are:

RTL	Resistor-Transistor Logic
DTL	Diode-Transistor Logic
TTL	Transistor-Transistor Logic
ECL	Emitter-Coupled Logic
DCTL	Direct-Coupled Transistor Logic
I^2L	Integrated Injected Logic

Integrated injected logic is another approach to constructing a logic gate. In this arrangement, a bipolar transistor is operated in the reverse mode, with the emitter as collector and collector as emitter. An advanced I^2L gate design is formed with the bipolar transistor constructed with a Schottky diode (sometimes called a clamp) between the base and collector. This arrangement results in a faster transistor, and hence a faster circuit.

Custom-semicustom logic. Using any of the logic gate approaches listed, hundreds of thousands of different logic circuits can be constructed. They vary from custom small-volume circuits to off-the-shelf standards. The bulk of logic circuits require some degree of customization. Several design and fabrication approaches are used to deliver custom and semicustom circuits to the customer at reasonable costs. The approaches are:

1. Full custom

2. Standard circuit—custom gate pattern

3. Standard circuit—selective wiring gate

4. Programmable array logic

Full custom. A full custom-designed logic circuit is specified by the customer who pays for design and mask-making fees along with the fabrication costs. This approach is expensive and lengthy, and is not geared for experimenting with different circuits in the design stage of a project. Custom-designed circuits are not cost-effective in quantities of less than 100,000.

Standard circuit—custom gate pattern. This fabrication process starts with a standard logic circuit design, but only the gates required for the particular application are formed during the fabrication process. The input, output, and other circuit sections are standard for a family of circuits.

Standard circuit—selective wiring gate arrays. This system is similar to the custom gate approach but based on a standard circuit design for most of the fabrication process. These circuits are built with a standard number of gates. This gate section is called the array and the circuit is known as a *gate array*. Working with the basic design, the customer can instruct the fabrication department to wire together only the gates required to produce their custom circuit logic function.

The result is faster turnaround time than full custom processes can achieve, and moderate cost. The cost per logic function of gate arrays is higher than that of a custom circuit produced in production quantities. The larger gate section required to allow many different circuit variations results in a larger chip. This larger chip size leads to a higher manufacturing cost per chip and/or a lower yield.

The wafers receive a common process up to the contact mask. The contact mask is customized to form contacts only to the gates required for the particular circuit. After metal deposition only the gates with contacts are wired into the circuit. A variation of this process is to open the contacts in all the gates but use a customized metal mask that wires in only the wanted gates.

Programmable array logic (PAL). Each of the three systems described requires the chip manufacturer to do the "customizing." This requirement can result in delivery or scheduling problems and generally forces the user to buy a minimum quantity of parts. Monolithic Memories, Inc., addressed this problem with the introduction of their PAL line of circuits in 1978. PAL stands for programmable array logic. MMI applied the programmable fuse technology used in memory products to logic circuits. The result was a field programmable (custom) logic circuit.

The concept is similar to the standard gate array, but within the circuit, a fuse connects each logic gate into the circuit. The user programs the circuit by blowing the fuses at the unneeded gates, thus removing them from the circuit. The customer buys a quantity of standard PAL circuits and can program them as needed for their particular applications. In his book *Soul of a New Machine*, Tracy Kidder related the story of a Data General project team's decision to design a 16-bit computer using the newly introduced MMI circuits. The ease of programmability and shortened delivery time is credited with allowing the computer to be brought to market ahead of another digital design team using custom circuits.

Memory circuits

Around 1960, industry forecasters began predicting that solid-state memory circuits would overtake the traditional core memory. The advantages of solid circuits were their reliability, small size, and faster speed. This prediction was made every year until the early 1970s, when solid-state memories finally did surpass core memory. The major factor which prolonged the life of core was the lower cost of core memory.

MOS is the favored transistor structure for memory circuits. During the 1960s, however, the cleanliness requirement for MOS processing was not reliably available. High-yield MOS processing also requires accurate alignment and clean thin gate oxides. These processes were not fully developed in those years. The resultant low process yields kept MOS memories more expensive than core memories.

With process improvements and improved MOS structures, MOS has not only become the memory method of choice but will pass bipolar circuits in volume for both memory and logic circuits. Some bipolar memories are favored for their fast speed and switching capabilities. While logic circuits can be (and are) made in MOS technology, most MOS circuits produced are memories, with the majority incorporated into computers. They are also used in microprocessor-based products, which require auxiliary memory chips. There are two principal types of memory circuits: volatile and nonvolatile (Fig. 17.6).

Nonvolatile memories. A nonvolatile memory device is one that does not lose its stored information when it loses its power. An example of this is a phonograph record, which is an information-storage device. If power to the record player is lost, the songs are not lost from the

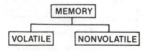

Figure 17.6 Memory circuit types.

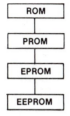

Figure 17.7   Nonvolatile memory types.

record itself. A number of nonvolatile memory circuits are listed in Fig. 17.7.

ROM. In integrated circuits, the ROM design is the principal nonvolatile circuit. ROM stands for *read only memory*. The sole function of this type of circuit is to give back prestored information. The information required in the circuit is specifically designed into the chip memory array section during fabrication. Once the chip is made, the stored information is a permanent part of the circuit.

Other memory types have *read and write* capability. That is, they can receive and store information from an input device (keyboard, magnetic tape, floppy disk, etc.). A phonograph record is a nonvolatile ROM device. A magnetic tape is an example of a nonvolatile device with both read and write capabilities because information can be erased and re-recorded.

In a calculator a lot of information is stored in the ROM section of the circuit. Every time you turn the calculator on, all the constants stored in ROM are still available. Internal instructions and constants required so that the circuit can function as a calculator are stored in a ROM section. ROM circuits, like logic circuits, number in the hundreds of thousands. Although there are many standard types, the industry also uses many custom ROM circuits. The choices offered to the user in selecting a standard or custom chip are similar to those available with logic circuits. The user can buy a standard circuit, specify a variation on a standard basic circuit, design a total custom circuit, or buy a PROM, EPROM, or EEPROM.

PROM. PROM stands for programmable read only memory. A PROM is the memory equivalent of a PAL. Every memory cell is connected into the circuit through a fuse. Users program the PROM to their own memory circuit requirements by blowing fuses at the unwanted memory cell locations. After programming, the PROM is changed to a

ROM, and the information is permanently coded in the chip, where it becomes a read only memory.

EPROM.   For some applications, it is convenient to change the information stored in the ROM without having to replace the whole chip. EPROM chips (erasable programmable ROM) are designed for this use. The erasable feature is built in with the use of MMOS (memory MOS) transistors detailed in Chap. 16. The transistors can be charged (programmed) selectively and they hold the charge for a long time. This structure has the ability to have the charge (erasing the memory) drained off by shining ultraviolet light on the chip. Reprogramming of the chip takes place by removing it from the circuit and putting in new memory information with an external programming machine. A typical EPROM can be reprogrammed up to ten times.

EEPROM.   The next level of convenience in memory design is the ability to program and reprogram the chip while it is in a socket in the machine. This convenience is available with the EEPROM, standing for electronically erasable PROM. Programming and erasing take place by pulses from the outside that place charges in selected memory cells or drain the charges away.

Bubble memory.   Bubble memories are solid-state devices which do not use a semiconductor transistor-based memory cell to store information. The substrate of a bubble memory (usually garnet) has hundreds of thousands of tiny magnetic regions called "bubbles." The bubbles can be programmed—directionally oriented—within the substrate by an electric current flowing in a surface circuit. Two orientation directions are possible: up and down, or north and south. (Remember that only two conditions are required for binary information coding and processing.)

The memory information is stored in the substrate as a string of north and south pole magnetized "bubbles." Retrieval occurs by moving the bubbles past a structure which senses the pole of each bubble. This is a *serial* format type of information retrieval. Gaining access to the first data put in memory requires printing out *all* of the information in memory. A phonograph record, without a mechanism to lift the tone arm, would also be a serial memory device. Listening to the last song recorded would necessitate listening to the entire record.

Bubble memories are capable of tremendous storage capability because of the small size of the individual bubbles. In fact, the storage capacity is similar to that of traditional bulk computer storage techniques such as disks and magnetic tapes. The major drawback of bubble memory use is the slowness of information retrieval. However, bubble memories are used where speed is not a critical factor. Telephone systems and lap computers both utilize these memories. Bubble memories are a nonvolatile, read-write memory. They retain the information without power, and new information can be entered into them for either permanent or temporary storage.

Volatile memories

Semiconductor circuit and computer design involves the constant evaluation of trade-offs. In the case of memory, nonvolatile memory provides protection against power loss, but these memories are frequently slow and not very dense (bits of storage per square centimeter). More important, none of the circuits described above (with the exception of bubble memories) has a write capability, an essential feature in operating a computer. New information, such as a change in pay status, must be conveniently entered into the computer and stored temporarily while the new check is being written. Memory must also be easily erasable so the computer can quickly process new information or accept a completely new program. There are several memory circuit designs used to produce fast and high-density memory circuits. Both are of the volatile type; that is, when power is lost to the chip all the stored information is lost; information presented on a computer screen, and not saved, is eligible for loss if the power to the computer goes off.

RAM. One type of circuit used for high-density memory (Fig. 17.8) storage is random-access memory, or RAM. "Random" refers to the ability of the computer to directly retrieve any information stored in the circuit. Unlike a serial memory, the RAM design allows the chip to give back the exact information asked for by the computer wherever it is located in the computer memory. This feature allows faster information retrieval and makes the RAM the principal memory circuit in computers.

Dynamic random access memory (DRAM). DRAMS come in two principal designs: dynamic and static (Fig. 17.8). A dynamic memory design called a DRAM, for dynamic RAM, is used in great quantities in computer memories. The memory cell design is based on only one transis-

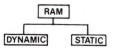

Figure 17.8 RAM cell designs.

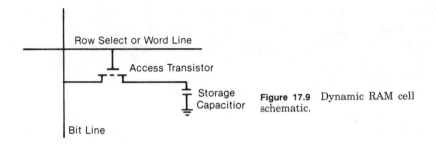

Figure 17.9   Dynamic RAM cell schematic.

tor and a small capacitor (Fig. 17.9). The information is stored in it by a charge built up in the capacitor. Unfortunately, the charge drains away very rapidly. To combat this problem, the memory information must be reinputted to the circuit on a constant basis. The term for this function is *refresh*. The refreshing of the circuit occurs many thousands of times per second. Dynamic RAMS are vulnerable to both power loss and interruption, or problems with the refreshing circuit.

The goal of DRAM design is small-cell design for high-density and closely spaced components with small and thin parts for speed. These requirements have driven DRAM design and processing to the highest levels of the technology. All the advantages available by advanced, state-of-the-art equipment and processing are applied to DRAM circuits. This fact make them the industry's leading-edge circuit.

Static random access memory (SRAM). Static random-access ram (SRAM) memories are based on a cell design that does not need a refresh function. Once the information is put into the chip, it will stay as long as the power remains on. This is accomplished with a cell containing several transistors and capacitors (Fig. 17.10). The information is stored as conditions of the transistors are alternately on or off. Information can be read and written with a SRAM cell much faster than with a RAM design since transistors can be switched faster than capacitors can be charged and drained. The penalty paid for this lesser degree of volatility and speed is loss of space. The larger cell design makes static memories less dense than DRAMs.

Memory capacity is measured by the number of bits that can be stored. A 1K RAM has a capacity of 1024 bits of information; 1024 is a power of 2. A 64K RAM actually has a capacity of 65,536 bits of information. RAM capacity is expanding rapidly, with one, four, and larger megabit memories (one million) expected to be produced in quantities with presently identified technology. Each step upward in RAM capacity places greater pressure on wafer processing and yield improvement. The nature of the semiconductor chip business is exemplified by the 64K RAM, introduced by IBM in 1977. The chips were

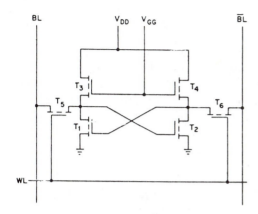

Figure 17.10   Static RAM cell schematic.

soon available in the merchant market, priced at over $100 per circuit. By 1985 competition and yield improvements had lowered the price to under $1 per circuit!

The Next Generation

The industry is moving toward a higher level of integration through the reduction of component size on the chip. A number of limits affect the size and levels which can be reached. First are the limits imposed by optical lithography. It is predicted that this technology will allow feature sizes down to the sub-1-μm dimension because of better process tools and control and improved clean rooms. Beyond that level x-ray and/or E-beam exposure systems will be used to produce geometries in the 0.1-μm range. The physical limit setting the bottom floor on device size and operation is 0.004 μm, the smallest size at which semiconductor devices will operate by the same physical laws as today's components. Below that size, conventional semiconductor physics do not apply.

Of more immediate concern are yield limits, especially as the individual circuits grow in size. Presently, the technology exists to cover an entire wafer with individual circuits. However, the yield of a whole wafer circuit would be virtually zero, owing to the presence of defects. Many circuits fail because of only one defect. A number of approaches are available to overcome the fatal influence of one or a few small defects on an otherwise functioning die.

Redundancy

Redundancy is the inclusion of extra circuit components in the design. If one or more of the components don't work, there are others available that do. The trade-off for redundancy is larger chip size. Also, extra circuitry is required within the main circuit to detect the functioning and nonfunctioning components and direct the selection of a functioning component. Although this approach to higher yield has been discussed for years, it hasn't yet become a mainstay of circuit design. This is due to the problem of locating the working and nonworking components and wiring the working ones into the circuit.

Wafer scale integration

Wafer scale integration (WSI) is a novel approach to integration aimed at fuller utilization of all the functioning areas on a wafer. The surface of a wafer using WSI is not covered by individual die. Instead, the wafer contains sections that are the functional subsegments of a circuit.

Failed chips on a conventional wafer usually fail from a single (or several) defect that prevents only one of the subsections of the circuit from performing. Even failed chips have many areas or subsections that *do* function. In the conventional chip approach, vast areas of the chip that contain working subsections are thrown away.

A WSI wafer is not covered by individual chips. Part of the wafer contains input sections, part has output sections, etc. At the end of the processing each of the subsections is electrically probed and tested. The locations of the working subsections are recorded. After testing, a "super" circuit is wired together from the working subsections. For example, consider a circuit board requiring 10 logic chips and 20 memory chips. The conventional approach requires manufacturing enough wafers to deliver the required number of individual chips.

With WSI, the wafer surface is covered with individual circuit subsections rather than complete circuits. In the supercircuit sample just given, the WSI wafer would have all the subsections plus extras required to make the 10 logic and 20 memory sections. The result is a supercircuit that uses all the functioning parts of the wafer surface. Other benefits include faster speed due to the lower number of package connections and lower cost resulting from eliminating individual packages. The drawbacks to WSI include longer processing time to form both logic and memory functions on the same wafer, reliability problems stemming from high heat levels which the more dense circuit generates, and the need for a more flexible metallization patterning than is currently available.

Key Concepts and Terms

Binary	ROM, PROM, EPROM, EEPROM
Memory/logic	Volatile, nonvolatile memories
RAM, DRAM, SRAM	Wafer scale integration

Review Questions

1. Name the two principal solid-state circuit types.
2. Name the two principal logic circuit types.
3. Give an example of a digital electrical device.
4. Which of the following describes a PAL circuit?
   a. Factory programmable
   b. Field programmable
   c. a and b
   d. Not programmable
5. How many times can a thin film fuse PROM be programmed?
   a. Once
   b. Five times
   c. Many times
6. What does RAM stand for?
7. What is the difference between a DRAM and a SRAM?
8. How does a serial memory device read back information? Give an example.
9. Why are DRAM circuits considered the leading-edge ULSI circuits?
10. How does the WFI approach to design and processing differ from conventional approaches?

18

Packaging

Overview

After wafer sort the chips on the wafer surface are completed and the electrically functioning ones identified. Most of the chips will be mounted onto the surface of a ceramic substrate as part of a hybrid circuit, connected directly onto a printed circuit board, or incorporated into a protective package. This chapter will describe the packages and processes used for chip protection.

Objectives

Upon completion of this chapter, you should be able to:

1. List the four functions of a semiconductor package.
2. List the five common parts of a package.
3. Recognize and identify the major package designs.
4. List and describe the major packaging process flows.

Introduction

After wafer sort, the chips on the wafer surface are completed and the electrically functioning ones identified. In most cases the chips will become part of a larger circuit. They may be mounted onto the surface of a ceramic substrate as part of a hybrid circuit, connected directly onto a printed-circuit board, or incorporated into a protective package (Fig. 18.1). In this chapter the packages used to protect the chip and the various packaging processes will be explained. This series of processes is known variously as *packaging, assembly,* or the *back-end process*. In the packaging process the chips are called *dies* or *dice* and

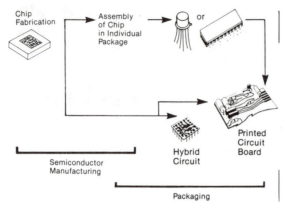

Figure 18.1 Chip packaging options.

in certain parts of the country they are referred to by the Texas Instruments' term, *bars*. In this chapter they will be referred to as *chips*
or *die*.

Over the years, semiconductor packaging has been considered to require a lower level of technology than the fabrication process because
at one time the packaging process was primarily a labor-intensive one.
The labor-intensiveness of packing was the reason why most packaging
facilities were established in countries where labor costs were lower. The
advent of the VLSI era in chip density has forced a radical upgrading
of chip packaging technology and production automation. Higher-
density chips result in more bonding pads, requiring more electrical
connections (Fig. 18.2) and cleaner packages and processes. And as
solid-state circuits have found more applications, the need for special
package designs has increased. The chips may end up in either consumer products or in the demanding and harsh environments of space
and military hardware. These harsh environments require special
packages, processing, and testing to ensure high reliability in the

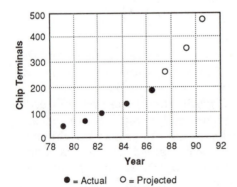

Figure 18.2 Projected chip terminal count.

field. These packages, processes, and tests are referred to as *hi-rel*. The other chips and packages are referred to as *commercial* or *commercial grade* parts.

No longer is packaging the stepchild of the semiconductor industry. Many feel that eventually packaging will be the limiting factor on the growth of chip size. For the time being, however, much effort is going into new package designs, new material development, and faster and more reliable packaging processes.

Chip Characteristics

Throughout this text many characteristics of discrete devices and integrated circuits have been mentioned. Several of them have a direct bearing on package design and the packaging processes (Fig. 18.3). The chip density (integration level) determines the number of connections required, with higher-density chips having larger surface areas and more bonding pads. The trend to larger chips has resulted in the need for thicker and larger-diameter wafers. These factors have caused changes in die separation processes, package design and the need for wafer thinning.

In previous chapters it was pointed out that the functioning of the chip components (transistors, diodes, capacitors, resistors, and fuses) can be altered by various influences. Primary among them are chemicals such as sodium and chlorine. Additionally, other chemicals can attack the chip layers, and environmental factors, such as particulates, humidity, and static can ruin chips or change their performance. Other concerns are the influence of light and radiation impinging on the chip surface. Some chips are extremely light or radiation sensitive, and this factor must be considered in the choice of package materials and processing. A dominant chip characteristic is the extreme vulnerability of its surface to physical abuse. The surface components are only a small distance down into the wafer surface and the surface wiring can be as thin as 1 μm.

These environmental and physical concerns are addressed in two

---

- Integration Level
- Wafer Thickness
- Dimensions
- Environmental Sensitivity
- Physical Vulnerability
- Heat Generation
- Heat Sensitivity

Figure 18.3 Chip characteristics affecting packaging process.

- Silox
- Vapox
- Pyrox
- Glassivation Layer
- PSG
- BSG
- PBSG

Figure 18.4   Passivation layer names.

ways. First is the passivation layer deposited near the end of the fabrication process. This layer is known by several different names as listed in Fig. 18.4. The two materials used for the passivation layer are silicon dioxide and silicon nitride. Often they are doped with boron, phosphorus, or both to increase their protective properties. The second method of protecting the chip is provided by the package itself.

Another chip characteristic of importance to the packaging process is heat generation. Chips used in high-power circuits and highly integrated circuits can generate enough heat to actually damage the circuit. This factor dictates the design of many packages. Heat is also an important parameter in packaging processes, where the vulnerability of the chip to high temperatures is of concern. The temperature of the chip cannot exceed 450°C. Above this temperature the aluminum and silicon contacts can form an alloy that can migrate down into the chip components and cause electrical shorts. This vulnerability is a constraining limit on the higher-temperature packaging operations of die attach and bonding.

Overview of Packaging Operations

In wafer fabrication the wafers pass many times in and out of four basic operations (layering, patterning, doping, and heat processing). In packaging also there are several basic operations (Fig. 18.5). As in fabrication, the exact order of the operations is determined by the package type and other factors. However, in packaging the flow is linear, that is, the chips proceed from operation to operation and they go through each operation only once. The operations are *backside preparation, die separation, die pick, inspection, die attach, wire bonding, preseal inspection, package sealing, plating, trim, marking*, and *final tests*. The details of the various operations and their optional processes are explained in more detail in the remainder of the chapter.

Backside preparation

At the end of the fabrication process some wafers have to be thinned (wafer thinning) to fit in the package or to remove backside damage or

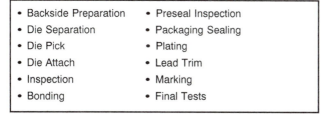

• Backside Preparation	• Preseal Inspection
• Die Separation	• Packaging Sealing
• Die Pick	• Plating
• Die Attach	• Lead Trim
• Inspection	• Marking
• Bonding	• Final Tests

Figure 18.5 Basic packaging operations.

junctions. Wafers whose die are going to be attached to the package by a gold-silicon eutectic will receive a deposited layer of gold (backside gold).

Die separation

The wafer is separated into individual die by sawing or scribe-and-break techniques.

Die pick and plate

The functioning die identified at wafer sort are picked from the separated wafer and placed in carriers (Fig. 18.6).

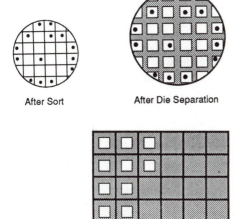

After Sort   After Die Separation

Plated Die

Figure 18.6 Die separation to plate.

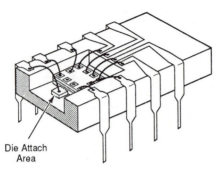

Die Attach
Area

Figure 18.7 Die attach area of package.

Inspection

The plated dies are optically inspected for edge integrity, contamination, and cosmetic defects.

Die attach

Each die is attached into the die-attach area of the package by either a gold-silicon eutectic layer or an epoxy adhesive material (Fig. 18.7).

Lead bonding

Thin wires are bonded between the chip bonding pads and the inner leads of the package.

Preseal inspection

The bonded die is optically inspected in the package. Criteria checked for are the alignment of the die in the package, proper bond placement, contamination, die-attach quality, and bonding quality.

Plating, trimming, and marking

The outer package leads are plated with an additional layer of conductive metal to improve their solderability into a printed-circuit board. Near the end of the process each package with outer leads will go through a trimming operation to separate the leads from supports. The marking operation is performed to code important information (Fig. 18.8) of the outside of the package enclosure.

Figure 18.8   Marked package.

Final test

A series of final tests, including electrical and environmental, are performed on each packaged chip to ensure quality. Some packages will receive a "burn-in" test to detect early failures.

Prepackaging Wafer Preparation

After the final passivation layer and an alloy step in wafer fabrication, the circuits are complete. However, one or two additional processes may be performed on the wafer before transfer to packaging. These steps, wafer thinning and backside gold, are optional, depending on the wafer thickness and the particular circuit design.

Wafer thinning

The trend to thicker wafers presents several problems in the packaging process. One of them is the effect of the thicker wafers on die separation and package design. Thicker wafers require the more expensive complete saw-through method at die separation. While sawing produces a higher-quality die edge, the process is more expensive in time and consumption of diamond-tipped saws. Thicker die also require deeper die attach cavities, resulting in a more expensive package. Both of these undesirable results are avoided by thinning the wafers before die separation. The thinning step may take place either before or after wafer sort.

Another situation requiring wafer thinning is electrical in nature. If the wafer backs are not protected as the wafers go through the dopant operations in fabrication, the dopants will form electrical junctions in the wafer back, which may interfere with good conduction in the back contact that is required for the circuit to operate correctly. These junctions may require physical removal by wafer thinning.

Three methods are used to thin the wafers. A mechanical method is back grinding on a precision grinder using an abrasive grit. Another method is a chemical-mechanical removal of the back by a process similar to the polishing of the starting wafers. A third technique is to dip the wafers in a tank of etchant and remove the required amount of material from the back.

Wafer thinning is a worrisome process. In back grinding there is the concern of scratching the front of the wafer and of wafer breakage. Since the wafer must be held down on the grinder or polishing surface, the front of the wafer must be protected, and once thinned, wafers are easier to break. In back etching there is a similar need to protect the front of the wafer from the etchant. The protection can be provided by spinning a thick layer of photoresist on the front side. Other methods include attaching adhesive-backed polymer sheets cut to the wafer diameter.

Backside gold

Another optional wafer process is adding a layer of backside gold. A layer of gold is required on wafers that are going to be attached to the package by eutectic techniques (see the die-attach section). The gold is usually applied in the fabrication area by evaporation or sputtering.

Cleanliness and Static Control

Contamination control is vital in the wafer fabrication process. However, the vulnerability of the chips to contamination exists during their entire lifetime. While assembly areas are not generally required to maintain the same cleanliness levels as fabrication areas, cleanliness is important. The table in Fig. 18.9 lists the common contamination control practices used in packaging areas. High-rel areas, particularly, demand higher cleanliness levels. In fact, many companies are finding that half-way contamination control programs are doomed to failure. Consequently, more assembly areas are practicing very stringent controls, especially for people-generated particles and chemicals.

An environmental danger that is most serious in packaging areas is static. In the fabrication clean rooms static is controlled primarily to prevent the attraction of particles to the wafer surface. This is also a concern in a packaging area. But the greatest concern is *electrostatic discharge*, or *ESD*. Static charge can build up to levels of tens of thou-

---

- HEPA Filters/VLF Air
- Smocks, Hats, Shoe Covers
- Finger Cots or Gloves
- Filtered Chemicals
- Sticky Floor Mats
- Static Control

Figure 18.9 Contamination control practices.

- Wrist Ground Straps
- Nonstatic Garments
- Antistatic Materials
- Grounded Equipment
- Grounded Work Surfaces
- Grounded Floors or Floor Mats

Figure 18.10 Static control practices.

sands of volts. If this voltage is suddenly discharged onto a chip surface, it can easily destroy a portion of the circuit. MOS gate structures
are particularly vulnerable to ESD.

Every packaging area making high-density chips should have an
active antistatic program (Fig. 18.10). Those assembling military
parts will need to have one as a condition of getting the contract. The
antistatic program is based on operators wearing grounding straps
and nonstatic smocks; the use of antistatic carrier materials; moving
work by lifting rather than sliding; and grounded equipment, work
surfaces, and floor mats. Static is also reduced by the placement of ionizers in nitrogen and air blow-off guns (Fig. 18.11) and in the path of
air coming out of Hepa filters.

Package Functions and Design

There are four basic functions performed by a semiconductor package.
They are to provide

1. A substantial lead system

2. Physical protection

3. Environmental protection

4. Heat dissipation

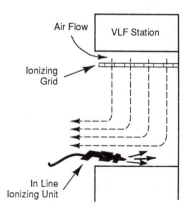

Figure 18.11 Static control techniques.

Substantial lead system

The primary function of the package is to allow connection of the chip to a circuit board or directly to an electronic product. This connection cannot be made directly due to the thin and fragile metal system used to interconnect the components on the chip surface. The metal leads are typically less than 1.5 µm thick and often only 1 µm wide. The thinnest wires available are 0.7 to 1.0 mils in diameter, which is many times larger than the surface wiring. This difference in wiring sizes is the reason that the chip wiring terminates in the larger bonding pads around the edge of the chip.

Even though the wires are larger, at 1 mil in diameter they too are very fragile. This fragility is overcome by a more substantial electrical lead system that serves as the connection of the chip to the outside world (Fig. 18.12). The lead system is an integral part of the package.

Physical protection

The second function of the package is the physical protection of the chip from breakage, particulate contamination, and abuse. Physical protection needs vary from low, as in the case of consumer products, to very stringent, as in the case of automobile circuits, space rockets, and military uses. The protection function is accomplished by securing the chip to a die-attachment area and surrounding the chip, wire bonds, and inner package leads with an appropriate enclosure. The size and eventual use of the chip dictate the choice of materials for the enclosure and the design and size of the package.

Environmental protection

Environmental protection of the chip from chemicals, moisture, and gases that may interfere with the chip functioning is provided by the package enclosure.

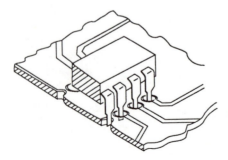

Figure 18.12 Dip through-hole assembly.

Heat dissipation

Every semiconductor chip generates some heat during operation. Some generate large quantities. The package enclosure materials serve to draw the heat away from most chips. Indeed, one of the factors in choosing a package material is its thermal dissipation property. The chips that generate large quantities of heat require additional consideration in the package design. This consideration will influence the size of the package and will often require the addition of metal heat-dissipating fins or blocks on the package.

Common package parts

The four functions of a chip package are accomplished through the use of a wide variety of package designs. However, all of the packages have five common parts: the die-attachment area, the bonding wires, the inner and outer leads of the metal lead system, and the enclosure.

Die-attachment area.   In the center of every package is an area for the chip. This area is where the chip is securely attached into the package. The die-attachment area may have an electrical connection that serves to connect the back of the chip to the rest of the lead system. A major requirement for this area is absolute flatness in order to intimately support the chip in the package (Fig. 18.7).

Bonding wires.   The bonding wires are actually added to the package assemblage during the packaging process. They are the electrical bridge between the chip bonding pads and the package lead system (Fig. 18.13).

Inner and outer leads.   The metal lead system chip is continuous from the die-attach cavity to the printed circuit board or electronic product. The system inner connections are called the *inner leads, bonding lead*

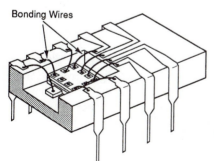

Bonding Wires

Figure 18.13   Bonding wires.

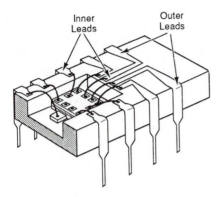

Inner Leads   Outer Leads

Figure 18.14 Inner and outer leads.

tips, or *bond fingers*. The inner leads are generally the narrowest portion of the lead system. The leads become progressively wider, finally ending outside the package. These portions of the lead system are called the *outer leads* (Fig. 18.14). Most of the lead systems are composed of one continuous piece of metal. One exception is the side-brazed package. In this package construction method, the outer leads are brazed onto the interior leads. Two different alloys are used for the outer lead system, either an iron-nickel alloy or a copper alloy. The iron-nickel alloy is desirable for its strength and stability, while copper is used for its electrical and heat-conduction properties.

Enclosures.   The die-attach area, bonding wires, and inner and outer leads constitute the electrical parts of the package. The other part is the *enclosure* or *body*. This is the part that provides the protection and heat-dissipation functions. These functions are achieved by several different techniques as described in the sealing section. The completeness of the seal falls into two categories: hermetic and nonhermetic (Fig. 18.15).

Hermetic sealing results in a package that is impervious to the penetration of moisture and other gases. Hermetic seals are required for chips operating in harsh and demanding environments such as in rockets and space satellites. Metal and ceramic enclosures are the preferred materials for hermetically sealed packages.

Nonhermetically sealed packages are adequate for most consumer applications such as computers and entertainment systems. This sealing system provides good and adequate environmental protection of the chip, except in the most demanding situations. A better term for this type of enclosure sealing method would be "less hermetic." Nonhermetic packages are composed of epoxy resins or polyimide materials and are generally referred to as "plastic packages."

Hermetic
• Metal
• Ceramic
Nonhermetic
• Epoxy Resins
• Polyimides

Figure 18.15   Package sealing designations.

Package Designs

Up to the early 1970s most chips ended up in either a metal package, known as a "can," or in the familiar dual in-line package (DIP). The trends in chip size and integration levels and newer methods of incorporating the chips into larger circuits have fostered a number of new package types. In this section the principal types are identified and discussed.

Metal cans

Metal cans are cylindrically shaped packages with an array of leads extending through the base (Fig. 18.16). The chip is attached to the base and wire-bonded to posts that are connected to the leads. The lid and base have matching flanges that are welded together to create a hermetic seal. These packages are designated by numbers, with the T0-3 and T0-5 being the most common. Metal cans are used to package discrete devices and small-scale integrated circuits.

Dual in-line packages (DIPs)

The DIP is probably the most familiar package design. It features a thick sturdy body with two rows of outer leads coming out of the side and bending downward. DIPs are constructed by three different tech-

Figure 18.16   Metal can.

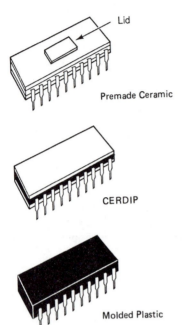

Lid

Premade Ceramic

CERDIP

Molded Plastic

Figure 18.17   DIP packages.

niques (Fig. 18.17). Chips designed for high-reliability use will be packaged in a premade ceramic DIP. The package is formed with a solid body of ceramic with the leads buried in the ceramic. The die-attachment area is a cavity recessed into the body. The hermetic seal is completed by a soldered metal lid or a glass-sealed ceramic lid.

Another approach to the DIP is the CERDIP, which stands for ceramic DIP. This type of package is composed of a bottom ceramic base with the lead frame held firm in a glass layer. The chip is attached to the base and wire-bonded to the lead frame. The hermetic seal is completed with a ceramic top glass sealed to the base. CERDIP construction is used for a number of package types. The vast majority of DIPs are made by the epoxy molding technique. In this technique the chip is attached to a lead frame and then wire-bonded. After bonding the frame is placed in a molding machine and the package is formed around the chip, wire bonds, and inner leads.

Flat packs

These packages are designed with flatter height profiles and have their leads bent out to the side of the package (Fig. 18.18). Flat packs are constructed by the same techniques used to form DIP packages.

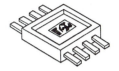

Figure 18.18   Flat pack.

They are used where package height and weight are considerations, as in airplanes and thin-profile calculators.

Pin grid arrays

Larger chips, with more leads, have outgrown the DIP configuration. The pin grid array is a ceramic package designed for larger chips. It features a premade ceramic "sandwich" with the outer leads coming out of the bottom of the package in the form of pins (Fig. 18.19). The chip is attached in a cavity that is formed in either the top or bottom of the body. The packages are hermetically sealed with a soldered metal lid.

Quad packages

While pin grid array packs are a convenient design for larger chips, their ceramic construction is expensive compared to molded epoxy packages. This consideration led to the development of the "quad" package. A quad (short for quadrant) pack is constructed by the epoxy-molded technique but has leads coming out of all four sides of the package (Fig. 18.20).

Surface-mount devices

There are two primary methods of attaching packages to printed circuit boards. The most familiar is by inserting and soldering the pack-

Figure 18.19   Pin grid array.

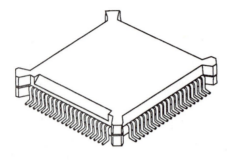

Figure 18.20   Quad package.

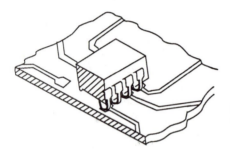

Figure 18.21   Surface-mount device.

age leads into holes in the board (Fig. 18.12). A newer method is *surface mount*, also referred to as *SMD*. This method features packages that have their leads bent into a J shape or bent outward to allow direct soldering to the board surface (Fig. 18.21).

Some SMD packages do not have leads—rather, they terminate in metal traces hugging the package body. They are known as *leadless packs*. For inclusion on a circuit board, they are inserted into chip carriers, which in turn have the leads that attach to the printed circuit board.

Blob top protection

Some chips will end up being bonded directly to the surface of a hybrid substrate or a printed circuit board. Others will be connected to a substantial lead system by the *tape automated bonding (TAB)* technique. After bonding, the chip still has to be protected from physical and environmental dangers. This protection is accomplished by covering the chip and bonds with a blob of epoxy resin material (Fig. 18.22). The material is similar in properties to that used to mold the plastic packages.

Package type summary

There are thousands of individual package types and there is no uniform system of identifying them. Some are named by their design

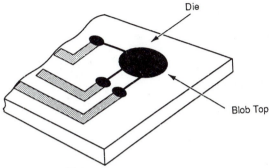

Figure 18.22   Blob top.

(DIP, flat packs, etc.), some are named by their construction technique (molded, CERDIP, etc.), and others are named by their use, such as SMDs. When trying to understand a package type, keep in mind the three considerations: design type, construction technique, and use.

Die Separation

The chip-packaging process starts with the separation of the wafer into individual dies. The two methods of die separation are scribing and sawing (Fig. 18.23).

Scribing

Scribing, or *diamond scribing*, was the first production die separation technique developed in the industry. It requires the alignment of the wafer on a precision stage, followed by dragging a diamond-tipped scribe through the center of the scribe lines. The scribe creates a shallow scratch in the wafer surface. The separation of the die is completed by stressing the wafer with a cylindrical roller. As the roller is moved over the surface, the wafer separates along the scribe lines. The breaks follow the crystal structure of the wafer, thus creating a

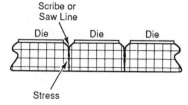

Figure 18.23   Scribe and saw separation.

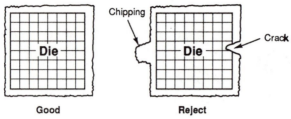

Figure 18.24 Die separation results.

right-angle edge on the die. Scribing becomes unreliable in wafers over 10 mils thick.

Sawing

The advent of thicker wafers has led to the development of sawing as the preferred die separation method. Wafer sawing is accomplished by two techniques. Both start with the passage of a rotating high-speed diamond-tipped saw over the scribe lines. For thinner wafers, the saw is lowered into the wafer surface to create a trench about one-third of the way through the wafer. The separation of the wafer into die is completed by the stress and roller technique used in the scribing method. The second sawing method is to separate the die by a complete saw-through of the wafer.

Often the wafers for complete saw-through are first mounted on a flexible plastic film. The film holds the die in place after the sawing operation and aids the die pick operation. Sawing is the preferred die separation method due to the cleaner die edges and the fewer cracks and chips left on the sides of the die (Fig. 18.24).

Die Pick and Plate

After sawing, the separated die are transferred to a station for selection of the functioning die (nonlinked). In the manual method, an operator will pick up each of the nonlinked dies with a vacuum wand and place it in a sectioned plate. Wafers that come to the station on the flexible film are first placed on a frame that stretches the film. The stretching separates the die, which aids the die pick operation.

In the automated version of this operation an automated vacuum wand moves over the surface of the separated wafer. The pick is directed to the functioning die by a memory tape or disk from the wafer sort operation. After picking up the die, the pick automatically places them in a sectioned plate for transfer to the next operation.

Die Inspection

Before being committed to the rest of the process, the die are given an optical inspection. Of primary interest is the quality of the die edge, which should be free of chips and cracks. This inspection also sorts out surface irregularities, such as scratches and contamination.

Die Attach

Die attachment has several goals: to create a strong physical bond between the chip and the package, to provide either an electrical conducting or insulating contact between the die and the package, and to serve as a medium to transfer heat from the chip to the package.

A requirement is the permanency of the die-attachment bond. The bond should not loosen or deteriorate during subsequent processing steps or when the package is in use in an electronic product. This requirement is especially important for packages that will be subjected to high physical forces such as in rockets. Additionally, the die-attach materials should be contaminant-free and should not out-gas during subsequent process heating steps. Lastly, the process itself should be productive and economical.

Eutectic die attach

There are two principal methods of die attach: *eutectic* and *epoxy adhesives*. The eutectic method is named for the phenomenon that takes place when two materials melt together (alloy) at a much lower temperature than either of them separately. For die attach, the two eutectic materials are gold silicon (Fig. 18.25). Gold melts at 1063°C, while silicon melts at 1415°C. When the two are mixed together, they

Conducting
- Gold/Silicon Eutectic
- Metal-Filled Epoxy
- Conducting Polyimide

Nonconducting
- Epoxy Adhesive
- Insulating Polyimide

Figure 18.25   Die attach material matrix.

start alloying at about 380°C. Gold is plated onto the die-attach area and alloys with the bottom of the silicon die when heated.

The gold for the die-attach layer is actually a sandwich. The bottom of the die-attach area is deposited or plated with a layer of gold. Sometimes a preformed piece of metal composed of a gold and silicon mixture is placed in the die-attach area. When heated, these two layers, along with a thin layer of silicon from the wafer back, forms a thin alloy layer. This layer is the actual bond forming the die-package attachment.

Eutectic die attach requires four actions. First is the heating of the package until the gold-silicon forms a liquid. Second is the placement of the chip on the die-attach area. Third is an abrasive action, called "scrubbing," that forces the die and package surfaces together. It is this action in the presence of the heat that forms the gold-silicon eutectic layer. The fourth and final action is the cooling of the system, which completes the physical and electrical attachment of the chip and package.

Eutectic die attach can be performed manually or by an automated machine which performs the four actions. Gold-silicon eutectic die attach is favored for high-reliability devices and circuits for its strong bonds, heat dissipation properties, thermal stability, and lack of impurities.

Epoxy die attach

The alternate die-attach process uses thick liquid epoxy adhesives. These adhesives can form an insulating barrier between the chip and package or become electrically and heat conducting with the addition of metals such as gold or silver.

The epoxy process starts with the deposit of the epoxy adhesive in the die-attach area by dispensing the adhesive with a needle or screen printing it into the die-attach area. The die, held by a vacuum wand, is positioned in the center of the die-attach area. The second action is to force the die into the epoxy to form a thin uniform layer under the die. The final action is a curing step in an oven at an elevated temperature that sets the epoxy bond.

Epoxy die attach is favored for its economy and ease of processing, in that the package does not have to be heated on a stage. This factor makes the automation of the process easier. When compared to gold-silicon eutectic die attach, epoxy has the disadvantage of potential decomposition at the high temperatures of bonding and sealing operations. Epoxy die-attach films also do not have the bonding power of the eutectics.

Regardless of the attachment method used, there are several marks of a successful die attach. One is the proper and consistent alignment of the die in the die-attach area. Proper placement pays off in faster and higher-yield automatic bonding. Another desired result is a solid, uniform, and void-free contact over the entire area of the chip. This is necessary for mechanical strength and good thermal conduction. One evidence of a uniform bond is a continuous joint or "fillet" between the die edge and the package. The final mark of a good die-attach process is a die-attach area free of flakes or lumps that can come loose during use and cause a malfunction.

Bonding

Once the die and package are attached, they go to the bonding process. This is perhaps the most critical of all the assembly operations. In wire bonding, up to hundreds of wires must be perfectly bonded from the bonding pads to the package inner leads. If the tape automated bonding (TAB) system is employed, the many bonding fingers of the lead frame must be precisely positioned on the chip bonding pads.

The wire bonding procedure is simple in concept. A thin (0.7- to 1.0-mil) wire is first bonded to the chip bonding pad and spanned to the inner lead of the package lead frame. The third action is to bond the wire to the inner lead. Last, the wire is clipped and the entire process repeated at the next bonding pad. While simple in concept and procedure, wire bonding is critical because of the precise wire placement and electrical contact requirements. In addition to accurate placement, each and every wire must make good electrical contact at both ends, span between the pad and inner lead in a prescribed arc without kinks and be at a safe distance from neighboring wires.

Wire bonding is done with either gold or aluminum wires. Both are highly conductive and ductile enough to withstand deformation during the bonding steps and still remain strong and reliable. Each has its advantages and disadvantages and each is bonded by different methods.

Gold wire bonding

Gold has several pluses as a bonding wire material. It is the best-known room-temperature conductor and is an excellent heat conductor. It is resistant to oxidation and corrosion, which translates into an ability to be melted to form a strong bond with the aluminum bonding pads without oxidizing during the process. Two methods are used for gold bonding. They are *thermocompression* and *thermosonic*.

Thermocompression bonding (also known as *TC bonding*) starts with the positioning of the package on the bonding chuck and the

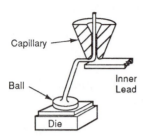

Capillary

Ball

Inner Lead

Die

Figure 18.26   Gold ball bonding.

heating of the chip and package to between 300 and 350°C. Chips that are going to be enclosed in an epoxy molded package are processed through die attach and bonded with the chip on the lead frame only. The bonding wire is fed out of a thin tube called a *capillary* (Fig. 18.26). An instantaneous electrical spark or small hydrogen flame melts the tip of the wire into a ball and positions the wire over the first bonding pad. The capillary moves downward, forcing the melted ball onto the center of the bonding pad. The effect of the heat (thermal) and the downward pressure (compression) forms a strong alloy bond between the two materials. This type of bonding is often called *ball bonding*. After the ball bond is complete, the capillary feeds out more wire as it travels to the inner lead. At the inner lead the capillary again travels downward to where the gold wire is forced by the heat and pressure to melt onto the gold-plated inner lead. The spark or flame then severs the wire, forming the ball for the next pad bond. This procedure is repeated until every pad and its corresponding inner pad are connected.

Thermosonic gold ball bonding follows the same steps as thermocompression bonding. However, it can take place at a lower temperature. This benefit is provided by a pulse of ultrasonic energy that is sent through the capillary into the wire. This additional energy is sufficient to provide the heat and friction to form a strong alloy bond.

The majority of production gold wire bonding is done on automatic machines that use sophisticated techniques to locate the pads and span the wire to the correct inner lead. The fastest bonding machines can perform thousands of bonds per hour. There are two major drawbacks to the use of gold bonding wires. First is the expense of the gold. Second is an undesirable alloy that can form between the gold and aluminum. This alloy can severely reduce the conduction ability of the bond. It forms a purplish color and is known as the "purple plague."

Aluminum wire bonding

Aluminum wire, while not having the conduction and corrosion-resistance properties of gold, is still an important bonding wire mate-

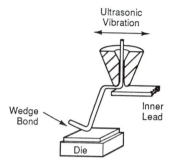

Wedge
Bond

Inner
Lead

Ultrasonic
Vibration

Die

Figure 18.27 Aluminum wedge bonding.

rial. A primary advantage is its lower cost. The second advantage is that the bond with the aluminum bonding pad is a monometal system and thus less susceptible to corrosion. Also, aluminum bonding can take place at lower temperatures than gold bonding, making it more compatible with the use of epoxy die-attach adhesives.

The bonding of the aluminum follows the same major steps as gold wire bonding. However, the method of forming the bond is different. No ball is formed. Instead, after the aluminum wire is positioned over the bonding pad a wedge (Fig. 18.27) forces the wire onto the pad as a pulse of ultrasonic energy is sent down the wedge to form the bond. After the bond is formed, the wire is spanned to the inner lead where another ultrasonic-assisted wedge bond is formed. This type of bonding is known variously as *ultrasonic* or *wedge bonding*.

After this bond, the wire is cut. At this point in the process a major difference between the bonding of the two materials occurs. In gold bonding the capillary moves freely from pad to inner lead, to pad, etc., with the package in a fixed position. In aluminum wire bonding the package must be repositioned for every single bonding step. The repositioning is necessary to line up the pad and inner lead along the direction of travel of the wedge and wire. This requirement places an additional difficulty on the designers of automatic aluminum bonding machines. Nevertheless, most production aluminum bonding is done on high-speed machines.

Tape automated bonding

Tape automated bonding (TAB) is used to wire up chips when extreme package thinness is required, such as credit-card-size radios. The TAB process starts with formation of the lead system on a flexible strip of tape. Various methods are used to form the lead system. The metal for the system is deposited on the tape by sputtering or evaporation. Formation of the lead system is either by mechanical stamping or patterning techniques similar to the fabrication patterning process. The

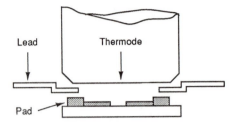

Figure 18.28 Tape automated bonding.

result is a continuous tape containing many individual lead systems. For the bonding operation (Fig. 18.28), the chip is positioned on a chuck and the tape is moved by sprockets until one of these lead systems is positioned exactly over the chip. In this position the inner leads of the system should be positioned over the bonding pads of the chip.

The bond is completed with a tool known as a *thermode*. The thermode is faced with a flat diamond surface and is heated. The thermode is moved downward, first contacting the inner leads. It continues downward with enough pressure to force the inner leads onto the bonding pads. The heat and the pressure are regulated to cause a physical and electrical bond between the two. Large chips require a larger TAB bonding area. For these chips the thermode is faced with a synthetic diamond.

TAB bonding is also used in conjunction with package bonding. The advantages of TAB are speed, in that all of the bonds to the chip are made in one action, and the ease of automation offered by the tape and sprocket system.

Preseal Inspection

An important step in the chip-packaging process is the preseal or precap inspection, sometimes called *third optical inspection*, which takes place after the bonding step. The inspection is performed to provide feedback on the quality of the operations already performed. It also is performed to reject packaged chips that may represent reliability hazards when the chip is in operation in the field.

While there are many levels of inspection criteria, they fall into two main categories: commercial and high-reliability. The commercial inspections are given to chips and packages destined for use in commercial systems and are derived from the internal quality levels of the producing company in conjunction with its experience and customer specifications. The high-reliability specifications are derived from a set of government standards identified as "Mil-Standard 883."

Commercial-level inspections screen the bonded chips for die-attach quality, correct placement of the bonds on the bonding pads and inner leads, the shape and quality of the ball or wedge bond, and the general condition of the chip surface in regard to contamination, scratches, etc.

Mil-Standard 883 covers the same general issues as the commercial inspections, but to more stringent requirements. In particular, this standard also specifies criteria for the chip surface, including pattern alignment, critical dimensions, and surface irregularities, such as small scratches, voids, and small defects. These criteria serve to reject bonded chips that may malfunction in the rigorous conditions encountered in space and military operation.

Sealing Techniques

After the bonded chip passes the optical inspection, it is ready for sealing in a protective enclosure. There are several methods used to achieve the enclosure of the chip. The method chosen depends on whether the seal must be hermetic or nonhermetic, and which type of package is to be used. The principal sealing methods use welded seals, soldered seals, glass-sealed lids, CERDIP package construction, and molded epoxy enclosures (Fig. 18.29).

Metal can

If the package is a metal can type, a hermetic seal is achieved by welding the flanged lid to a matching flange on the base of the package.

Premade packages

Premade ceramic packages are sealed by one of two methods, metal or ceramic lids (Fig. 18.30). Packages made for metal lids have a ring of gold around the top of the die-attach cavity, called the *seal ring*. Placed on top of the seal ring area is a preformed piece of gold-tin solder. The metal lid is clamped in position over the seal ring and placed on the belt of a conveyor furnace. As the clamped package passes

---

Hermetic
- Welding
- Soldered Lid
- Glass-Sealed Lid or Top

Nonhermetic
- Epoxy Molding
- Blob Top

Figure 18.29 Package sealing methods.

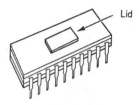

Figure 18.30   Premade ceramic package.

through the furnace, the lid and package are soldered together to form a hermetic seal. The sealing takes place at a temperature range of 320 to 360°C in a pure nitrogen atmosphere.

If the package is to receive a ceramic lid, the procedure is similar. The part of the ceramic lid that contacts the base, outside the die-attach area, is coated with a layer of low-melting-point glass. The hermetic seal is completed as the package passes through a conveyor furnace. This sealing takes place at a temperature of about 400°C in an atmosphere of clean dry air.

CERDIP packages

A completed CERDIP package results in a hermetic seal around the chip and bonding wires. This seal is accomplished with glass, similar to the ceramic seal on the premade packages. In the case of the CERDIP package, the inner metal lead system is buried in a glass layer. The ceramic top of the package system has a cavity (Fig. 18.31). The underside of the lid, outside the cavity area, is coated with a layer of low-melting-point glass. The lid is placed over the base and clamped. The seal is accomplished as the assembly is passed through a conveyor furnace or placed in an oven. In the furnace or oven, the glass melts, fusing the base and top together. The CERDIP glass sealing system is used to seal DIP and flat packs. These latter packages are known as *Cerpacks* and *Cerflats*.

Molded epoxy enclosures

The fourth major method of enclosure, *epoxy molding*, produces the plastic package (Fig. 18.32). The resultant seal, while protecting the die from moisture and contamination, is not classified as hermetic. However, there exists a considerable amount of research into the development of improved epoxy materials to create better enclosures. The major advantages of epoxy molded enclosures are weight, low material cost, and manufacturing efficiency.

This sealing method follows a different process flow. The die is attached and bonded to a lead frame containing a number of lead sys-

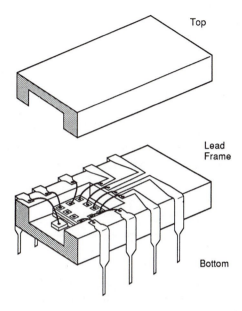

Top

Lead
Frame

Bottom **Figure 18.31** CERDIP parts.

Molded Plastic

Figure 18.32 Molded CERDIP.

tems (Fig. 18.33). After the preseal inspection, the lead frames are transferred to the molding area. The frames are placed on a mold mounted in a transfer molding machine. The molding machine is in turn charged with pellets of the epoxy material, which have been previously softened by a radiofrequency heater. Inside the pellets are forced by a ram into a liquid state. The ram then forces the liquid around the die on the lead frames, forming an individual package around each lead frame. After the epoxy sets in the mold, the frames are removed and placed in an oven for final curing.

Lead Plating

An important feature of the completed package is the finish on the package leads. Most package leads are coated with lead-tin solder, tin plate, or gold plate. The plating serves several important functions.

Figure 18.33   Lead frame.

First is the solderability of the leads into a circuit board. The additional metal finish improves the lead solderability, resulting in a more reliable electrical connection of the package and the printed circuit. The second benefit of the lead finish is that it protects the leads from oxidation or corrosion during periods of storage prior to mounting on the circuit board. The third benefit of lead plating is the protection of the leads from corrosive agents in the packaging and printed-circuit-board mounting processes. These agents include solder flux, corrosive cleaners, and even tap water. The plating continues to protect the leads during their lifetime of use.

Electrolytic plating

Plated layers, such as gold and tin, are applied by electrolytic procedures. The packages are mounted on racks with each lead connected to an electric potential. The racks are placed in a tank containing a plating solution. Next, a small current is passed between the packages and an electrode in the tank. The current causes the particular metal in the solution to plate onto the leads.

Tin-lead solder

Tin-lead solder layers are applied either by dipping the packages into a pot of the molten solder or by a wave soldering technique. This latter technique offers the advantage of good control of the layer thickness and provides a shorter exposure of the package to the molten solder.

Plating process flows

Metal cans, side-brazed DIP packs, and pin grid arrays have their leads plated before starting into the packaging process. CERDIP and plastic packages go through the plating process after the sealing steps.

Lead Trimming

One of the last steps in the package assembly process is trimming away excess material from the leads. The outer leads of DIP and flat-pack packages are made with a tie-bar (Fig. 18.34). This bar keeps the leads from becoming bent during the packaging process. At the end of the process, the package goes through a simple trimming machine that simultaneously trims away the tie-bar and trims the leads to the proper length.

Plastic package lead frames have an additional piece of material. It is a bridge of metal close to the package body that functions as a dam to prevent the liquid epoxy material from running into the lead area (Fig. 18.35). The dam is cut away from the frame with a series of precise cutting tools. After the dam is cut away, the packages move to another station on the cutter where the frame is separated into indi-

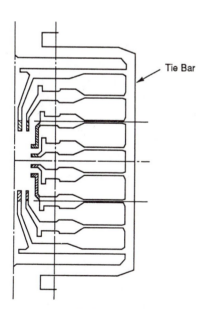

Figure 18.34 Tie-bar.

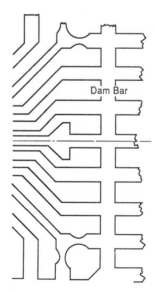

Figure 18.35 Lead frame dam.

vidual packages. If the package is a surface-mount type, the leads will be bent into the required shape.

Deflashing

Plastic packages receive an additional process, called *deflashing*, which is required to remove excess molding material from the package enclosure. Deflashing is done by either dipping the packages in a chemical bath followed by a rinse or by a physical abrasion process. Physical deflashing is done in a machine similar to a sand blaster which uses plastic beads as the abrasive.

Package Marking

Once completed, a package must be identified with key information. Typical information coded on the package is the product type, device specifications, date, lot number, and where it was made. The main methods of marking are ink printing and laser inscription. Ink printing has the advantage of good adherence to all of the package materials. The composition of the ink is chosen for permanence in the eventual operating environment of the device. The ink is applied by an offset printer followed by a curing step. Curing is done by oven drying, room temperature air drying, or by ultraviolet light.

Laser printing is especially suited for plastic packages. The mark is permanently inscribed in the package surface and can provide good contrast on the dark packages. Additionally, laser marking is fast and noncontaminating since no foreign material is added to the package surface and no curing step is required. A drawback to laser marking is the difficulty of changing the mark if a wrong code was used or the status of the device is changed. Regardless of the marking method, all marks must meet the requirements of legibility, especially on smaller packages, and permanenance when exposed to harsh environments.

Final Testing

At the conclusion of the packaging process the completed package is put through a series of environmental, electrical, and reliability tests. These tests vary in type and specifications, depending on the customer and use of the packaged devices. The tests may be performed on all of the packages in a lot or on selected samples.

Environmental tests

Environmental tests are performed to weed out leaking and defective packages. Defects detected are loose chips, contaminants and particles in the die-attach cavity, and faulty bonding. This testing series starts with a stabilization bake to drive off any volatile substances in the package. A typical bake is at 150°C for 24 hours.

The first environmental test is *temperature cycling*. The packages are loaded into a chamber and cycled between two temperature extremes. The number of cycles may reach several hundred. The high and low temperatures of this test vary with the device use. Commercial parts receive a narrower temperature range than hi-rel parts. The hi-rel cycle range is −25 to 125°C. During the cycling any weakness in the seal, die attachment, or bonding will be aggravated and be detected in later electrical tests.

A second environmental test is *constant acceleration*. In this test the packages are accelerated in a centrifuge (Fig. 18.36) that creates a force as high as 30,000 times the pull of gravity on the earth (30,000 g's). During the acceleration, loose particles in the package, poorly attached dies, and weakly attached bonds are stressed so that they will be detected at the final electrical tests.

Leaks in the package enclosure are detected by two tests. *Gross leak* testing (Fig. 18.37) is conducted by submerging the packages in a hot liquid. The heated liquid raises the temperature of the package and forces trapped gases in the cavity to escape. The escaping gases are observable as bubbles rising in the liquid. The chamber has a transparent side, allowing an operator to observe the bubbles. Smaller (or fine) leaks are detected by using tracer gases. For this check, helium is pumped under pressure into a chamber containing the packages. If the package has small leaks, the gas will be pumped into the package cavity. The gas is detected as it escapes through the small leaks by a

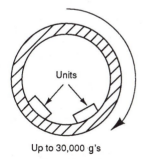

Units

Up to 30,000 g's

Figure 18.36 Acceleration test.

Figure 18.37 Gross leak bubbles test.

machine known as a mass spectrometer, which can identify the escaping gas. An alternate fine leak test uses the radioactive gas krypton-85. It too is pumped through any leaks into the package under pressure. Detection of any krypton-85 in the package is by a device similar to a geiger counter.

Electrical testing

The purpose of the wafer fabrication and packaging processes is to provide to the customer a specific semiconductor device that performs to specific parameters. Thus, one of the last steps is an electrical test of the completed unit to verify that it performs to the specifications. The tests are similar to the wafer sort tests. The overall objective is to verify that the good chips identified at wafer sort have not been compromised by the packaging process.

First there is a series of parametric tests. These electrical tests check the general performance of the device or circuit and ensure that it meets certain input and output voltage, capacitance, and current specifications. The second part of the final test is called the *functional test*. This test actually exercises the specific chip functioning. Logic chips are put through logic tests, memory chips are exercised in their data storage and retrieval capabilities.

The equipment used to conduct the final test is electrically similar to that used in the wafer sort operation. The electrical tests are performed by a computer-controlled tester that directs the sequences and levels of the parametric and functional tests. The packages are connected to the tester through sockets; the socket unit is known as the *test head*. The packages are inserted into the test head manually or by an automatic unit known as a *handler* (Fig. 18.38). This handler may be mechanical or robotic, depending on the speed and complexity of the operation.

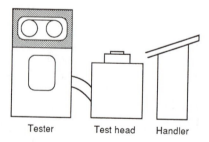

Figure 18.38 Final test.

Tester   Test head   Handler

Burn-in tests

The last of the tests is the optional burn-in test(s). The reason it is optional is that although it is required for all high-reliability device lots, it may or may not be performed on commercial devices. The test requires the insertion of the packages in sockets and mounting in a chamber with temperature-cycling capability. During the test, the circuits are temperature-cycled while under an electrical bias.

The burn-in test is intended to stress the electrical interconnection of the chip and package and drive any contaminants in the body of the chip into the active circuitry, thus causing failure. This test is based on data that indicate that chips prone to these types of failures actually malfunction in the early part of their lifetime. By conducting burn-in tests, the early failures are detected. The devices passing the test are statistically more reliable.

Package Process Flows

It should be obvious that there is no universal packaging process flow. The package construction technique and lead-plating requirements dictate the flow. The table in Fig. 18.39 illustrates typical flows for three of the more common package types.

Process Step	Premade Ceramic	CERDIP	Epoxy Molded
Die Separation	X	X	X
Pick and Plate	X	X	X
Die Attach	X	X	X
Wire Bonding	X	X	X
Preseal Inspect	X	X	X
Lid Sealing	X	X	
Epoxy Molding	X		
Deflashing	X		
Lead Plating and Trim		X	X
Marking	X	X	X
Environmental Test	X	X	X
Final Electric Test	X	X	X
Burn-In	X	X	X

Figure 18.39   Table of packaging process flows.

Key Concepts and Terms

Bonding	Package designs
Cleanliness requirements	Package functions
Die attach	Package parts
Final test(s)	Pick and plate
Hermetic/nonhermetic seals	Sealing techniques

Review Questions

1. Name the four functions of a chip package.
2. What is the function of marking dies with ink dots?
3. Name the five parts of a package.
4. Define a hermetic package.
5. Give an example of a hermetic and nonhermetic package.
6. List in order the six major steps in the packaging process.
7. Describe the process and material used for eutectic die attachment.
8. Describe the principal thermosonic and ultrasonic bonding techniques.
9. What is a DIP package and how is it formed?
10. Name two methods of package marking.

Glossary

acceptor An impurity that causes semiconducting materials to accept valence electrons, thereby leaving "holes" in the valence band. The holes act like carriers of positive charge, referred to as P type.

aligner A system which transfers an image from a mask to a wafer.

alignment Refers to the positioning of a mask or reticle with respect to the wafer.

alloy (1) A compound composed of two metals. (2) In semiconductor processing, the alloy step causes the interdiffusion of the semiconductor and the material on top of it, forming an ohmic contact between them.

aluminum (Al) The metal most often used in semiconductor technology to form the interconnects between devices on a chip. It can be applied by evaporation or sputtering.

amorphous Materials with no definite arrangement of atoms, e.g., plastics are amorphous.

angstrom A unit of length, an angstrom (Å) is one ten-thousandth of a micron (10^{-4} μm) or 100,000,000 Å = 1 cm.

anistrophic An etch process that exhibits little or no undercutting.

anneal A high-temperature processing step (usually the last one), designed to minimize stress in the crystal structure of the wafer.

antimony (Sb) A Group V element that is an N-type dopant in silicon. It is often used as the dopant for the buried layer.

arsenic A group V element that is an N-type dopant in silicon.

assembly The series of operations after fabrication in which the wafer is separated into individual chips and mounted and connected to a package.

atmospheric oxidation A process of oxidation of silicon carried out at atmospheric pressure. The equipment used for thermal oxidation is the same as that used for thermal diffusion. It is composed of four subassemblies: a reactant source cabinet, a reaction chamber, a heating source, and a wafer holder.

Source: Glossary terms are abstracted from Beverly Griggs, Anne Miller, and Peter Van Zant, *Semiconductor Terminology—Graphic Glossary of Terms*, Semiconductor Services, Redwood City, Calif., 1989.

atomic number   A number assigned to each element equal to the number of protons (therefore the number of electrons) in the atom.

atomic particles   The parts of an atom: electrons, protons, and neutrons.

base   (1) The control portion of an NPN or PNP junction transistor. (2) The P-type diffusion done using boron that forms the base of NPN transistors, the emitter and collector of lateral PNP transistors, and resistors.

binary notation   A way of representing any number using a power of 2 (the digits 0 and 1).

bipolar transistor   A transistor consisting of an emitter, base, and collector, whose action depends on the injection of minority carriers from the base by the collector. Sometimes called NPN or PNP transistor to emphasize its layered structure.

boat   (1) Pieces of quartz joined together to form a supporting structure for wafers during high-temperature processing steps. (2) A Teflon or plastic assemblage used to hold wafers during wet processing steps.

boat puller   A mechanical arrangement to push a boat loaded with wafers into a furnace and/or withdraw it at a fixed speed.

BOE   See **buffered oxide etch**.

bonding pad   The relatively rectangular or square areas of metallization on a die that are the sites for electrical testing (probing) and are utilized to electrically attach the chip lead system to its package.

boron (B)   The P-type dopant commonly used for the isolation and base diffusion in standard bipolar integrated circuit processing.

boron trichloride (BCl_3)   A gas that is often used as a source of boron for doping silicon.

bubbler   An apparatus in which a carrier gas is "bubbled" through a heated liquid, causing portions of the liquid to be transported with the gas, e.g., a carrier gas (nitrogen or oxygen) is bubbled through deionized water at 98 to 99°C on its way to the oxidation tube.

buffered oxide etch   A mix of hydrogen flouride (HF) and ammonium flouride (NH_4F) used to allow oxide etching to occur at a slow, controlled rate.

buried layer   The N+ diffusion in the P-type substrate done just prior to growing the epitaxial layer. The buried layer provides a low-resistance path for current flowing in a device. Common buried layer dopants are antimony and arsenic.

can   A metal package used for connecting a chip to a printed circuit board with from three to five leads.

capacitor   A discrete device which stores electrical charge on two conductors separated by a dielectric.

capacitance   Electrical charge storage capability.

capacitance-voltage plot (C/V plot) Plot that provides information on the amount of mobile ionic contamination present in the oxide.

carrier gas An inert gas which will transport atoms or molecules of a desired substance to a reaction chamber.

centistokes Units used to measure viscosity, centipoise divided by density.

channel A thin region of a semiconductor that supports conduction. A channel may occur at a surface or in the bulk, essential for the operation of MOSFETs and SIGFETs. In cases where channels are not part of the circuit design, their presence may indicate contamination problems or incomplete isolation.

channeling A phenomenon in which an ion beam will penetrate into the crystal planes of the wafer. Preventing channeling is accomplished by cutting the wafer "off orientation." The effect is to tilt the crystal planes relative to the beam direction.

charge carrier A carrier of electrical charge within the crystal of a solid-state device, such as an electron or hole.

chemical etching Selective removal of material by means of liquid reactants. The precision of the etch is controlled by the temperature of the etchant, the time of immersion and the composition of the acid etchant.

chemical vapor deposition See **CVD**.

chip Die or device, one of the individual integrated circuits or discrete devices on a wafer.

chrome A metal often used in mask fabrication to form the layer in which the circuit pattern is generated.

circuit board See **printed circuit board**.

circuit layout The calculation of the physical device dimensions required to produce the required electrical parameters. Vertical dimensions determine CVD and doping thickness specifications. Horizontal dimensions determine the wafer pattern dimensions and are the basis for a scaled drawing of the finished circuit (composite drawing).

class number Number of contaminant particles in a cubic foot of air.

clean room An area in which semiconductor device fabrication takes place. The cleanliness of the room is highly controlled in order to limit the number of contaminants to which the semiconductor is exposed.

clear field mask A mask on which the pattern is defined by the opaque areas.

CMOS (complementary field-effect transistor) N- and P-channel MOS transistors on the same chip.

collector Along with the emitter and base, one of the three regions of the bipolar type of transistor.

collimated light Light in which the rays are parallel; used for gross visual inspection of surfaces.

composite drawing   A scaled drawing of the finished circuit.

conductivity   The ability of materials to conduct electricity (measured in siemens for conductance or ohms for resistance).

conductor   A material which has low resistivity and high conductivity.

contact   The regions of exposed silicon that are covered during the metallization process to provide electrical access to the devices.

contact mask   The step at which holes are put into the wafer layers to allow the metal layer to reach down to the doped silicon substrate.

contamination   A general term used to describe unwanted material that adversely affects the physical or electrical characteristics of a semiconductor wafer.

critical dimensions (CDs)   The widths of the lines and spaces of critical circuit patterns as well as the area of contacts.

cryogenic pump   A vacuum pump that can produce a vacuum to the 10^{-10}-torr range, the same level as the vacuum of space. It does not require forepumps or cold traps and is faster than other types of vacuum pumps.

crystal   A material in which the atoms are arranged in structured groups called unit cells.

crystal defects   Vacancies and dislocations in a crystal which influence the electrical performance of a circuit.

crystal planes   The planes in the semiconductor crystal structure along which the die must be aligned in order to prevent "ragged" die edges when the wafer is separated into individual die.

CUM yield   See **fabrication yield**.

current   A measure of the number of charged particles passing a given point per unit time.

curve tracer   A piece of electrical test equipment that displays the characteristics of a device visually on a screen.

CVD (chemical vapor deposition)   A method for depositing some of the layers which function as dielectrics, conductors, or semiconductors. A chemical containing atoms of the material to be deposited reacts with another chemical, liberating the desired material, which deposits on the wafer while by-products of the reaction are removed from the reaction chamber.

DI water   Deionized water; purity of this water is measured by its resistivity, with the standard being 18 MΩ.

dark field mask   A mask on which the pattern is defined by the clear portion of the mask.

deep ultraviolet (duv)   A light wavelength often used to expose photoresist which has the advantage of an ability to produce smaller image widths.

dehydration baking   A heating process by which wafer surfaces are restored to a hydrophobic condition by baking. Surface water is evaporated from the wafer at elevated temperatures.

depletion layer The region in a semiconductor where essentially all charge carriers have been swept out by the electric field which exists there.

deposition Process in which layers are formed as the result of a chemical reaction in which the desired layer material is formed and coats the wafer surface.

develop inspection The first inspection in the photomasking process consisting of measurement of critical dimensions and inspection for defects. It is done after development or development and hard bake if an automatic baking system is used.

development A photoresist processing step in which photoresist is removed from areas defined by the masking and exposure step of wafer fabrication.

developer Chemical used to remove areas defined in the masking and exposure step of wafer fabrication.

device A single-function component such as a transistor, resistor, or capacitor or an integrated circuit.

diborane (B_2H_6) A gas that is often used as a source of boron for doping silicon.

die One unit on a wafer separated by scribe lines; after all of the wafer fabrication steps are completed, die are separated by sawing; the separated units are referred to as *chips*.

die bonding Assembly step in which individual chips are attached to the package with conductive adhesives or metal alloys.

die sort See **wafer sort**.

dielectric A material that conducts no current when it has a voltage across it. Two dielectrics encountered in semiconductor processing are silicon dioxide and silicon nitride.

diffusion A process used in semiconductor production which introduces minute amounts of impurities (dopants) into a substrate material such as silicon or germanium and permits the impurity to spread into the substrate. The process is very dependent on temperature and time.

diffusivity The rate of movement or diffusion of dopants in a semiconductor.

diode Device which enables current flow in one direction but not in another.

DIP (dual in-line package) A rectangular circuit package, with leads coming out of the long sides and bent down to fit into a socket.

discrete device A circuit having a single electrical function. Discrete devices include capacitors, resistors, transistors, diodes, and fuses.

dislocation A discontinuity in the crystal lattice, a type of crystal defect.

DMOS (diffused MOS) A transistor structure which features a narrow (channel length) separation between the source and drain. The channel length is created by two sequential diffusions through the same hole.

donor   An impurity that can make a semiconductor N-type by donating extra "free" electrons; electrons carry a negative charge.

dopant   An element that alters the conductivity of a semiconductor by contributing either a hole or electron to the conduction process. For silicon, the dopants are found in Groups III and V of the periodic table.

doping   The introduction of an impurity (dopant) into the crystal lattice of a semiconductor to modify its electronic properties—for example, adding boron to silicon to make the material P type.

DRAM (dynamic random access memory)   Memory device for the storage of digital information. The information is stored in a "volatile" state.

drain   Along with the source and gate, one of the three regions of a unipolar or field-effect transistor (FET).

drive-in   Stage in diffusion where the dopant is driven deeper into the wafer.

dry etching   A process resulting in the selective removal of material achieved by use of gas.

dry ox   The growth of silicon dioxide using oxygen and hydrogen, which form water vapor at process temperatures, rather than using water vapor directly.

dry oxide   Thermal silicon dioxide grown using oxygen.

E-beam (electron beam)   An exposure source which allows direct image formation without a mask. An E-beam can be deflected by electrostatic plates and therefore directed to precise locations, resulting in the generation of submicron-size patterns.

E-beam evaporation (electron beam evaporation)   Phase change that uses the energy of a focused electron beam to provide the required energy to change solid metal or alloys from solid to gas.

E-beam exposure system   A machine in which the image pattern is stored in a computer memory and used to control the electrostatic plates that in turn direct the E-beam, resulting in the generation of patterns without the use of reticles or photomasks.

edge die   The incomplete die located on the edge of the wafer.

electromigration   The diffusion of electrons in electric fields set up in the lead while the circuit is in operation. It occurs in aluminum and is exhibited as a field failure, not as a process defect. The metal thins and eventually separates completely, causing an opening in the circuit.

electron   A charged particle revolving around the nucleus of an atom. It can form bonds with electrons from other atoms or be lost, making the atom an ion.

ellipsometer   Instruments that use laser light sources to measure thin film thickness.

emitter (1) The region of a transistor that serves as the source or input end for carriers. (2) The N-type diffusion usually done using phosphorus, which forms the emitter of NPN transistors, the base contact of PNP transistors, the N+ contact of NPN transistors, and low-value resistors.

epitaxial (Greek for "arranged upon") The growth of a single-crystal semi-conductor film upon a single-crystal substrate. The epitaxial layer has the same crystallographic characteristics as the substrate material.

EPROM (erasable programmable read-only memory) Device that allows stored information to be erased; erasing is typically accomplished with ultra-violet light.

etch A process for removing material in a specified area through a wet or dry chemical reaction or by physical removal, such as by sputter etch.

evaporation A process step that uses heat to change a material (usually a metal or metal alloy) from its solid state to a gaseous state with the result of the source being deposited on wafers. Both electron beam and filament evap-oration are common in semiconductor processing.

exposure Method of defining patterns by the interaction of light or other form of energy with photoresist that is sensitive to the energy source.

fabrication Integrated circuit manufacturing processes.

fabrication yield The percent of wafers arriving at wafer sort compared with the number started into the process.

FET (field-effect transistor) A transistor consisting of a source, gate, and drain, whose action depends on the flow of majority carriers past the gate from the source to drain. The flow is controlled by the transverse electric field un-der the gate. See **unipolar transistor**.

field oxide The region on an electrical device where the oxide serves the function of a dielectric.

final test The final assembly step in which the packaged die is put through its last electrical test.

flat zone The highly temperature controlled region of a tube furnace.

four-point probe A piece of electrical test equipment used to determine the sheet resistivity of a wafer.

furnace A piece of equipment containing a resistance-heated element and a temperature controller. It is used to maintain a region of constant tempera-ture with a controlled atmosphere for the processing of semiconductor devices.

fuse A circuit component which can be blown to allow a desired memory cell or logic gate to be programmed.

gallium arsenide (GaAs) Most common of compound semiconductor materi-als. It has the advantage of producing higher-speed devices than those pro-duced using silicon as a substrate.

gate Along with the source and drain, one of the three regions of the unipolar or field-effect transistor (FET or MOS).

gate array Type of integrated circuit made up of an arrangement of inter-connected gates used to provide custom functions.

gate oxide (gate ox) The thin oxide which causes the induction of charge, creating a channel between source and drain regions of an MOS transistor.

germanium Semiconducting material used in the manufacture of crystal diodes and of early transistors.

Hepa filter (high-efficiency particulate attenuator) A filter constructed of fragile fibers in an accordion-folded design which allows a larger filtering area at an air velocity low enough for operator comfort. This filter permits a filtering efficiency of 99.99 percent.

hexamethyldisilizane (HMDS) Primer used to promote photoresist adhesion.

high-pressure oxidation Oxidation carried out at high pressure (10 to 20 atm) in order to reduce the amount of heat or time required. The reaction chamber for this process must be constructed of stainless steel to safely contain the pressure.

hole (1) The absence of a valence electron in a semiconductor crystal. Motion of a hole is equivalent to motion of a positive charge. (2) A "hole" in a surface layer created by the photomasking process.

hybrid integrated circuit A structure consisting of an assembly of one or more semiconductor devices and a thin-film integrated circuit on a single substrate, usually of ceramic.

hydrofluoric acid (HF) An acid used to etch silicon dioxide; often diluted or buffered before it is used.

hydrogen (H$_2$) A gas used in semiconductor processing primarily as a carrier gas for high-temperature reaction steps such as epitaxial silicon growth.

hydrophilic Affinity toward water (water-loving); a hydrophilic surface is one that will allow water to spread across it in large puddles.

hydrophobic Aversion to water; a hydrophobic surface is one that will not support large pools of water. The water is pulled into droplets on the surface. These surfaces often are termed "dewetted."

hydroscopic Attracts and absorbs water.

integrated circuit A circuit in which many elements are fabricated and interconnected on a single chip of semiconductor material, as opposed to a "nonintegrated" circuit, in which the transistors, diodes, resistors, etc., are fabricated separately and then assembled.

intrinsic semiconductor An element or compound that has four electrons in its outer ring (i.e., elements from Group IV of the periodic table or compounds of Group III and V).

ion   An atom that has either gained or lost electrons, making it a charged particle (either negative or positive).

ion beam milling   A dry etching method which uses an ion beam. Argon atoms are ionized and accelerated toward a wafer. The exposed areas are removed through a sputtering action.

ion implantation   Introduction of selected impurities (dopants) by means of high-voltage ion bombardment to achieve desired electronic properties in defined areas.

isolation diffusion   Diffusion step resulting in P-N junctions surrounding the areas to be separated.

isotrophic etching   Refers to the etching of the photoresist both downward and to the side.

interconnect   See **lead**.

JFET (junction field-effect transistor)   Device in which voltage is applied to a terminal to control current between the source and drain regions.

junction   The interface at which the conductivity type of a material changes from P type to N type or vice versa.

layering   A process by which thin layers of different materials are grown on, or added to, the wafer surface.

lead   A metal strip on the wafer surface.

LED (light-emitting diode)   A semiconductor device in which the energy of minority carriers in combining with holes is converted to light. Usually, but not necessarily, constructed as a P-N junction device.

light field mask   See **clear field mask**.

lithography   Process of pattern transfer; when light is utilized, it is termed photolithography; and when patterns are small enough to be measured in microns, it is referred to as microlithography.

LSI (large-scale integration)   Refers to chips with between 5000 and 100,000 components each.

majority carrier   The mobile charge carrier (hole or electron) that predominates in a semiconductor material—for example, electrons in an N-type region.

mask   A glass plate covered with an array of patterns used in the photomasking process. Each pattern consists of opaque and clear areas that respectively prevent or allow light through. Masks are aligned with existing patterns on silicon wafers and used to expose photoresist. Mask patterns may be formed in emulsion, chrome, iron oxide, silicon, or a number of other opaque materials.

masking   See **patterning**.

memory   The storing of data or information.

metal mask   The step at which an island of conducting material is left on the wafer surface.

micrometer   One-millionth of a meter (10^{-6} m); symbol is μm.

micron   Same as micrometer.

microchip   See **chip**.

minority carrier   The nonpredominant mobile charge carrier in a semiconductor—for example, electrons in a P-type region.

MMOS (memory MOS)   A nonvolatile memory device structure that enables information to be retained during power shutdown.

molecule   Smallest quantity of a substance that retains the properties of that substance.

monochromatic light   Light of a single wavelength.

MOSFET   A field-effect transistor containing a metal gate over thermal oxide over silicon.

MSI (medium-scale integration)   Refers to chips with between 50 and 5000 components each.

negative resist   Photoresist that remains in areas that were not protected from exposure by the opaque regions of a mask while being removed by the developer in regions that were protected. A negative image of a mask remains following the develop process. A "clear" or "light" field mask is most often used with negative resist.

nitric acid (HNO$_3$)   A strong acid often used to clean silicon wafers or etch materials.

nitrogen (N$_2$)   A gas that seldom reacts with other materials. It is often used as a carrier gas for chemicals in semiconductor processing.

NMOS   N-channel MOS; type of MOSFET in which the channel is negative during conduction.

NPN transistor   A transistor which has a base of P-type silicon sandwiched between an emitter and a collector of N-type silicon.

N-type   A semiconductor material in which the majority of carriers are electrons and therefore negative. N-type dopants in silicon are Group V elements, in which the fifth outer electron is free to conduct current.

oil diffusion pump   A type of high-vacuum pump that uses evaporated hot oil particles to "push" chamber particles out of the system.

Ohm's law   A relationship between resistance, voltage, and current: $R = V/I$.

overall yield   The percentage of functioning packaged chips from a wafer related to the number of die mapped onto the wafer. Overall yield is the product of fabrication yield, sort yield, and assembly yields.

oxidation The growth of oxide on silicon when exposed to oxygen. This process is highly temperature dependent.

oxidation reaction chamber A chamber in which oxidation takes place. Quartz or silicon carbide tubing is used to make an oxidation reaction chamber due to their thermal resistance and purity.

oxide See **silicon dioxide**.

oxide etching An etching process which uses acid [usually hydrofluoric acid (HF)]. The acid must be buffered in order for the reaction to proceed at a rate slow enough to be controlled. Buffered oxide etch (BOE) is often used.

oxygen (O_2) Gas used to combine with silicon to form silicon dioxide.

package Protective container for semiconductor chip having electrical leads.

passivation Sealing layer added at the end of the fabrication process to prevent deterioration of electronic properties through chemical action, corrosion, or handling during the packaging processes. The passivation layer, usually silicon dioxide or silicon nitride, protects against moisture or contamination.

patterning A process in which the pattern in a reticle or photomask is transferred to a wafer resulting in the identification of areas to be doped or selectively removed.

pellicle A thin film of an optical-grade polymer that is stretched on a frame and secured to a mask or reticle. This solves the problem of airborne dirt collecting on the mask and acting as an opaque spot. During the exposure, the dirt is held out of the focal plane and does not "print" onto the wafer.

phosphine (PH_3) A gas that is often used as a source of phosphorus for doping silicon.

phosphorus (P) The N-type dopant commonly used for the sinker and emitter diffusions in standard bipolar integrated circuit technology.

phosphorus oxychloride ($POCl_3$) A liquid that is often used as a source of phosphorus for doping silicon.

photomasking See **patterning**.

photoplate A coated mask blank before imaging.

photoresist The light-sensitive film spun onto wafers and "exposed" using high-intensity light through a mask. The exposed (or unexposed depending on its polarity) photoresist is dissolved with developers, leaving a pattern of photoresist which allows etching to take place in some areas while preventing it in others.

pinhole A small undesired hole in the photoresist or in the opaque region of a mask or reticle.

planar structure A flat-surfaced device structure fabricated by diffusion and oxide masking, with the junctions terminating in a single plane.

plasma   High-energy gas made up of ionized particles.

plasma etching   An etching process similar to the chemical etching which uses an etching gas instead of a wet chemical.

PMOS (P-channel MOS)   Type of MOSFET in which the channel is positive due to conduction achieved by holes.

PNP   Semiconductor crystal structure consisting of an N-type region sandwiched between two P-type regions, as commonly used in bipolar transistors.

polymer   A complex organic chemical compound made up of repeating units.

polycrystalline silicon (Poly)   Silicon composed of many crystal unit cells randomly arranged.

positive resist   Photoresist that is removed in areas that were not protected from exposure by the opaque regions of a mask while remaining after develop in regions that were protected from exposure. A positive image of the mask remains following the develop process. A "dark field" mask is used most often with positive resist.

predeposition (predep)   The process step during which a controlled amount of a dopant is introduced into the crystal structure of a semiconductor.

primer chemical   A chemical which enhances the adhesion of a desired layer (in semiconductor technology, the layer is usually photoresist).

programmable read-only memory   See **PROM**.

projection alignment   An exposure system in which the image on the mask is projected onto the wafer. This results in little mask or photoresist damage and has about the same productivity as the contact method. For LSI and VLSI, projection alignment is standard.

PROM (programmable read-only memory)   A technology in which fuses are used in every memory cell and selected fuses are blown in order to program the chip with user-specified information.

P-type   Semiconductor material in which the majority carriers are holes and therefore positive. P-type dopants in silicon are Group III-A elements.

quartz   Commercial name for silicon dioxide formed into glass products. Because of its high temperature resistance, quartz is used in many processing steps in integrated circuit fabrication.

RAM (random access memory)   Device that temporarily stores digital information.

RCA clean   A multiple-step process to clean wafers before oxidation; named after RCA, the company that developed the procedure.

reactive ion etching (RIE)   An etching process that combines plasma and ion beam removal of the surface layer. The etchant gas enters the reaction chamber and is ionized. The individual molecules accelerate to the wafer surface. At the surface, the top layer removal is achieved by the physical and chemical removal of the material.

reactor   (1) A piece of equipment used for the deposition of a layer of material used in semiconductor processing. Common types of reactors are epitaxial reactors, vapox reactors, and nitride reactors. (2) See **plasma etcher.**

resistivity   A measure of the resistance to current flow in a material. A function of the attraction between the outer electrons and inner protons of a material. The more tightly bound the electrons, the greater the resistivity.

reticle   A reproduction of the pattern to be imaged on the wafer (or mask) by a step-and-repeat process. The actual size of the pattern on the reticle is usually several times the final size of the pattern on the wafer.

rinse   The removal of wet etchants or developers from the wafer. This process results in stopping the etching or developing processes and removing the active chemical from the surface. There are several different methods of rinsing: overflow rinsing, spray rinsing, dump rinsing, and spin-rinse dryers.

ROM (read-only memory)   Device in which information is permanently stored.

scanning electron microscope (SEM)   Microscope used to magnify images as much as 50,000 times by means of scanning with an electron beam. The impinging electrons cause electrons on the surface to be ejected. The ejected electrons are collected and translated into a picture of the surface.

scribe lines   Lines used to separate die on a wafer. The wafer will be sawed along the scribe lines, resulting in individual chips.

semiconductor   An element such as silicon or germanium, intermediate in electrical conductivity between the conductors and the insulators, in which conduction takes place by means of holes and electrons. Common single-element semiconductors are Si (silicon) and Ge (germanium); a common compound semiconductor is GaAs (gallium arsenide).

sheet resistance   A measurement of resistance with dimensions of ohms per centimeter squared that shows the number of N-type or P-type donor atoms in a semiconductor.

silicon (Si)   The Group IV element used for fabricating diodes, transistors, and integrated circuits.

silicon dioxide (SiO$_2$)   A nonconducting layer that can be thermally grown or deposited on silicon wafers. Thermal silicon dioxide is commonly grown using either oxygen or water vapor at temperatures above 900°C.

silicon nitride (Si$_3$N$_4$)   A nonconductive layer chemically deposited on wafers at temperatures between 600 and 900°C. When it is the final layer in the process, it protects devices against contamination.

single crystal   Refers to substances which have all unit cells arranged in a definite and repeated fashion as opposed to polycrystalline materials, which have unit cells randomly arranged.

slope etching   Controlled undercutting; an etch strategy in which the sides of the contact holes are purposely overetched so as to reduce the shadow effect of the sidewall and the resultant thinning of the film.

soft baking   A heating process used to evaporate a portion of the solvents in resist. The term "soft" describes the still soft resist after baking. The solvents are evaporated to achieve two results: to avoid retention of the solvent in the resist film and to increase the surface adhesion of the resist to the wafer.

solid-state electronics   Designation used to describe devices and circuits fabricated from solid materials such as semiconductors, ferrites, or thin films, as distinct from devices and circuits making use of electron tube technology.

source   Along with the gate and drain, one of the three regions of a unipolar or field-effect transistor (FET).

sputtering   A method of depositing a thin film of material on wafer surfaces. A target of the desired material is bombarded with radiofrequency-excited ions which knock atoms from the target; the dislodged target material deposits on the wafer surface.

spectrophotomoter   An analytical instrument used to collect interference measurements which are used to calculate film thickness.

spinning   A technique in which the photoresist is spun onto the wafer, resulting in a typical photoresist layer 0.5 μm thick with an allowable thickness variation of 10 percent.

spreading resistance   A technique used for measuring the dopant concentration profile in a wafer.

SSI (small-scale integration)   Refers to chips with between 2 and 50 components each.

static RAM (static random access memory)   Fast read-write memory cell based on transistors.

steam oxide   Thermal silicon dioxide grown by bubbling a gas (usually oxygen or nitrogen) through water at 98 to 100°C.

step and repeat   An operation in which the pattern on the reticle is transferred to the mask or wafer. The photoresist-coated mask blank (chrome, emulsion, or iron oxide) or wafer is placed on an x-y stage and the reticle pattern is repeatedly imaged until the entire surface is filled with the reticle pattern.

stepper   A machine which steps a reticle directly onto the wafer. A reticle can be produced to lower defect level and with tighter dimensional control than an entire mask, resulting in wafer images having fewer defects. Alignment of reticle to wafer is accomplished by reflecting a laser beam through a special reticle pattern (alignment target) and off a corresponding pattern on the wafer.

step coverage   The ability of new layers to evenly cover steps formed in the existing wafer layers.

stripping   Removal process; usually refers to photoresist.

subcollector   See **buried layer**.

substrate The underlying material upon which a device, circuit, or epitaxial layer is fabricated.

sulfuric acid (H_2SO_4) A strong acid often used to clean silicon wafers and to remove photoresist.

susceptor The flat slab of material (usually graphite) on which wafers are held during high-temperature deposition processes such as epitaxial growth or nitride deposition.

target The material to be sputtered during the sputtering process.

test die Die on a wafer that appear to have a different pattern from most others. These contain test devices created by the same processes as the regular die; however, the devices on these die are designed on a larger scale to allow in-process quality control.

TCE (trichloroethylene) A solvent used for wafer and general cleaning.

thermal diffusion A process by which dopant atoms diffuse into the wafer surface by heating the wafer in the range of 1000°C and exposing it to vapors containing the desired dopant.

thermal oxide On silicon semiconductor devices, an oxide fabricated by exposing the silicon to oxygen at high temperatures. The resulting interface has low levels of ionic impurities and defects (surface states).

thermocouple A device to measure the temperature in a furnace of a reactor. It is made by welding two wires together at a point. Heat generates a voltage between the two materials that is proportional to the temperature.

torr Pressure unit; international standard unit replacing the English measure, millimeters of mercury (mmHg).

transistor A semiconductor device that uses a stream of charge carriers to produce active electronic effects. The name was coined from the electrical characteristic of "transfer resistance."

tube (1) See **furnace**. (2) A cylindrical piece of quartz with fittings on one or both ends. It is placed in a furnace to provide a contamination-free and controlled atmosphere.

undercutting See **isotrophic etching.**

ULSI (ultra-large-scale integration) Refers to chips with over 1,000,000 components each.

ultraviolet A light wavelength that is commonly used to cause a response in a photoresist (photosolubilization or photopolymerization).

unipolar transistor A transistor such as an FET whose action depends on majority carriers only.

vacancy (1) A position in the crystal for an atom which is empty. (2) A type of crystal defect.

vacuum A low-pressure condition.

vapor priming A technique in which primer is applied in a vapor state such that the wafer never comes in contact with any possible contamination in the liquid or, in the case of HMDS, any particles of hydrolyzed HMDS.

via Vertical opening filled with conducting material used to connect circuits on various layers of a device to one another and to the semiconducting substrate; serves same purpose as "contacts."

viscosity The qualitative measure of liquid flow. Viscosity measurements are made by measuring the force required to move an object through the liquid. It is a measurement of "internal friction."

VLF hood A work station with vertical laminar airflow to keep particulate levels low.

VLSI (very-large-scale integration) Refers to chips with between 100,000 and 1,000,000 components.

voltage The force applied between two points causing charged particles (and hence current) to flow.

wafer A thin, usually round slice of a semiconductor material, from which chips are made.

wafer fabrication The series of manufacturing operations in which the circuit or device is put in and on the wafer.

wafer flat Flat area(s) ground onto the wafer's edges to indicate the crystal orientation of the wafer structure and the dopant type.

wafer sort The step after wafer fabrication during which the electrical parameters of integrated circuits are tested for functionality. Probes contact the pads of the circuit to conduct the test leading to the name "prober" for the equipment that performs electrical tests on each die site of completed wafers.

wafer sort yield The number of functioning die at wafer sort as compared to the total number of die started; typically, the lowest major yield point for integrated circuits.

wire bonding An assembly step in which thin gold or aluminum wires are attached between the die bonding pads and the lead connections in the package.

x-ray exposure system Imaging system using x rays as the exposure source. Due to their short wavelengths, x rays exhibit no detrimental diffraction effects.

yield A percentage used in the semiconductor industry which indicates the amount of finished products leaving a process as compared to the amount of product entering that process.

Index

ABOUT THE AUTHOR

Peter Van Zant's career spans twenty-five years in the
semiconductor industry, including positions in process
development and engineering, engineering management,
college instruction, consulting, marketing, and training. In
1981 Van Zant founded Semiconductor Services, Inc., a
training and information company serving the
semiconductor industry. Today, Semiconductor Services is a
leader in supplying training to integrated circuit
manufacturers and to equipment and material suppliers.
Mr. Van Zant has also held positions at Monolithic
Memories, National Semiconductor, Texas Instruments,
and the IBM Semiconductor Process Laboratory. He has
written a number of books on microchip technology,
including *Chip Packaging Manual* and *Wafer Fabrication,*
and was a contributing technical editor for Time-Life
Books' *The Chip Makers.* Mr. Van Zant earned his B.S. in
physics at Marist College.